'Magic is No Magic'
The Wonderful World of
Simon Stevin

WITPRESS

WIT Press publishes leading books in Science and Technology.
Visit our website for the current list of titles.
www.witpress.com

WIT*eLibrary*

Home of the Transactions of the Wessex Institute, the WIT electronic-library provides the international scientific community with immediate and permanent access to individual papers presented at WIT conferences. Visit the WIT eLibrary at
http://library.witpress.com

The only portrait of Stevin was probably painted shortly after he died. In the top left and right corners are the dates and places of his birth and death.

Leiden, University Library, Special Collections

'Magic is No Magic'
The Wonderful World of
Simon Stevin

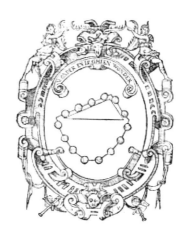

Authors:

Jozef T. Devreese
University of Antwerp, Belgium

Guido Vanden Berghe
Ghent University, Belgium

WITPRESS Southampton, Boston

Jozef T. Devreese
University of Antwerp, Belgium

Guido Vanden Berghe
Ghent University, Belgium

Translated by Lee Preedy
Cover concept: Sara Van Gucht
Graphics: Vladimir Gladilin
Design: Keith Godwin
Cover photo: © Davidsfonds Publishers (Daniël Geeraerts).

Published by

WIT Press
Ashurst Lodge, Ashurst, Southampton, SO40 7AA, UK
Tel: 44 (0) 238 029 3223; Fax: 44 (0) 238 029 2853
E-Mail: witpress@witpress.com
http://www.witpress.com

For USA, Canada and Mexico

WIT Press
25 Bridge Street, Billerica, MA 01821, USA
Tel: 978 667 5841; Fax: 978 667 7582
E-Mail: infousa@witpress.com
http://www.witpress.com

British Library Cataloguing-in-Publication Data

A Catalogue record for this book is available
from the British Library

ISBN (hardback) 978-1-84564-092-6
ISBN (paperback) 978-1-84564-391-1

Library of Congress Catalog Card Number: 2006934780

No responsibility is assumed by the Publisher, the Editors and Authors for any injury and/or damage to persons or property as a matter of products liability, negligence or otherwise, or from any use or operation of any methods, products, instructions or ideas contained in the material herein.

© WIT Press 2008

Printed in Great Britain by Biddles Ltd.

All rights reserved. No part of this publication may be reproduced, stored in a retrieval system, or transmitted in any form or by any means, electronic, mechanical, photocopying, recording, or otherwise, without the prior written permission of the Publisher.

Dedications

Dedicated by Jozef T. Devreese to:

Rose-Marie, Jeroen,

Gert, Veronique, Sarah, Charlotte, Jorik,

Annemie, Jacques, Jessica, Stefanie, Jasper, Jana

Dedicated by Guido Vanden Berghe to:

Magda,

Kristof, Beatrijs, Hanne, Lore, Quinten, Tine,

Stefaan, Caroline, Machteld, Michiel, Marthe, Hermelien,

Jeroen, Griet, Jona,

Elke, Peter, Lotte, Robin,

Maaike, Stijn,

Maarten,

Hannes

Contents

Foreword	xiii
Preface	xv
Acknowledgements	xix
Note to the reader	xxi
Stevin's works mentioned in the text	xxiii
Prologue. The Low Countries in Stevin's day	xxix
Chapter One. Simon Stevin and the Renaissance	1
1. Middle Ages, Renaissance and the Modern Age	1
1.1. The transition from the Middle Ages to the Renaissance was a gradual one	1
2. Printing	1
2.1. From the Officina Plantiniana in the Netherlands scientific knowledge spread across the whole of Europe	1
3. The universities	3
3.1. From 1426 Leuven was in the ascendant in the Southern Netherlands. 150 years later, Leiden became a Calvinist bastion in the north	4
4. The Renaissance in art, literature, science and technology	5
4.1. A new lease of life for ancient texts	5
4.2. The Renaissance was the prelude to the great scientific revolutions of the seventeenth century	7
5. Simon Stevin: Scholar and Engineer	8
Box 1 : How did Stevin know the works of the ancient Greeks?	9
Chapter Two. Simon Stevin, religious exile?	17
1. The Bruges period	17
1.1. The birth of a natural child in Bruges	17
Box 1 : Documents in the Bruges city archives	19
1.2. Burgomasters and aldermen on the spear side	22
1.3. Magistrates and merchants on the distaff side	23
Box 2 : Calvinism in Bruges between 1550 and 1584	25

Box 3: Important dates in Simon Stevin's life	28
1.4. Coming of age at twenty-eight	32
1.5. Simon Stevin's travels	33
2. Dedication to the Republic	34
Box 4: The University of Leiden	35
2.1. The young Stevin is enormously productive. He also applies his theoretical research in practice	40
Box 5: A letter to the town council of Gdánsk	41
2.2. Friendship with Maurice of Nassau ensures Stevin's future in the northern Netherlands	42
2.3. The scholar from Bruges becomes involved with the politics of the northern Netherlands	43
2.4. Stevin's mathematical approach to fortification-building is translated into German, French and English	44
2.5. Stevin's sailing chariot	45
Box 6: Maurice of Nassau	47
3. Married life and descendants	49
3.1. Three natural children	50
3.2. Stevin's death	51
3.3. Stevin's descendants	52

Chapter Three. The man who 'invented' the decimal system — 55

1. Decimal fractions for elementary calculations	55
2. Translations of Stevin's *De Thiende*	55
3. Uniform ways of counting and measuring	57
Box 1: The decimal system before and after Stevin	59
3.1. Stevin begins with a definition of the concept of the *Thiende* – the Tenth, or *Dime*	62
Box 2: An addition of decimal fractions	64
Box 3: Extracting roots	66
Box 4: Of weights and measures	68
Box 5: A chaotic monetary situation in the Low Countries	70
Box 6: Shakespeare and *De Thiende*	74
Box 7: The dollar and the disme	75

Chapter Four. Engineer and Inventor — 79

1. Patents	79
1.1. Raising water	79
Box 1: Patent granted by the States of Holland on 17 February 1584	80
1.2. A roasting spit	81
2. Mills	84
2.1. Stevin turned his mills by theory	84
Box 2: The IJsselstein mill	86

2.2. A long tradition of impoldering	87
3. Sluices	88
3.1. Of sluices and locks	89
4. Hydraulic designs	91
Box 3: Extract from *Van den handel der Waterschuyring onses Vaders Simon Stevin*	93
4.1. Stevin's son Hendrick and the impoldering of the Zuiderzee	95
5. Navigation	96
5.1. The straight and curved tracks…	96
5.2. Stevin replaces Plancius's four agonic lines with six	98
6. The Art of War (military science)	99
6.1. Stevin's presence in military camps	100
6.2. A proposal for army administration	101
7. Stevin as an architect	105
7.1. Town Planning	107
7.2. House building	107

Chapter Five. Economist *avant la lettre*   111

1. Introduction	111
1.1. Stevin and the reorganization of the princely lands	112
2. The *Nieuwe Inventie*	113
2.1. Is *Nieuwe Inventie* Stevin's first publication?	113
Box 1: Double-entry or Italian bookkeeping	115
3. The *Tafelen van Interest*	118
Box 2: *Le grand parti*	119
3.1. Demystifying the calculation of interest	120
3.2. Examples to illustrate the theory of simple and compound interest	121
4. Bookkeeping	122
4.1. Bookkeeping for merchants	122
4.2. Bookkeeping for a prince and his lands	126

Chapter Six. *Wonder en is gheen wonder*   131

1. Introduction	131
1.1. The negation of perpetual motion applied in both statics and hydrostatics	133
1.2. Practical problems provide the impetus for studies on weighing and hydrostatics	133
1.3. Definitions, axioms and propositions: the axiomatic approach	135
2. The structure of *De Beghinselen der Weeghconst*	136
Box 1: The *clootcrans*	136
2.1. The centre of gravity of a solid body	139
2.2. The lever	139

2.3. From the inclined plane to the parallelogram of forces: the *clootcrans*	140
3. *De Weeghdaet*	145
3.1. The *snaphaen*: a didactic gem	145
3.2. Cranes and windlasses according to Memling and Stevin	147
3.3. Stevin's 'Almighty'	149
3.4. The *Byvough der Weeghconst*	151
3.5. Dynamics	152
4. *De Beghinselen des Waterwichts* and the *Anvang der Waterwichtdaet*	154
4.1. Two lines in the historical development of hydrostatics	155
4.2. Stevin was the first to make any headway in hydrostatics since the time of Archimedes	155
4.3. The structure of *De Beghinselen des Waterwichts*	156
Box 2: Stevin was there before Pascal	157
4.4. Generalized derivation of Archimedes's Principle	162
4.5. Stevin's hydrostatic paradox	163
Box 3: The pressure of water on vertical and inclined walls	170
Box 4: Why do ships float?	172
4.6. Situating Stevin's work on hydrostatics	175
5. Summary: the significance of Stevin's work in physics	177
5.1. Stevin was the first to use the negation of perpetual motion in deriving a natural law	177

Chapter Seven. The link between Italian and French algebra — **181**

1. Introduction	181
1.1. Stevin and the sixteenth-century notion of numbers	182
Box 1: Quadratic, cubic and quartic equations	182
Box 2: Completing a square with an example given by Al-Khwarizmi (*c*.800)	186
2. Geometrical numbers	187
2.1. Geometrical numbers and geometrical figures	188
3. Algebraic numbers and the representation of polynomials	191
3.1. Stevin's notation	191
4. Equations	193
4.1. Introduction	193
4.2. The quadratic equation	195
4.3. Cubic equations	197
4.4. Quartic equations	198
4.5. A numerical method	198
5. The Rule of Algebra	199
5.1. …the Inexhaustible fountain of infinite Arithmetical Theorems…	200

Chapter Eight. Stevin's contribution to the Dutch language   **201**

   1. Stevin and the building of the Dutch language   201
   2. Early printed editions of works on medicine and
        the natural sciences   202
   3. Stevin's presence in the *Woordenboek der Nederlandse Taal*   203
   4. Neologisms and semantic neologisms   204
        4.1. A comparison of Stevin's work with the dictionaries
            compiled by Plantin and Kiliaan   204
        4.2. Arithmetic books in Dutch from 1445   205
        4.3. What can be learned from
            the *Woordenboek der Nederlandse Taal*?   207
        4.4. Stevin's ideas about Dutch   207
        4.5. The interest in the vernacular as an international phenomenon   211

Chapter Nine. Perspective   **213**

   1. The historical context and pioneering aspect of
        Stevin's *Vande Verschaeuwing*   213
        1.1 Early perspective   213
        1.2. Stevin and Guidobaldo del Monte   214
   2. A short description of *Vande Deursichtighe*   215
        2.1. The six theorems   217
        2.2. The inverse problem   221
   3. Maurice, Stevin and Dürer   222
   4. Stevin's followers: De Vries, 's Gravesande…   224
   5. Impact   225

**Chapter Ten. An extraordinary didactic genius:
Stevin's 'visible language'**   **227**

   1. Introduction   227
   2. Stevin's information graphics   228
        2.1. The *snaphaen* brings the concept
            of 'equal apparent weight' to cunning life   229
   3. *Vanden Hemelloop*   230
        3.1. The exposition of Ptolemy and Copernicus   231
        3.2. A didactic jewel   233
        3.3. Show and explain   236

Chapter Eleven. An unfinished composition   **239**

   1. Introduction   239
        1.1. In historical terms, Stevin's rational mathematical theory
            represents a transitional stage   240

2. Stevin divides the octave into 12 equal intervals	241
2.1. Experiments had a different status in Stevin's day	245
Box 1: Explanation of the most important ideas in the *Singconst*	247
3. 'Now in order to divide the monochord geometrically...' and the 'Arithmetical Division of the Monochord'	249
4. Temperament in Stevin's time and the placing of the frets on the neck of a lute	255
4.1. Stevin's practical knowledge was probably confined to the lute	255
5. What Verheyen had to say ...	256
6. The historical place of Stevin's tone system	256

Chapter Twelve. The resonance of Simon Stevin and his work   261

1. Introduction	261
Box 1: The statue of Simon Stevin in Bruges	261
Box 2: Stevin's books in the antiquarian book world	263
2. Stevin's influence on some of his contemporaries	267
2.1. Adriaan van Roomen. 'Ad Staticem Stevinii quod attinet, non facile in latinam linguam vertetur…'	267
2.2. William Gilbert and Stevin's *De Havenvinding*	269
2.3. Willebrord Snel van Royen (Snellius), who translated Stevin's *Wisconstige Gedachtenissen* into Latin	271
2.4. Isaac Beeckman and his *Journal*. References to and transcriptions of Stevin's work	273
2.5. Christaan Huygens	275
2.6. Nicolaes Witsen. 'I am more convinced by the discourse of the renowned Mr S. Stevin…'	278
Box 3: Simon Stevin: an innovator and polymath	278
3. Later reactions of eminent scholars	281
3.1. Joseph Louis Lagrange. 'Cette démonstration de Stevin… très ingénieuse…'	282
3.2. Ernst Mach. 'Die Stevin'sche Ableitung ist eine der werthvollsten Leitmuscheln in der Urgeschichte der Mechanik…'	282
3.3. Richard Feynman. '…because the chain does not go around…'	283

Epilogue   **285**

Bibliography   **287**

Index of names   **293**

Index of subjects   **303**

Foreword

The decimal number system, using a decimal place followed by digits ranging from 0 to 9, in order to characterize any number with any desired accuracy, is so familiar to most of us that one might forget that it actually had to be invented by someone. This someone also had to find out how to add, subtract, multiply and divide such numbers. His name is Simon Stevin, an ingenious physicist and mathematician born in Bruges in the sixteenth century. The word 'dime' stems directly from 'disme', an English translation of 'thiende', or 'tenth', as introduced in Stevin's work. That and many other aspects of his many-sidedness and versatility can be found in this book.

Stevin's contributions cover a range of disciplines: mathematics and physics, engineering and technology, navigation, financial theory, fortifications and city planning, linguistics, theory of music, etc. He was one of the very first 'Copernicans', while at the same time he took part in the ethical discussions of his time, advocating tolerance and civic spirit, and furthermore he created an Engineering School. Stevin's books, several of them published by the humanist master printer Christopher Plantin and his son-in-law, use strong and innovative visual representation as a tool of generating and disseminating new science. In the present book this spectrum of endeavours by Stevin and their historical setting in the Renaissance is presented for a broad audience.

The authors, Jozef T. Devreese and Guido Vanden Berghe, profoundly studied the works of Stevin and their impact. As physicists the authors based their study on the objective analysis of existing material. Interpreting the objective data they concluded that Stevin's work merits to be known, not only by specialists but also by a wide public.

The Scientific Revolution is sometimes considered to have been initiated in 1543 with Copernicus's '*De Revolutionibus*', even when the heliocentric world picture was known to Aristarchos around 300 BCE. Stevin is often cited as one of the links in the chain of progress leading from Copernicus via Kepler, Gilbert and Galileo to Newton and Huygens. As scientists we stress the significance of Stevin's explicit requirement of the combination of theory and experimental verification, in his words '*Spiegheling en Daet*'. It is not commonly realised that Stevin, in his treatise on hydrostatics was one of the first to stress this point. It is this systematic testing of theory using experiment that constitutes a key difference between the science of the ancient Greeks and the modern scientific method. One of the reasons why Greek science came to a halt was that the ideas of leading philosophers were accepted as decisive, while empirical tests were barely pursued. Historians of science have devoted extensive studies to this problem of the decline of Greek science after

Archimedes. As many as 1,800 years passed before first Simon Stevin and subsequently Galileo, who went a step further, succeeded to open new avenues in mechanics, beyond the achievements of the Greeks.

Stevin should be credited for his recognition of the need for standard units (of length, weight, mints, etc) and his making of a plea for those units, in a time when a great variety of different systems were used leading to great confusion. The introduction and use of standard units is a very hard task. It took till the French Revolution before the metric system took off. Even today miles, inches, and pounds are still being used in most parts of the Anglo-Saxon world.

An example of a fascinating theme touched upon in the present book concerns the *perpetuum mobile.* The idea has it roots in the Orient. In the Middle Ages and during the Renaissance period, there was widespread interest in the *perpetuum mobile* (that we would call today 'of the first kind'). Leonardo da Vinci had explored the concept of perpetuum motion machines; he made some sketches of imaginary machines, and took a sceptic attitude regarding their practical realization. Stevin formulated the negation of the possibility of a *perpetuum mobile* as a law of nature and from it he drew several conclusions, such as the law of the inclined plane. He came close to the crucial concept of energy in his famous thought experiment which he devised for the inclined plane. Subsequently Galileo and Huygens, in their works, also used this negation of the *perpetuum mobile.*

Stevin's main contributions to physics were realized in a short period between 1581 and 1586, a period in which he already devoted considerable efforts to engineering. After 1586 he expanded his activity to many fields as described in several chapters in this book.

The reader will find a very accessible account of Stevin's multiple activities and findings, and how those efforts fit into the broad picture of progress during the Renaissance. *'Magic is No Magic'. The Wonderful World of Simon Stevin* also incorporates the new insights that came to light after earlier bibliographic works on Stevin had appeared (1943-1966).

I am convinced the reader will find this book about Simon Stevin enjoyable and accessible.

Prof. Dr. Gerard 't Hooft,
Nobel Laureate for Physics (1999)

Preface

Wonder en is gheen wonder – **'Magic is No Magic'** – expresses Simon Stevin's vision of science. According to Stevin a natural phenomenon is only 'magic' as long as it is not understood. As soon as it is grasped, it is **'no magic'**. This is a very pragmatic motto that appears on the title pages of many of Stevin's works, and with which Stevin perhaps contrasted his approach with the scholastic.

Dozens of times each day the drivers of the horse-drawn carriages that convey tourists around Bruges extol Simon Stevin as they pass his statue: 'He was a great mathematician, physicist and engineer, inventor of the decimal system and,' they sometimes add, 'of the sailing chariot'. 'Mathematician and physicist', we read, on the sign that bears the name of the square where his statue stands, the Simon Stevinplaats.

Numerous societies in Flanders and the Netherlands are named after him, including sailing clubs, organizations concerned with the conservation of fortifications, schools, observatories and student associations. His name is attached to important prizes, journals, streets, sluices and many more and diverse things. Less well known to the general public is the fact that Stevin was an intellectual all-rounder who made significant contributions in numerous disciplines.

Internationally, Stevin is chiefly known in academic and scientific circles of various specialities. His name is to be found in select lists of the seven or so pioneers in the history of the computer and is included on the famous 'Printing and the Mind of Man' list, which highlights books which have had a significant impact on history in general.

Stevin was already renowned in his own day and men of learning in various disciplines were well aware of his work. Yet it was to be a long time before the Bruges scholar and his manifold contributions received the public recognition they merited. In recent decades there has been a growing interest in Stevin's work, but even today the resonance of his name is not commensurate with his achievements.

Physics provides a case in point. As is underlined in this book, Stevin's name should be associated with at least two important laws. In hydrostatics some of the credit that is rightly Stevin's is given to Pascal, while the composition law of forces has become such a fundamental fact that no individual scholar's name is associated with it, although – for the general case – it was actually propounded by Stevin. And though Newton is indisputably the creator of modern physics, Simon Stevin was one of those who paved the way for the modern science.

The purpose of this book, however, is not the scientific study and definition of Stevin's priority in discoveries and inventions, though modern insights into these things are covered. Its main aim is to introduce Simon Stevin and his work to the widest possible readership. In doing so, the authors have kept to established historical facts and critically underpinned scientific insights into Stevin and his work.

To date, apart from short essays on single aspects of his work, two standard works on Stevin have appeared: *Simon Stevin* by Eduard Jan Dijksterhuis (1943) and *The Principal Works of Simon Stevin,* published in five volumes between 1955 and 1966, an initiative of the Koninklijke Nederlandse Academie van Wetenschappen. *The Principal Works* reproduces most of Stevin's works, with an English translation and introductory commentary. These standard references are extremely valuable, and are essential sources for those wanting to study Stevin. They are aimed particularly at specialists in one or other of the many disciplines Stevin practiced. In addition, since 1943, and respectively in 1966, our picture of Stevin has been filled out with many new insights and discoveries. Thus our purpose here is to introduce Stevin's oeuvre in an accessible way with today's insights taken into account.

Particular attention is devoted to the more recent insights into Stevin's work, which appear in virtually every chapter. As a rule these are not gone into in detail: often they are situated succinctly and the interested reader is referred to the appropriate source. For instance, a lengthy chapter could have been written about Stevin's posthumously published work on architecture, *De Huysbou*. Such a chapter could have been based partly on the research of Charles van den Heuvel. Instead, when preparing the original, Dutch version of this book, we opted for a brief reference to this work, as in 2003 that author's extensive work on the *Huysbou* was still in preparation.

For each of the subjects discussed the background, the significance, the impact and the status of Stevin's contribution are analysed and interpreted.

Archive-browsing has resulted in the discovery of additional references to Stevin's presence in the army camps during Maurice of Nassau's campaigns (Chapter Four). Also in the present volume we have included reproductions of Stevin autographs of the *Singconst* which, as far as we know, have hitherto only appeared as transcriptions in the literature (Chapter Eleven).

The polymathic nature of the Bruges scholar is immediately striking. Stevin dealt in a profound and creative way with a multiplicity of subjects, often using innovative didactic design. This is reflected in the structure of this book, with chapters devoted to extremely diverse subjects and specialities. After situating Stevin as a true *homo universalis* and looking at his life and family, we examine his original and innovatory contributions chapter by chapter. The reader will get to know the man behind the introduction of decimal fractions, the engineer and inventor with his numerous patents, the economist *avant la lettre*, the brilliant physicist who also contributed to the mathematics of his day. Subsequently we look at Stevin's contribution to his own Dutch language, and at his crystal clear contribution to the theory of perspective and music. A special chapter deals with Stevin's exceptional pedagogic genius and his 'visible language', which is receiving much attention today. Finally the resonance of Stevin's work is discussed.

It will be obvious to the reader that given the way science works today it would be impossible for a single individual to make meaningful contributions in so many areas.

We have done our utmost to ensure that every chapter can be enjoyed by everyone. None the less, Chapters Six and Seven will perhaps be appreciated more by those who did not altogether abandon science in secondary school. As Menaechmus reputedly pointed out to Alexander the Great when the ruler asked for a shortcut in geometry, there is no royal road; everyone must travel the same way.

This fact was also recognized by Maurice of Nassau, Prince of Orange, who made great efforts to master the science and technology of his day, and who chose Stevin as his permanent tutor and councillor. The relationship between Stevin and Maurice is not explored in depth in this book. In 2000 the excellent *Maurits, Prins van Oranje* (Kees Zandvliet, 2000) appeared, in which the interaction between the polymath and the prince is discussed in detail. However, the important new insights into Stevin's status at Maurice's court that follow from this publication are explained in the present volume.

For the authors, both of whom have their roots in Bruges, an interest in Simon Stevin has grown gradually, as a sort of hobby. From 1990, separately at first and then jointly, both organized events based on Stevin and his work: lectures, short publications, exhibitions on Stevin, and, in 1998, together with Dr Charles van den Heuvel, a symposium in Bruges. The wealth of Stevin's world of ideas, the multifaceted nature of his approach and the subjects he dealt with, his probing and aesthetic didactic designs, and the situating of Stevin in the Renaissance period, made this exploration a unique and most fascinating experience.

We hope that we can pass on to the reader something of this contact with the Bruges scholar and his world in **'Magic is No Magic'. The Wonderful World of Simon Stevin.**

The Authors

Acknowledgements

Simon Stevin's multifaceted nature is also reflected in the many contacts we have had in the creation of this book, and it is a great pleasure now to express our thanks. In the first place we thank Dr Charles van den Heuvel, art historian, Senior Researcher Koninklijke Nederlandse Academie van Wetenschappen and, in connection with Stevin, specialist in the areas of architecture (theory), town planning, fortification and civil technology. He has a particular interest in Stevin's manuscripts and transcriptions thereof, and in the reception of this written heritage. We had several fruitful exchanges of ideas with Dr van den Heuvel. He provided us with numerous interesting references, documents and data, and was kind enough to read an early version of our manuscript.

We are grateful to Sara Van Gucht, whose suggestions contributed significantly to the content and presentation of the book. As a high-school teacher of History and English, her grasp of the historical and didactic aspects of the text and their resonance in the Anglo-Saxon world not only enhanced the style, clarity and accessibility of the original Dutch version, but also set out a sound basis for the English translation.

We are indebted to Dr Koos van de Linde, an authority on the organ, for rewarding conversations regarding the history of tuning theory.

We also thank those who through discussions, critical commentary and the provision of data or assistance have contributed to the creation of this book:

Ludo Vandamme, scientific associate, Historical Collections of the City Library (Biekorf) Bruges; Professor Dr Carl Reyns, Universiteit Antwerpen, Former Rector Universiteit Antwerpen (UFSIA), professor of financial accounting; Dr Marjolein Kool, Hogeschool Domstad, Utrecht, teacher of arithmetic/mathematics and pedagogics; Dr Nicoline van der Sijs, University of Utrecht, etymologist and publicist on language; J.P. Puype, conservator, Army Museum, Delft; Professor Dr Fons Brosens, Universiteit Antwerpen, physicist; Christiane Broos, teacher of physics; Professor Dr Guido de Baere, Universiteit Antwerpen, professor of religious hermeneutics; Professor Emeritus Dr Ludo Simons, Universiteit Antwerpen, professor of book and library science at the Universiteit Antwerpen and Katholieke Universiteit Leuven, Honorary Chief Librarian of the Universiteit Antwerpen; Dr M. De Reu, keeper of manuscripts and rare books, Universiteit Gent; Professor Dr Pierre Delsaerdt, professor of book and library history at the universities of Antwerp and Leuven, Head of the Department of Library and Information Sciences at the Universiteit Antwerpen; Julien van Borm, Universiteit Antwerpen, Chief Librarian of the Universiteit Antwerpen; Francis Wray MA, PhD (Cantab), Technology Transfer Consultant; Dr M.A. Mooijaart, editor-in-chief *Woordenboek der Nederlandsche Taal*, Instituut voor Nederlandse Lexicologie (Leiden); Professor Dr Josse Van Steenberge, Universiteit Antwerpen, President of the Association of the University - Higher Education Colleges Antwerp, Former Rector Universiteit Antwerpen (UIA), professor of social security law, Universiteit Antwerpen; Professor Dr Francis Van Loon, Rector Universiteit Antwerpen, professor of statistics; J. Dhondt (City Archive, Bruges); Professor Emeritus Dr Roland Baetens, Universiteit Antwerpen, historian; Professor Dr Vladimir Fomin and Dr V. Gladilin who also contributed to the logistics of the editing process, physicists, Universiteit Antwerpen and Moldova State University; Yola de Lusenet, Edita of the Koninklijke Nederlandse Academie van Wetenschappen (Amsterdam); Dr iur. P. Van Remoortere, former President of the Universiteit Antwerpen (UIA); Professor Dr Georges Wildemeersch, professor of Dutch Literature, Universiteit Antwerpen; Dr Ike van Hardeveld, Poppel (Belgium); N. Blontrock (VRT and press journalist); Professor Dr E. Coppens (University of Nijmegen); J. Kok (Elsevier Science); Professor Dr G. Van Hooydonck, Universiteit Gent; Professor Dr H. Symoens, Universiteit Gent; the late Professor Dr Jozef Brams, Institute of Philosophy, Katholieke Universiteit Leuven; Mrs C. De Smedt, Merchtem; Professor Dr R. Beyers, Universiteit Antwerpen, Department of Romance Languages and Latin; P. Donche, Universiteit Antwerpen; Professor Dr Alain Verschoren, President Universiteit Antwerpen, Former Rector Universiteit Antwerpen (RUCA), professor of mathematics.

We thank the collaborators of the Davidsfonds and WIT Press Publishing Houses for their contributions to the design of the Dutch and English editions of this book, respectively.

The authors are pleased to express special thanks to Lee Preedy for her outstanding translation from Dutch to English of our original book *'Wonder en is gheen wonder'. De geniale wereld van Simon Stevin 1548-1620* (Davidsfonds, Leuven, 2003 - ISBN 90-5826-174-3). Ms Preedy's contribution was not limited to the translation of the original text, but also included the discussion and refinement of several points, their historical context and factual accuracy.

The publication of the original Dutch edition of this book, *'Wonder en is gheen wonder'. De geniale wereld van Simon Stevin 1548-1620,* was made possible through the support of a number of individuals, companies and institutions, to whom we express our thanks. Professor Dr Ir. Jef Roos, Honorary Chairman VEV, Chairman UGINE & ALZ-Belgium NV. Dr Ludo Verhoeven, Past President and Chairman of the Board of AGFA-GEVAERT NV, Honorary Chairman of the VEV. AGFA's support made possible the inclusion of the colour reproductions in the original Dutch edition of this book. The Fonds voor Vlaanderen who sponsors projects on the theme of 'Fleming, who are you? Highs and Lows in our History'. The Marnixring Internationale Service Club VZW who encourages the Dutch language irrespective of political and philosophical borders. Ir. Herman Deroo, retired General Chairman of the KVIV. The Koninklijke Vlaamse Ingenieursvereniging (Royal Flemish Society of Engineers) supported the publication of the original Dutch edition of this book to mark its sevety-fifth anniversary. The KVIV regards Simon Stevin as the first university trained engineer and as a pioneer for the many who follow in his footsteps today. The Laboratory for Theoretical Solid State Physics (TFVS) of the University of Antwerp.

Note to the reader

Virtually every subject that Stevin turned his attention to resulted in a written work. His output was prolific and many of his texts were published, either by himself, by his son Hendrick, or by others. Many of his works were also translated into European vernaculars or Latin.

Even though the Dutch of Stevin's day will probably be unfamiliar to many readers it was decided to leave the titles of Stevin's works in their original languages. Where a title appears for the first time a translation follows in brackets. A list of titles with an English translation will be found on page numbers xxiii-xxviii.

Simon Stevin's works were frequently translated. In *Les Œuvres Mathematiques de Simon Stevin de Bruges, Le tout reveu, corrigé, & augmenté par Albert Girard Samielois* Girard, an exiled French Huguenot, provides a translation of much of Stevin's mathematical work.

Leiden, University Library, 252 A 36

Stevin's works mentioned in the text

Below is an overview of Simon Stevin's published works. Only the first editions are mentioned, with the name of the printer and the year and place of publication. A more extensive list including reprints, translations and so on can be consulted in E. Dijksterhuis (1943) and E. Dijksterhuis (1955).

Nieuwe Inventie van Rekeninghe van Compaignie, inde welcke verclaert worde een zekere corte ende generale reghele om alle Rekenynghe van Compagnie ghewislic ende lichtelic te solveren, Gheinventeert ende nu eerst int licht ghegheven door Simonem Stephanum, Ghedruckt tot Delft, by Aelbert Hendricz woonende aent Marct-veldt, Anno 1581.

New Invention for Company Accounting, in which is explained a certain short and general rule for correctly and easily solving all business calculations. Invented and now published for the first time by Simonem Stephanum, Printed at Delft by Aelbert Hendricz. living on the market place, Anno 1581.

It is not absolutely certain that Stevin was the author of this work. This is discussed at length in Chapter Five.

Tafelen van Interest, Midtsgaders De Constructie der selver, ghecalculeert door Simon Stevin Brugghelinck (1582) door Christoffel Plantijn in den gulden Passer te Antwerpen.

Tables of Interest, Together with the construction of same, calculated by Simon Stevin of Bruges (1582) by Christopher Plantin in the Golden Compasses at Antwerp.

The Golden Compasses was the name of Plantin's Antwerp printing house.

Problematum Geometricorum In gratiam D. Maximiliani, Domini a Cruningen ect. editorum Libri V. Auctore SIMONE STEVINIO Brugense (1583) door Johannes Bellerus te Antwerpen.

Problematum Geometricorum In gratiam D. Maximiliani, Domini a Cruningen ect. editorum Libri V. Auctore SIMONE STEVINIO Brugense (1583) by Johannes Bellerus at Antwerp.

Dialectike ofte Bewysconst, Leerende van allen saecken recht ende constelic oirdeelen; oock openende den wech tot de alderdiepste verborghentheden der Natueren, Beschreven int Neerduytsch door SIMON STEVIN van Brugghe, (1585) door Christoffel Plantijn van Leyden.	*Dialectics or the Art of Demonstration*, Teaching to judge of all things rightly and aptly; also opening the way to the deepest mysteries of nature, Written in Dutch by Simon Stevin of Bruges, (1585) by Christopher Plantin of Leiden.
De Thiende Leerende door onghehoorde lichticheyt allen rekeningen onder den Menschen noodich vallende afveerdighen door heele ghetalen sonder ghebrokenen. Beschreven door Simon Stevin van Brugghe, (1585) door Christoffel Plantijn te Leyden.	*The Disme: the Art of Tenths*, Teaching how to perform all computations whatsoever, by whole numbers without fractions. Written by Simon Stevin of Bruges, (1585) by Christopher Plantin at Leiden
L'Arithmetique de SIMON STEVIN de Bruges: Contenant les computations des nombres Arithmetiques ou vulgaires: Aussi l'Algebre, avec les equations de cinq quantitez. Ensemble les quatre premiers livres d'Algebre de Diophante d'Alexandrie, maintenant premierement traduicts en François. Encore un livre particulier de la Pratique d' Arithmetique, contenant entre autres, Les Tables d'Interest, La Disme; Et un traicté des Incommensurables grandeurs: Avec l'Explication du Dixiesme Livre d'Euclide, (1585) door Christophle Plantin A Leyde.	*Arithmetic by SIMON STEVIN of Bruges*: Containing the computations of arithmetic or common numbers. Also Algebra, with equations of the fourth degree. Together with the first four books of the Algebra of Diophantos of Alexandria, now translated for the first time in French. Also another book, the Practise of Arithmetic, containing Tables of Interest, The Disme; And a treatise on Incommensurable quantities: With the explanation of the tenth book of Euclid, (1585) by Christopher Plantin at Leiden.
De Beghinselen der Weeghconst, Beschreven duer Simon Stevin van Brugghe, (1586), Tot Leyden, Inde Druckerye van Christoffel Plantijn, By Françoys van Raphelinghen	*The Elements of the Art of Weighing*, Written by Simon Stevin of Bruges (1586), [published] at Leiden in the printing house of Christopher Plantin, by Frans van Ravelingen

The *Weeghconst* is prefaced by

Uytspraeck vande Weerdicheyt der Duytsche Tael	*Discourse on the Worth of the Dutch Language*

Published together with the *Weeghconst* were:

De Weeghdaet	***The Practice of Weighing***
De Beghinselen des Waterwichts	***The Elements of Hydrostatics***

***Vita Politica, Het Burgherlick Leven**, Beschreven deur Simon Stevin (1590), uitgegeven te Leyden door Franchoys van Ravelenghien.*

***Vita Politica, Civic Life**, Written by Simon Stevin (1590), published at Leiden by Frans van Ravelingen.*

***Appendice Algebraique**, de Simon Stevin de Bruges, contenant regle generale de toutes equations, (1594)*

***Algebraic Appendix**, by Simon Stevin of Bruges, containing a general rule for all equations, (1594)*

Published at Leiden by Frans van Ravelingen.

***De Stercktenbouwing**, beschreven door Simon Stevin van Brugghe, (1594) uitgegeven door Françoys van Ravelenghien te Leyden.*

***The Art of Fortification**, written by Simon Stevin of Bruges, (1594), published by Frans van Ravelingen at Leiden.*

***De Havenvinding** (1599), Tot Leyden, In de Druckerye van Plantijn, By Christoffel van Ravelenghien, Gesworen drucker der Universiteyt tot Leyden.*

***The Haven-Finding Art** (1599), At Leiden, in the printing house of Plantin, by Christopher van Ravelingen, official printer to the University at Leiden.*

***Wisconstige Gedachtenissen**, Inhoudende t'ghene daer hem in gheoeffent heeft den Doorluchtichsten Hoochgheboren Vorst ende Heere, Maurits, Prince van Oraengien, Grave van Nassau, Catzenellenbogen, Vianden, Moers &c. Marckgraaf van der Vere, ende Vlissinghen, &c. Heere der Stadt Grave ende S'landts van Cuyc, St. Vyt, Daesburch &c. Gouverneur van Gelderlandt, Hollant, Zeelant, Westvrieslant, Zutphen, Utrecht, Overyssel &c. Opperste Veltheer vande vereenichde Nederlanden, Admirael generael van der Zee &c. Beschreven deur Simon Stevin van Brugghe (1608), gedrukt in de Druckerye van Jan Bouwensz. te Leyden.*

***Mathematical Memoirs**, containing that which has been practiced by our Excellent Sovereign and Lord, Maurice, Prince of Orange, Count of Nassau, Catzenellenbogen, Vianden, Moers, etc. Margrave of Veere and Flushing, etc. Lord of The Hague and the Lands of Cuyc, St. Vyt, Daesburch etc. Governor of Guelders, Holland, Zeeland, Westfrisia, Zutphen, Utrecht, Overijssel, etc. Commander-in-chief of the United Netherlands, Admiral-general of the Sea, etc. Written by Simon Stevin of Bruges (1608), printed in the printing house of Jan Bouwensz. at Leiden.*

On the title page the date of publication is given as 1608; many parts of this work had already been published in 1605, however. Stevin's vast opus has a complicated arrangement. The principal division is into five parts, with further complex subdivisions. Only those writings mentioned in the present volume are listed here:

i. *Vant Weereltschrift*	i. *Cosmography*
i.2. *Vant Eertclootschrift*	i.2. *Geography*
i.2.2. *Vant Stofroersel des Eertcloots*	i.2.2. *Geomorphology*
i.2.4. *Vande Zeylstreken*	i.2.4. *The Sailings*
i.2.6. *Vande Spiegheling der Ebbenvloet*	i.2.6. *The Theory of Ebb and Flow*
i.3. *Vanden Hemelloop*	i.3. *The Heavenly Motions*
ii. *Van de Meetdaet*	ii. *The Practice of Measuring*
iii. *Van de Deursichtighe*	iii. *Optics*
iii.1. *Vande Verschaeuwing*	iii.1. *Perspective*
iii.2. *Vande Beginselen der Spieghelschaeuwen*	iii.2. *The Elements of Catoptrics*
iii.3. *Vande Wanschaeuvving*	iii.3. *Refraction*
iv. *Van de Weeghconst*	iv. *The Art of Weighing*
iv.7. *Byvough der Weeghconst*	iv.7. *Supplement to the Art of Weighing*
iv.7.1. *Van het Tauwicht*	iv.7.1. *Of the Cord Weight*
iv.7.2. *Van het Catrolwicht*	iv.7.2. *Of the Pulley Weight*
iv.7.3. *Vande vlietende Topswaerheyt*	iv.7.3. *Of the Floating Top-heaviness*
iv.7.4. *Vande Toomprang*	iv.7.4. *Of the Pressure of the Bridle*
v. *Van de Ghemengde Stoffen*	v. *Miscellaneous Subjects*
v.2.1. *Coopmans Bouckhouding op de Italianse Wijze*	v.2.1. *Mercantile Bookkeeping in the Italian Manner*
Van Compaignieslot	*On Closing Company Activities*
v.2.2. *Vorstelicke Bouckhouding op de Italianse Wijze*	v.2.2. *Princely Bookkeeping in the Italian Manner*
v.2.2.1. *Bouckhouding in Domeine*	v.2.2.1. *Bookkeeping of Lands*
v.2.2.3. *Bouckhouding in Finance Extraordinaire*	v.2.2.3. *Bookkeeping of Extraordinary Financial Affairs*

Castrametatio, Dat is Legermeting and *Nieuwe Maniere van Sterctebou, door Spilsluysen* (1617)

Castrametatio, That is the Marking Out of Army Camps and the *New Manner of Fortification by means of Pivoted Sluice-Locks* (1617)

Published at Rotterdam by Jan van Waesberghe

A number of works were published posthumously. In some of these, earlier publications were reprinted and supplemented with new material taken from manuscripts retained by Stevin's wife. Below is a brief summary of these works.

Les Œuvres Mathematiques de Simon Stevin de Bruges. Ou sont insérées les Memoires Mathematiques Esquelles s'est exercé le Tres-haut & Tres-illustre Prince Maurice de Nassau, Prince d'Aurenge, Gouverneur des Provinces des Païs-bas unis, General par Mer & par Terre, &c. Le tout reveu, corrigé, & augmenté par ALBERT GIRARD Samielois, Mathematicien, (1634) uitgegeven door Bonaventure en Abraham Elzevier te Leiden.

The Mathematical Works of Simon Stevin of Bruges. In which are included the Mathematical Memoirs practiced by the Most High and Most Illustrious Prince Maurice of Nassau, Prince of Orange, Governor of the United Provinces, General of Sea and Land, etc. Revised, corrected and augmented by ALBERT GIRARD Samielois, Mathematician, (1634) published by Bonaventure and Abraham Elzevier at Leiden

Materiae Politicae, Burgherlicke Stoffen and *Verrechting van Domeine mette Contrerolle en ander behouften van dien (1649)*

Materiae Politicae, Civic Matters and *Administration of Domains with Control and other requirements of the matter, (1649)*

Compiled by Hendrick Stevin and published by Justus Livius at Leiden. Both parts are usually found in a single volume. In Section VIII of *Civic Matters* we find Stevin's non-mathematical writings on warfare under the title:

Van de Crijchspiegeling	*On the Theory of War*
Vande Spiegeling der Singconst and *Vande Molens*	*On the Theory of the Art of Singing* and *On Mills,*

Two unedited treatises, reprinted by Dr D. Bierens de Haan, (1884)

Some of Stevin's works are known only from the publications of other authors. Below are two of the principal works to contain extracts from Stevin's work.

Journal tenu par Isaac Beeckman de 1604 à 1634 publié avec une introduction et des notes par C. de Waard	*Journal kept by Isaac Beeckman from 1604 to 1634 published with an introduction and notes by C. de Waard* (The Hague 1942).

Appendix I contains extracts from the following:

Huysbou	*House building*
De Spiegheling der Singconst	*Theory of the Art of Singing*
Cammen ende Staven	*Cogs and Staves*
Watermolens ende Cleytrecking	*Water Mills and the Drawing of Clay*
Waterschueringh	*Waterscouring*
Van de Crijchconst	*On the Art of Warfare*

Wisconstich Filosofisch Bedryf, Van Hendric Stevin, Heer van Alphen, van Schrevelsrecht, &c. Begrepen In veertien Boeken. (Leiden 1667).	*Mathematical Philosophical Transactions* By Hendrick Stevin, lord of Alphen, of Schrevelsrecht, etc. In fourteen volumes (Leiden 1667).

The following volumes contain parts of Simon Stevin's work:

Boek VI: *Van den handel der cammen en staven onses Vaders, als bewegende oirsaec van dese*	Vol. VI: *Essay on Cogs and Staves by our Father Simon Stevin*
Boek X: *Van den handel der Watermolens onses Vaders Simon Stevin*	Vol. X: *Essay on Water Mills by our Father Simon Stevin*
Boek XI: *Van den handel der Waterschuyring onses Vaders Simon Stevin*	Vol. XI: *Essay on Waterscouring by our Father Simon Stevin*

Prologue

The Low Countries, an area that more or less corresponds to the present-day Benelux and part of northern France, have a complex history. It was at no time more complex than in Simon Stevin's day, when this mercantile hub of northern Europe was convulsed with religious and political strife (see fig. 1). At the time of Stevin's birth the Low Countries were in process of being forged into a unified whole within the Habsburg empire. By the time of his death they had split into two fundamentally opposed halves, and the half in which he had chosen to live was well on the way to becoming a colonial power in its own right.

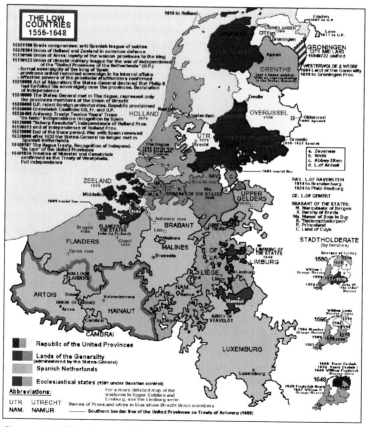

Figure 1: Political map of the Low Countries in the period 1556-1648, Courtesy of Mr J. de Salas Vara de Rey.

The Seventeen Provinces

When Simon Stevin was born in 1548 the Low Countries were ruled by the Habsburg emperor, Charles V. Born in Ghent, Charles reigned over an empire that stretched from the Andes to the Balkans. From his father, Philip the Fair, he had inherited the possessions of Burgundy and Habsburg, from his mother, Joanna of Castile, the Spanish kingdoms of Castile and Aragon and their territories in the Americas and southern Italy. As Holy Roman Emperor his dominion extended over present day Germany, Austria, the Czech lands and northern Italy.

In the Low Countries Charles completed the conquests begun by his ancestors, the Valois dukes of Burgundy, who had ultimately come to rule not only the duchy of Burgundy and the Franche-Comté (both eventually lost to France) but had also acquired in succession Flanders, Artois, Namur, Brabant, Limburg, Holland, Zeeland, Hainaut and Luxemburg. To these Charles added the episcopal cities of Tournai and Cambrai, Friesland, Utrecht, Overijssel, Groningen, Drenthe, Guelders and Zutphen. Having brought all of the Low Countries except the prince-bishopric of Liège under his control he took steps to ensure that they would descend to his heir entire. In 1548 he designated his Low Countries territories as the 'Burgundian Circle', a consolidated part of his empire. The Pragmatic Sanction of 1549 established the Seventeen Provinces of the Netherlands as a separate entity, a personal union of principalities, and decreed that a single heir would inherit them. When Charles abdicated in 1555 this fledgling state passed to his son, Philip II of Spain.

In Stevin's day the Low Countries were among the richest lands in Christendom. There was immense wealth in the great trading metropolis of Antwerp, where Stevin was to work for a time as a bookkeeper and cashier, and cities such as Ypres, Ghent, Brussels, Mechelen and Stevin's birthplace, Bruges, still dazzled with their splendour. There was considerable artistic and mercantile activity: a vast output of painting and sculpture, a flourishing textile, tapestry and damask industry, capacity shipbuilding, cannon and bell-founding, beer-brewing, coalmining and agriculture. There was bustling maritime business, with major ports at Antwerp, Flushing and Rotterdam. There was also a vigorous intellectual life. In the sixteenth century the language of scholarship and letters was Latin, but there was a growing tendency toward the use of the vernacular. It was already employed in the literary activities of the chambers of rhetoric, amateur societies set up in most towns in the Low Countries to promote the practice of rhetoric through theatrical performances and poetry competitions and to provide edifying entertainment. Simon Stevin was very likely a member of the Bruges chamber known as 'The Holy Ghost', and his interest in his own Dutch language and his desire to promote its use is ever more explicitly expressed in his own works.

New religions

By the sixteenth century there were many who deplored the laxity and excessive display of the Catholic Church and sought an alternative. The Council of Trent (1545-1563) gave structure and direction to the process of change within the Church, but Protestant reform had already taken root in the Low Countries. Martin Luther found followers, particularly in merchant cities trading with Germany, such as

Antwerp. From shortly after 1530 the Anabaptists rapidly gained support in the northern provinces. But the most significant reformist movement as far as the history of the Low Countries is concerned was Calvinism. By the end of Charles V's reign John Calvin's teachings had permeated the Low Countries and the Calvinists soon became the largest and best-organized Protestant group. There is reason to suppose that Stevin's mother was a Calvinist, and some have even conjectured that Stevin himself left his home town of Bruges for reasons of faith, though evidence to support this idea is lacking.

The government made powerful efforts to suppress these reformist movements. Charles V imposed more stringent heresy laws in the Low Countries than anywhere else in his vast empire and Philip II zealously enforced them. Heretics were persecuted and brought to judgement before a range of tribunals, civic, royal and ecclesiastical.

Repression, resistance, Revolt

At the chapter of the Order of the Golden Fleece held at Ghent in 1559 Philip II, the duke of Alva, William of Orange and the counts Egmond and Horne sat together in apparent amity as fellow knights of the Fleece. Philip then returned to Spain, where he remained. He appointed his half-sister Margaret of Parma as regent, though her authority was nominal and real power lay in the hands of Cardinal Granvelle, primate of Mechelen, much to the ire of other councillors of state who felt excluded from their rightful share in the country's government. Granvelle pursued a rigorous policy of religious repression, driving Protestants to desperation and affronting the civic magistracies who saw inquisitors encroaching on their jurisdictions. A group of noblemen led by William of Orange and counts Egmond and Horne petitioned for Granvelle's removal, to which Philip agreed in 1564. In 1566 Egmond was sent to Madrid with the States General's appeal for the suspension of the heresy laws and devolution of more power to the Council of State. This Philip refused, and the first signs of organized resistance began to appear in the Low Countries. It took shape at the highest level. On 5 April 1566 a group of 200 nobles presented the regent with a similar plea. A scoffing courtier referred to the petitioners as *gueux* – beggars – a name that would become a rallying cry. None the less, the heresy laws were slightly relaxed and Calvinists began to meet more openly, sometimes coming together in their thousands to hear field-preachers. In July Margaret banned such gatherings; the following month Calvinist frustration came to a head in the 'Iconoclastic Fury'. In churches throughout the Low Countries travelling bands of image-breakers smashed or defaced carvings and paintings, and burned books and vestments. Leading Catholics were threatened or assaulted and in several places Catholic services were suspended.

Eventually, order was restored. Margaret demanded a new oath of obedience from the nobles and few hesitated to take it, destroying hopes of a united resistance to royal policy. Though protesting his loyalty to Philip, William of Orange did refuse the oath and prudently withdrew to the family castle at Dillenburg in Nassau. But even as the commotion was abating, Alva was advancing with an army, sent by Philip to reassert control over his rebellious Seventeen Provinces.

In 1567 Margaret retired and Alva assumed the position of governor-general of the Low Countries. Philip had invested him with unlimited powers for the extirpation of heretics and he at once set up an emergency tribunal called the Council of Troubles, soon to be popularly known as the Council of Blood. Egmond and Horne were among the council's first victims, their right to trial before their peers as knights of the Golden Fleece brushed aside. They were executed on the Grand' Place in Brussels, along with some twenty other nobles, in July 1568, a deed that sent a shock wave throughout the Low Countries. Calvinist chroniclers attribute tens of thousands of executions to Alva. This is an exaggeration, but thousands of death sentences were passed before he was recalled to Spain in 1573.

By the mid-sixteenth century repression and persecution had already caused a certain number of Protestants to emigrate. The reprisals that followed the Iconoclastic Fury brought about a massive exodus of émigrés who left the Low Countries for England, Germany and France.

When William of Orange returned to his German estates it was to assemble troops and supporters to continue the resistance to Spanish rule. He invaded the Low Countries in 1568 and won the first battle of the Revolt on 23 May, though he was soon forced to withdraw. This marked the beginning of the Eighty Years' War, in whose shadow Stevin was to live his entire adult life. As a speedy means of creating a navy Orange issued letters of marque to Calvinist privateers, enabling them to attack Spanish shipping under his princely authority. Between 1568 and 1572 the 'Sea Beggars' played a significant role in the rebellion. Operating from East-Friesland, from English ports and the Huguenot sea town of La Rochelle, they made the seas unsafe for the Spanish. In April 1572 they seized Den Briel, then Flushing, using them as bases from which to take a number of other towns. By the end of the year most of the towns of Holland and Zeeland had declared for Orange and were in rebel hands, often thanks to the help of their inhabitants and to the 'Wood Beggars', commando groups of land-based Calvinists.

In July 1572, William of Orange called together the States of Holland at Dordrecht and formed an embryonic government of the rebel provinces. He was acknowledged as stadholder and captain-general of Holland, Zeeland and Utrecht, albeit nominally on behalf of the king. In the autumn of that year Alva sent a punitive expedition northwards: Mechelen, Zutphen, Naarden and Haarlem fell and were brutally plundered, their inhabitants massacred. But at Alkmaar Alva's troops were repulsed, and in the Zuiderzee his navy was defeated by the Sea Beggars. Leiden determinedly held out against besieging Spanish forces for almost a year, until Orange finally forced the royal troops to withdraw by the desperate expedient of breaching the dikes. In 1575 Orange founded Holland's first university there, to be a spiritual and intellectual bastion of Calvinism. Simon Stevin and Orange's son, Maurice of Nassau, were both among its earliest students; their long association may date from this time.

Reconciliation? The Pacification of Ghent

The Revolt took its toll on rebel and loyalist alike. Economic depression set in. Those who remained loyal to the crown saw the decline in trade and were dismayed by seemingly unreasonable orders from Spain. They turned envious eyes on Holland and Zeeland's self-government and began to urge negotiations. Don Luis de Requesens had taken Alva's place as governor-general in 1573, but it was a further two years before Philip authorized him to open official negotiations with the rebel northern provinces. The formal talks that took place in Breda in the spring of 1575 made clear the irreconcilability of king and rebels on two essential points – religion and form of government. The rebels declared that they had no wish to be anything other than the Philip's loyal subjects, but he must allow Protestantism and give the States a share in government. Philip would concede neither of these conditions. Negotiations collapsed and battle was rejoined.

In 1576 the Revolt entered a new phase. The Spanish treasury was bankrupt and the army in the Low Countries could not be paid. Widespread mutiny followed, and the Spanish soldiery became the most serious danger to the provinces they had been sent to protect, sacking and looting in an orgy of violence. Requesens's sudden death amid these difficulties added to the crisis. Left without a head, the Council of State was powerless.

The States of Brabant now took autonomous action and called for a convention of the States General – an unlawful deed, for only the sovereign had the right to do this. The assembled States immediately agreed an armistice with the Sates of Holland and Zeeland and entered negotiations which, while not repudiating royal authority, were intended to lead to the removal of the hated Spanish troops from the land and the restoration of peace in the Seventeen Provinces. Ghent was chosen as a venue. At the end of October 1576 agreement was reached and a cease-fire or 'pacification' was declared. By its terms the loyalist southern provinces agreed to join with Orange and the States of Holland and Zeeland in driving out the mutinying Spanish troops and setting up a provisional government under a single States General. William of Orange's position as stadholder of Holland and Zeeland was confirmed. As to the vexed issue of religion, things would remain as they were: public practice of Protestantism would be allowed in Holland and Zeeland, the other provinces would remain Catholic, though royal edicts concerned with the suppression of heresy were suspended and private Protestant practice was now allowed. With the Pacification of Ghent it seemed as if the civil war was over at last and harmony was restored to the Low Countries.

The parting of the ways
Although the rebel and loyalist provinces were now ostensibly in agreement there was still an immense gap between them. The southern provinces were willing to accept Philip's new governor-general, Don John of Austria, provided that he sent away the Spanish troops and upheld the Pacification of Ghent. Don John agreed, and signed the 'Eternal Edict' with the States General in February 1577. But the edict specified that Catholicism was to be maintained throughout the country and made no guarantees for the Reformed Church. Holland and Zeeland promptly withdrew from the States General. The reconciliation was over.

The already complicated situation became more involved still. By the autumn of 1577 a complex power play was in progress between Don John, seeking to rebuild royal authority, the States General, willing to settle with the king so long as Philip made some political concessions and kept the Spanish soldiery away, and William of Orange with Holland and Zeeland. Added to this was the growing split between moderates and radicals in the south, and the emergence of a group of 'Malcontents' led by Catholic nobles who, while not willing to be reconciled with the king, were opposed to the States regime. Don John set up his headquarters in Namur and recalled the Spanish troops, pushing the States General towards alliance with Orange. In September William of Orange entered Brussels and was enthusiastically received. In October opponents of Orange in the States General brought a young Habsburg prince and nephew of Philip of Spain, Archduke Matthias, to the Low Countries and appointed him 'governor-general'. In the following months there were a number of Calvinist and Orangist coups in the towns and cities of the Low Countries, and fresh outbursts of iconoclasm.

In January 1578 Alexander Farnese, prince of Parma, and nephew of Philip of Spain, arrived in the Low Countries at the head of a new army. He succeeded Don John as governor-general and began his campaign to bring the Low Countries to heel.

In January 1579 Arras, Hainaut and Gallicante Flanders signed the Union of Arras, upholding the spirit of the Pacification of Ghent but declaring for Catholicism and obedience to the king's authority. Holland and Zeeland, together with Utrecht, Friesland, Guelders, the Ommelanden and the towns in Flanders and Brabant that had come under Calvinist control the previous year, such as Ghent, Ypres, Antwerp, Bruges and Mechelen, responded with the Union of Utrecht, which was intended to forge a close confederacy of States without splitting with the States General. Not all who signed did so without misgivings, for it was regarded as anti-Catholic and as such was likely to drive non-signatories into the arms of Spain. Even so, the creation of the Union of Utrecht marked the real beginning of the Dutch Republic.

In a move to make the northern provinces more accommodating to Catholicism as well as securing outside help against Parma, Orange urged the States General to offer sovereignty, albeit with constitutional conditions, to the duke of Anjou, younger brother of the king of France. A treaty between Anjou and the States General was signed in September 1580 and Anjou was proclaimed 'prince and lord of the Netherlands' in January 1581. In acquiring a new sovereign it was necessary to be rid of the old one: by July the States General had agreed the text of the Act of Abjuration, by which Philip II and his heirs were repudiated forever. The Union of Arras responded by recognizing in full the king's sovereignty and Parma's authority as governor-general. Philip attainted William of Orange and put a price on his head. In July 1584, at his quarters in Delft, he was assassinated.

Meanwhile, Parma's reconquests continued and the territory of the Union of Utrecht steadily shrank. Town after town was taken. With the fall of Antwerp in Augusts 1585 the whole of the southern Low Countries was back in Philip's hands. Protestants were allowed a period of grace in which to consider their position: either they could return to the Catholic fold or they could leave the country. Many chose to do the latter, to the great impoverishment of the south. Tens of thousands, both of the intellectual and financial elite and the artisanry, took refuge in the provinces of the Union, taking with them their capital and international commercial contacts, their specialized crafts and skills. Almost half Antwerp's population left. The great trading metropolis lost its leading mercantile role to Amsterdam, and in the smaller towns trade dwindled to next to nothing.

The Spanish Netherlands and the Dutch Republic

As the new sovereign of the United Provinces Anjou had found himself frustratingly circumscribed. He attempted to seize greater power through a military coup, failed, and left the country. The rebel regions now turned to England for help. Elizabeth declined sovereignty but sent military assistance and nominated the Earl of Leicester to be 'governor-general' and commander of these forces. In 1586 Leicester established himself in Utrecht, but his actions and the conduct of the English troops caused considerable friction and the following year he returned to England. The provincial States of Holland, Zeeland, Utrecht, Guelders, Overijssel, Friesland and Groningen decided to have done with foreigners and declared themselves the bearers of sovereignty within their own borders. France and England signed treaties with the United Provinces, recognizing them as an independent state. The Dutch Republic had come into being.

Parma's northwards campaign was brought to a halt by Philip's preparations for an invasion of England, and he moved instead on Ostend and Sluis. The defeat of the Spanish Armada should have allowed him to turn his attention back to the rebel northern provinces, but now Philip ordered him to France, to support the Catholic opposition to Henry IV, and he met his death there in 1592.

In 1598 Philip II of Spain died. Shortly before his death he bequeathed the 'obedient provinces' of the Netherlands to his daughter Isabella and saw her married to Archduke Albert of Austria, who had assumed the governor-generalship of the Low Countries two years earlier. The two, referred to as the 'Archdukes', were to rule jointly, though the Spanish army, paid for by the Spanish treasury, was to remain and its commanders were to be appointed by the king of Spain. It was further stipulated that the Catholic faith must be defended and heretics extirpated, and that the Netherlands would revert to the Spanish crown if either Albert or Isabella died without producing a legitimate heir, which turned out to be the case. Thus the southern Netherlands continued to be perceived as the 'Spanish Netherlands'.

William of Orange's second son, Prince Maurice of Nassau, had been appointed stadholder of Holland and Zeeland in 1585. In 1590 and 1591 the States of Utrecht, Guelders and Overijssel also named Maurice stadholder and captain-general of their provinces. He had already proved himself a competent military leader. Now he became commander-in-chief of the Republic's army, and in the course of the next few years regained many of the losses in the northern Netherlands, driving out the last royalist garrisons in 1597. In 1600 Maurice invaded Flanders. Having taken a number of towns along the coast with little opposition he clashed with crack Spanish forces led by Archduke Albert at Nieuwpoort. Though his tactical skill carried the day his campaign was brought to a halt: both sides hailed the Battle of Nieuwpoort as a victory. The war reached a stalemate. None the less, in the following years the Republic's army was tripled in size and top military engineers – including Simon Stevin, who had by now become Maurice's quartermaster – devised highly sophisticated fortifications.

By 1604 both France and England had made peace with Spain, leaving the Republic without strong allies. But the treasuries of Spain and the Spanish Netherlands were exhausted and the Spanish army in the Low Countries was subject to recurrent mutinies. In 1607 an armistice was agreed and talks began. Yet once again the Republic's demand to be recognized as an independent sovereign state and Spain's demand for Catholic equality in the Republic seemed insuperable obstacles. Moreover, Spain demanded that the Republic put an end to its profitable trading with the East Indies. Finally, in 1609, a Twelve Years' Truce was signed, Spain grudgingly agreeing to treat the Republic 'as if' it was a sovereign state. The rest of Europe and the Ottoman court regarded the Truce as the legitimization of the Dutch Republic. The sundering of the Low Countries of Stevin's childhood was effectively complete.

Golden Age

The years of the Truce allowed both sides to recover and rebuild. The economy of the Spanish Netherlands regained momentum and trade picked up again. A creative army of artists, architects and specialized craftsmen of all kinds was recruited to make good the damage caused by war and iconoclasm. The Jesuits built churches and opened schools throughout the country, and the issue of poverty was addressed. Under the Archdukes the early seventeenth century was an artistic golden age.

But the recovery of the south did not begin to rival the meteoric rise of the north. The Dutch Republic was entering its own golden age of mercantile predominance, artistic splendour and scientific achievement. The influx of refugees from the Spanish Netherlands in the previous decades had given the commercial, artistic and intellectual life of the break-away states a massive boost. Amsterdam had replaced Antwerp as the greatest mercantile metropolis in Europe. Fleets had set out in search of new markets and by the time of Stevin's death in 1620 the Dutch Republic would be trading with the East and West Indies, the Far East and the Arctic North. The United East India Company – the Vereenigde Oost-Indische Compagnie, or VOC for short – provided the foundation of Dutch colonial power in East Asia. There was a vigorous artistic life. Literature flourished. New seats of learning opened and scientific enquiry pushed forwards the frontiers of knowledge. In this the polymath Simon Stevin of Bruges played a leading role, as we shall see in the following pages. His life and the Twelve Years' Truce came to an end more or less at the same time, but it can be argued that as Maurice of Nassau's friend and mentor he contributed directly not only to many branches of science but to the establishing of the Dutch Republic itself.

Chapter One

Simon Stevin and the Renaissance

The Renaissance was a high point in the history of visual arts, architecture and literature. In science, however, it was a transitional period. Increasing awareness of the knowledge of classical antiquity gave rise to important new impulses and inspirations. Advances resulting from the interaction between scholars and practicians provided the foundation for the scientific revolution of the seventeenth century. The work and character of Simon Stevin were entirely in tune with this prelude to the Modern Age.

1. Middle Ages, Renaissance and the Modern Age...

1.1. The transition from the Middle Ages to the Renaissance was a gradual one

The transition from the Middle Ages to the Renaissance, like that from the Renaissance to the Modern Age, is often described as a steady and continuous process. We see a gradual transformation of the characteristics that dominated the society and its working. Expanding scales and changing worldviews were two of the catalysts that precipitated the shift from the Middle Ages to the Renaissance. The invention of printing midway through the fifteenth century, the voyages of discovery of the fifteenth century and the economic prosperity that went hand in hand with the great population increases of the sixteenth century were all factors that contributed to the functioning of society on a larger scale. The growth of towns and cities, the emergence of regional centres of power, the rise of new monarchies and the development of universities were likewise important elements of change.

2. Printing

2.1. From the Officina Plantiniana in the Netherlands scientific knowledge spread across the whole of Europe

The impact of printing can scarcely be overestimated. Block printing had existed in China since 770, and it was from that country, too, that the knowledge of paper-making spread. The invention of printing brought several techniques together – the oil paint used by the Flemish Primitives provided the basis for printing ink, for instance. In the mid-fifteenth century Johannes Gutenberg, Johann Fust and Peter Schöffer, experimenting with a printing press and loose type whose characters were separately cast in lead and set in a forme, overcame the final hurdle. The invention of printing with moveable type brought about a revolution in communications as significant as the invention of writing itself – it changed the world.

The Netherlands were at the forefront in the development of printing. Indeed, the spread of knowledge in the Low Countries was largely due to the printers' industry. In the first half of the sixteenth century, Antwerp became the most important centre for book production in the Netherlands, with 66 printers producing 2,254 publications. Jan and Hendrik Van der Loe set up their printing office a little before Christopher Plantin. Here, between 1552 and 1578, seven works by Rembert Dodoens were printed. Later, this botanist would call on the skill of Plantin.

Christopher Plantin (Saint-Avertin [Tours] *c.* 1520–Antwerp 1589) was not a scholar himself, yet he was responsible for spreading throughout the whole of Europe, and beyond, the scientific and scholarly work produced in the Low Countries. In 1549, he settled in Antwerp. His printing house, the Officina Plantiniana, became the largest typographic enterprise in Europe; in its most prosperous period, at least 16 presses were in operation. In his 34-year career, Plantin published some 1,500 books. He was renowned as the printer of the monumental *Biblia Regia*, an extensively glossed eight-volume edition of the Bible in five languages: Aramaic, Greek, Latin, Hebrew and Syriac. Plantin was appointed *architypographus* to Philip II, king of Spain, and obtained a monopoly on the production of all liturgical and devotional works to be sold in Spain and her vast colonies. Within a few years, thousands of liturgical works destined for Spain were printed in the Antwerp Officina Plantiniana.

In the aftermath of the Spanish Fury – the sacking of the city by mutinying Spanish troops – and the subsequent formation of a Calvinist town council, Plantin's printing office went into a decline. The lucrative export to Spain came to a standstill and the printer looked about him for another base of operations. In the spring of 1583, on the advice of his close friend, the humanist scholar Justus Lipsius, he moved to Leiden where he was appointed as the newly founded university's printer. In Antwerp his sons-in-law, Jan I Moretus (Jan Moerentorf, 1543–1610) and Franciscus Raphelengius (Frans van Ravelingen, 1539–1597), kept the Officina Plantiniana going and in the two years that Plantin remained in Leiden, 120 publications were printed on its presses. When Antwerp fell to the Spanish governor-general Alexander Farnese, Plantin returned to his adopted city, where he would die. Frans van Ravelingen, a convert to Calvinism, took over the Leiden branch. After Plantin's death, the Antwerp printing house continued to operate under Jan I Moretus. Here, mostly liturgical and devotional works were printed while scientific, scholarly and classical works issued from the Plantin press in Leiden (fig. 1).

Figure 1: The invention of printing in the mid-fifteenth century led to an unprecedented diffusion of ideas and knowledge. Jan van der Straet (Johannes Stradanus, 1523–1605) designed this print as one of a series devoted to contemporary inventions. It was engraved by Theodoor Galle.
Bruges, Municipal Museums, Steinmetz Cabinet, © Reproductiefonds.

3. The universities

Although the schools of Athens and Alexandria are sometimes described as universities, it is generally held that universities as we know them today originated in Europe. They evolved from medieval schools that were open to scholars from all over Europe and were truly cosmopolitan. Initially the universities were private associations. In the thirteenth century the more structured *universitas studii* came into being, and in the fourteenth century the university became an officially recognized corporation of teachers and scholars, with charters and privileges, known as a *studium generale*. The University of Bologna, regarded as the oldest, dates back to the end of the twelfth century while Paris, Oxford and Cambridge came into being in the thirteenth century. Prague, Heidelberg, St Andrews, Marburg and Uppsala are likewise among the earliest universities.

The history of the university is complex. The modern university is probably more closely related to the medieval school with its faculty-based teaching than to the university of the Renaissance. Humanism made its entrée into these centres of learning only after it had developed outside the university walls. The careers of Copernicus, Kepler, Stevin and Gilbert show that scientific discovery could take place outside of universities just as well.

The universities suffered greatly during the Reformation, which went hand in hand with considerable dissension. In that period they lost a great deal of their international character.

3.1. From 1426 Leuven was in the ascendant in the Southern Netherlands. 150 years later, Leiden became a Calvinist bastion in the north

What was the situation as regards the universities in the Low Countries? In the southern provinces, Leuven was the most important institution. With the bull *Sapientiae immarcescibilis* of 9 December 1425, Pope Martin V founded the first university in the Low Countries at Leuven at the request of the city's magistracy and the chapter of St Peter's, supported by Duke John IV of Brabant. The privileges he granted to the *studium generale* ensured an almost total independence from the temporal authorities. The university at Leuven acquired five faculties: arts, medicine, civil law, canon law and (from 1432) theology. Courses began on 2 October 1426. Over the next 150 years the *studium lovaniense* became the most eminent scientific and scholarly institution in Europe. Leuven's university had a veritable monopoly on education.

Entry to the higher faculties was via the faculty of arts. Students arrived at the university at around the age of 15, after they had completed Latin school, and were matriculated into the arts faculty. Subsequently, they went on to a superior faculty – either theology, where their studies lasted 10 or 11 years, or medicine or law, each requiring 6 years of study. Teaching was given in four colleges: *Porcus* (the Pig), *Falco* (the Falcon), *Castrum* (the Castle) and *Lilium* (the Lily). Four public chairs were added to the existing programme: rhetoric, ethics, dialectics and mathematics. The Lily was the first to embrace humanism. Here, the predilection for ancient languages and literature was stimulated by Desiderius Erasmus. The Rotterdam-born scholar stayed twice in Leuven, from 1502 to 1504 and again from 1517 to 1521, during which period the *Collegium Trilingue* was created for the teaching of Greek, Latin and Hebrew. The study of ancient languages and literature was seen as a necessary introduction to the study of not only theology but also medicine and the sciences.

Many French-speaking students went to universities in France. To achieve the dual purpose of halting this drain from the Spanish Netherlands and controlling a French-language centre from which to combat Calvinism, Philip II founded a university at Douai (a part of Flanders at that time) in 1562. However, this university never attained the same lofty reputation as that enjoyed by Leuven.

The first university in the northern Netherlands was established in Leiden, which owed its selection as a university town to its favourable strategic location. Students now had no need to go to the Catholic Leuven or other foreign seats of learning. Leiden's university, founded on 8 February 1575, was intended to be a spiritual bastion of Calvinism in the northern provinces.

4. The Renaissance in art, literature, science and technology

The tension between a religious focus on the life to come, propagated by the Church in the Middle Ages, and a secular focus on the present and an interest in the natural world, was a characteristic of the Renaissance. This latter viewpoint was inherent to humanism, one of the essential factors of the Reformation. The two contrasting worldviews collided. The printed word played a crucial role in the argument – though that does not alter the fact that the conflict between the ideologies of Church and Reformation was still largely fought out on the battlefield.

The Renaissance used to be seen as the revival of culture following the 'dark' Middle Ages, an image that has subsequently undergone some adjustment. The Renaissance evolved organically from the achievements of the Middle Ages, with their sophisticated intellectual environment and network of developing towns and cities on the one hand, and from intense contacts with the Islamic world via the crusades and the kingdoms of Spain on the other.

Paradigms that characterize the Renaissance include

> 'the State as a work of art'
> 'the development of the individual'
> 'the discovery of the world and of man'
> 'the rediscovery of a body of knowledge from classical antiquity', via Arabic translations of Greek works, for instance.

4.1. A new lease of life for ancient texts

In painting and sculpture a style developed – initially in Italy in the fourteenth century and later in the north – that was typified by a new handling of space and human proportion. Leonardo da Vinci, Michelangelo, Raphael, El Greco, Rubens and many others were exponents of this style.

Literature provided a tremendous impetus to the development of the Renaissance. Fortunately a number of important classical literary and scientific – especially mathematical – texts had survived the Middle Ages, mostly in the libraries of monasteries and cathedrals. But there was a far greater acceleration in the 'rediscovery' of the texts and ideas of classical antiquity in the Renaissance than during the Middle Ages.

In the fourteenth century Petrarch laid the basis for the classical humanities. He believed that to understand classical civilization it was necessary to study the languages in which Plato and Socrates, Apollonius and Archimedes, or Cicero and Virgil had expressed their thoughts. To Petrarch and the humanists, 'rhetoric' had a pivotal role. Petrarch's pursuit of *eloquentia et sapientia* influenced the humanists in their setting up of study programmes. They were inspired by the system of the seven *artes liberales* that originated in Rome in the fifth century. The introduction of the 'humanities' was coupled with the revaluation of original Greek and Latin texts. Here too the invention of printing had a great impact. Whereas previously one or, at best, two or three copies of the texts of the ancient Greeks and Romans were kept in a library, during the Renaissance, they could be made more widely available.

The Italian Renaissance produced new literary methods and models. In addition to Petrarch, Machiavelli and Guicciardini were among the most original Renaissance thinkers.

Remarkably, the Renaissance, which so often sought its inspiration in classical antiquity and in Latin and Greek, also went hand in hand with the rise of the vernacular; thus, Renaissance culture renewed contact with its roots. The combination of new literary forms with popular tradition resulted in the brilliant works of writers such as Rabelais, Cervantes and Shakespeare.

Renaissance ideals and educational models spread northwards beyond the Alps. From the second half of the fifteenth century cities such as Paris, Antwerp, Augsburg and London grew by leaps and bounds. Erasmus was one of the most influential humanists of his time. He was an early advocate of religious tolerance. With his ideal of forbearance, so strikingly interpreted in *Praise of Folly* (1509) with its play of gentle humour and biting satire, he was too far ahead of his time. The same can be said of the ideas his friend Thomas More expressed in *Utopia*.

Tolerance was not a hallmark of the reaction to the Reformation. Emanating chiefly from Spain religious wars and inquisitions were let loose upon Europe and outbreaks of witch-hunting and torture would make the Flanders of Pieter Bruegel into a kind of Vietnam of Europe.

The turning of the Renaissance to classical antiquity was a boon to science. Renaissance scholars studied and cultivated the brilliant contributions made to mathematics, statics and hydrostatics by the great Greek mathematicians – Thales of Miletus, Euclid, Pappus of Alexandria, Apollonius of Perga, Menelaos and, above all, Archimedes. Their surviving texts provided the base for the advances made by Tartaglia, Cardano, Stevin and others. What a catastrophe it would have been for our civilization had the contributions of Greek mathematicians – conics, for instance – been lost! Would this knowledge have evolved anew? That in 1544 Renaissance scholars translated and published a number of hitherto unknown works by Archimedes has had a significant impact on the dissemination of knowledge.

The principal contribution to cosmology during the Renaissance was Copernicus's *De Revolutionibus Orbium Coelestium* (Nuremberg, 1543). Those who study this pioneering work are struck by its concordance with Greek ideas. Copernicus's formulations of problems and working methods are those of classical Greece. His basic principles – that the motion of the heavens could be explained with the help of uniform circular orbits, for instance – are also Greek concepts. In cosmology Copernicus took up where Ptolemy left off, just as Stevin carried on from Archimedes in mechanics and hydrostatics.

De Revolutionibus Orbium Coelestium was a milestone in the history of science. Kepler would build on it and discover mathematical regularity in the orbits of the planets. Kepler was also involved in another important Renaissance reform, that of the calendar. Although a Protestant he publicly supported the reform of Pope Gregory XIII. In Graz, where he held the post of *Landschaftsmathematicus* or district mathematician, he was required to produce a calendar (with a horoscope!) every year. In 1595 two of his prognostications were borne out. We know that the winter of 1595 was exceptionally cold from a painting by Pieter Bruegel the Younger (fig. 2), and Kepler's prediction of a renewed Turkish offensive also proved to be accurate.

4.2. The Renaissance was the prelude to the great scientific revolutions of the seventeenth century

In addition to Copernicus and Kepler, mathematicians such as Nicolo Tartaglia, Giovanni Battista Benedetti and Girolamo Cardano were also among the creative spirits who laid the foundations for the later scientific revolutions. In mechanics the Spanish Dominican Domingo de Soto and Simon Stevin in particular contributed significant insights. Somewhat earlier, Renaissance artists such as Piero della Francesca, Leon Battista Alberti and Leonardo da Vinci had also contributed to the development of science. The

Figure 2: Pieter Bruegel the Younger (c. 1564–1637/38) painted the bitter winter of 1595, which was forecast by Kepler. The painting 'Return from the Inn' (c. 1620) is now in *The Montreal Museum of Fine Arts*, Gift of the Maxwell family in memory of Mrs Edward Maxwell. Photo: *The Montreal Museum of Fine Arts*, Christine Guest.

geographer and cartographer Gerardus Mercator should also be mentioned here. And in addition to the mystic elements in his work, Nicholas of Cusa (Nikolaus von Kues or Cusanus) also propounded progressive scientific ideas. Gilbert, Beeckman, Descartes, Galileo and Huygens, who would take the next steps in the development of science, already belong to the modern period. The actual scientific revolution took place only after the Renaissance, in the seventeenth century. Newton's seminal work *Philosophiae Naturalis Principia Mathematica* (1687) was undoubtedly pivotal in that development. Yet, without printing and the spread of ideas on an unprecedented scale, the scientific revolution, with its international interaction between numerous scholars, would have been impossible.

Technological developments also accelerated the advancement of science during the Renaissance. In the fifteenth and sixteenth centuries technicians and engineers made remarkable strides, even though their perceptions were still rooted in the Middle Ages. The technicians' practice fertilized science and vice versa. In this respect, the development of the sciences during the Renaissance differs from its growth since classical antiquity. Praxis was conveyed and expanded by artists who, like Leonardo da Vinci, were sometimes also engineers, or by architects who were also involved with town planning, hydrology or fortifications. Ever greater demands were made of the instruments required in building, navigation, warfare, astronomy, voyages of discovery, the performance of music and so on. Gradually, the need was felt to provide a sound scientific foundation for the empirical knowledge of the *ingenieri* and the craftsmen. This was seen first in mathematics.

Contemporary historians have meanwhile come to agree that the work of 'practicians' and naturalists such as Leonardo da Vinci and Stevin was as important to the ideas of the Renaissance as the contributions of the humanist men of letters.

5. Simon Stevin: Scholar and Engineer

Undoubtedly, the Renaissance represented a high point in the fields of visual arts, architecture and literature. One has only to recall the names of visionary painters and sculptors such as Michelangelo, Raphael, Bernini, Van Eyck, Bruegel and Dürer, of an architect such as Brunelleschi, of poets and writers like Dante and Shakespeare. In science, on the other hand, the Renaissance was a transitional period. A broadening awareness of the body of knowledge of classical antiquity gave rise to significant new impulses and inspiration. The interaction between scholars and practicians brought about advances that would make the scientific revolution of the seventeenth century possible.

The work and character of Simon Stevin were entirely in tune with this period leading up to the Modern Age. He was very familiar with the literature of classical antiquity and was one of the first, after Leonardo da Vinci, to read and comprehend the works of Archimedes, Diophantos and others (box 1). Just as Copernicus had taken up the thread of the Greek astronomy of Ptolemy in his pioneering work *De Revolutionibus Orbium Coelestium*, so Stevin exploited new areas of statics and hydrostatics, expanding on the work of Archimedes. Stevin was also an exponent of the interaction between 'scholars' and *ingenieri*. In 1586, he authored the works *De Beghinselen der Weeghconst, De Weeghdaet, De Beghinselen des Waterwichts, Anvang der Waterwichtdaet (The Elements of the Art of Weighing, The Practice of Weighing, The Elements of Hydrostatics, Preamble of the Practice of Hydrostatics)*, which opened up new vistas. In these, he coupled physics and mathematics with the praxis of mechanics, building construction, seafaring and so on. Moreover, through the multiplicity of his contributions, Stevin matches the ideal Renaissance image of *homo universalis*, of which Leonardo da Vinci is regarded as the prototype. Stevin himself specifically referred to the importance of combining theory and practice in his *Spiegheling en daet* (see e.g. K. van Berkel (1985)).

> **Box 1: How did Stevin know the works of the ancient Greeks?**
>
> We still possess original clay tablets on which mathematical symbols can be seen just as they were inscribed some 3,700 years ago by Egyptians and the Babylonians of Mesopotamia. Classical Greek mathematics developed as early as *circa* 600 BCE. Thales of Miletus was one of its earliest practitioners. It might therefore be expected that original Greek texts would be found as well. Unfortunately, this was not the case. The works of Thales of Miletus (*c.* 600 BCE), Pythagoras (*c.* 540 BCE), Euclid (*c.* 300 BCE), Apollonius of Perga (*c.* 262–190 BCE), Archimedes (287–212 BCE) and others have not come down to us directly. This is because from 450 BCE onwards, the Greeks wrote on papyrus. Papyrus, already in use since 3000 BCE, was very vulnerable and rotted easily. Scrolls could survive only in a very dry environment.
>
> Fortunately, those works of Greek scholars that were thought important were copied over and over again. In the process of copying, however, the original text may have undergone changes. For instance, a copyist with no technical knowledge of the material being copied can easily make mistakes. If, on the other hand, he is very familiar with the topic he might be inclined to add elements based on later knowledge that the writer of the original text did not possess. In either case, the parent text is not passed on to us. Codices – manuscripts gathered together in a binding – were not in common use until the fourth century (though some date from the second century) (figs. 3a and 3b).

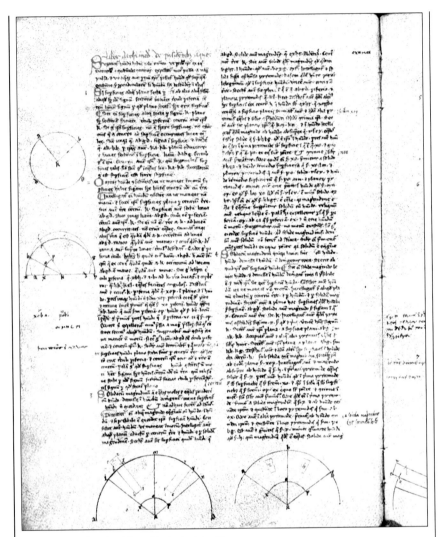

Figure 3a: *Liber Archimedis de insidentibus aque...*
First page of William of Moerbeke's Latin translation of Archimedes's *On Floating Bodies*. William of Moerbeke (c. 1215–1286), a Flemish Dominican, was a prolific and accurate translator of philosophical, medical and scientific texts from Greek to Latin. His translations contributed to the beginnings of the Renaissance movement. In 1269, while at the papal court at Viterbo, he translated works by Archimedes with commentaries by Eutocius (480–540). From 1277 until his death, William of Moerbeke was the archbishop of Corinth.
© Biblioteca Apostolica Vaticana (Vatican). Ottobonianus latinus 1850, fol. 55^V (ektachrome: JTD).

Figure 3b: *Recta portio rectanguli...*
Here, in the same translation by William of Moerbeke, as seen in fig. 3a, are Archimedes's theorems describing the equilibrium of a floating boat. He proposes that the section of the boat has a particular shape (namely that of a straight segment of a paraboloid). The translations of classical Greek works made by William of Moerbeke, Clavius (1538–1612), Federigo Commandino (1509–1575) and others provided a treasury of ideas in mathematics and the sciences for Renaissance scholars and engineers such as Leonardo da Vinci and Simon Stevin.
© Biblioteca Apostolica Vaticana (Vatican). *Ottobonianus latinus 1850, fol. 57ʳ* (ektachrome: JTD).

The oldest surviving complete copy of Euclid's *Elements* dates from 888 CE. This text was copied by the scribe Stephanus for Arethas, the bishop of Caesarea, in Cappadocia. Older fragments of the *Elements* do exist, some dating back to 225 BCE. Remarkably enough, the first versions of the *Elements* to appear in Europe in the Middle Ages were not Latin translations of the aforementioned Greek texts. At that point, the only known texts were those that had been translated into Arabic.

So how did Simon Stevin come to know the works of the ancient Greeks, especially those of Archimedes? Here, another Fleming, the Dominican William of Moerbeke (c. 1215–1286) played a crucial role. Between 1268 and 1280 he worked at the papal court at Viterbo, where he translated several Greek works, especially the writings of Aristotle and Archimedes, into Latin. Knowledge of Latin was fairly common in those days, at least among intellectuals. Familiarity with Greek, on the other hand, was quite exceptional. For his translation of Archimedes's works, William of Moerbeke used the so-called codex A. This codex was compiled in the ninth century in Constantinople by Leo of Thessaloniki, a great scholar of the Byzantine school. Even though few significant mathematical creations originated from Byzantium, there was great interest there in the important mathematical and scientific contributions of the past. Archimedes sent his mathematical findings to Erathostenes (c. 230 BCE) in Alexandria, among others. From the works of Heron (c. 75 CE?), Pappus (c. 300) and Theon of Alexandria (c. 390), it appears that more of Archimedes's works were known in the third and fourth centuries CE than are known today. Byzantine scholars clearly had a keen interest in Greek works, especially in those of Archimedes. The sixth-century mathematician Isodorus of Miletus, for instance, studied and taught these works.

In addition to codex A, we also know of codex DD and codex C (this terminology was introduced at the beginning of the twentieth century by the Dane, Johan Ludwig Heiberg, who made a profound study of Greek manuscripts). Codex DD, also from Constantinople, contains works by Archimedes on mechanics and optics, and was used in the thirteenth century for Latin translations. Codex C is the famous manuscript that was discovered at the end of the nineteenth century. The history of these codices is interesting. Codex A, which had already been translated by William of Moerbeke in 1269, seems to have been lost between 1550 and 1564. Codex A and codex DD had arrived in Europe in the twelfth century via Sicily. Around 1266, both codices came into the possession of the pope. There is evidence that William of Moerbeke also made use of the Byzantine codex DD. Sometime after 1311, when it was still in the Vatican, codex DD vanished without trace. Although the original codex A has been lost there are several Greek copies of it. There is another Latin translation of codex A in addition to that of William of Moerbeke. It dates from 1450, and is the work of Jacob of Cremona. Thanks to Johannes Regiomontanus, a reworked version of this last translation found its way to Germany in 1468 or thereabouts.

At the time of Stevin's researches in physics, Latin translations of various works of Archimedes had already appeared in print – by Thomas Gechauff in Basle (1544), and by Luca Gaurico (1503), Nicolo Tartaglia (1543), Curtius Trojanus (1565) and Federigo Commandino (1558, completed in 1565) in Venice. Below, we take a brief look at the fascinating history of codex C.

The adventures of codex C

Codex C refers to a work by Archimedes that came to light only in the nineteenth century, when it was discovered in the library of the Metochion of the Holy Sepulchre in Constantinople (Istanbul). The first hint of its existence appeared in 1846 in a travel book by the German biblical scholar Constantine Tischendorf, who had visited the library in Constantinople and found nothing of particular interest 'apart from a palimpsest dealing with mathematics' (a page of which he took away with him).

A palimpsest is a manuscript or piece of parchment from which the original writing has been scraped or washed so that it can be reused. The Archimedes text had originally been transcribed in the mid-tenth century in Constantinople. In the second half of the twelfth or perhaps in the first half of the thirteenth century, an attempt was made to efface the original writing so that the leaves could be used for liturgical texts. Fortunately, the attempt was not entirely successful (figs. 4a and 4b).

In 1899 A.I. Papadopoulos-Kerameus published a description of the volume in his catalogue of the library's manuscripts. That description was brought to the attention of the Danish classicist Johan Ludwig Heiberg, an expert in the history of Greek science. Heiberg examined the manuscript at Constantinople in 1906 and again in 1908 and, using only a magnifying glass, was able to decipher a good deal of the text. In addition to known writings by Archimedes, the palimpsest contained a work that was believed to have been lost. All that had remained were references to it in other texts. This 'new' work is the Εφόδιου or *Method of Mechanical Theorems*. The Εφόδιου gives us much information regarding the Greek mathematician's method of solving problems. Significantly, codex C proved to be independent of codex A and other works that William of Moerbeke had used for his translation.

What had happened to the codex after Heiberg's second examination is unclear. In the 1920s the library was dissolved and somehow the manuscript came into private hands. In 1998 it reappeared as lot no. 1 at a Christie's sale in New York. On 29 October of that year the Archimedes palimpsest was sold for over two million dollars to an anonymous collector who subsequently deposited the codex at The Walters Art Museum, Baltimore, for conservation and scholarly study. With the use of modern digital and optical techniques it is hoped that Heiberg's findings on Archimedes's thought processes can be completed.

Figure 4a: This is what the Archimedes palimpsest looks like today. In the twelfth century a religious text (the horizontal lines) was written over the original Greek text (the vertical lines that can be made out beneath). This palimpsest was discovered in 1899.
New York, Christie's Images Ltd. (ektachrome: JTD).

The study of and quest for old manuscripts are fascinating pursuits. We know that a considerable number of works by the ancient Greeks have been lost, including at least six by Archimedes. One of these is *On Balances and Levers*, a title mentioned by Pappus.

That a number of works from antiquity are still known today is thanks in no small measure to the school of Byzantine mathematicians and the Greek Orthodox Church. Monks and scribes copied the texts. They were driven by their interest in learning and in practical applications like architecture. Sometimes, they passed on the works to us unintentionally, as in the case of codex C.

But the Byzantine route was not the only path by which these Greek works were transmitted; they came via the Arabian world as well. The Arab mathematician Thabit ibn Qurra, who came from Mesopotamia and was born around 830 CE, translated Greek works – including those of Archimedes – into Arabic. In the seventeenth century, thanks to Borelli (1608–1679) in Florence and others, several of Thabit ibn Qurra's works were translated into Latin.

Stevin had access to a number of printed translations of works by the great Greek mathematicians. He mentions Federigo Commandino, for instance, when he refers to Archimedes in his study of the centres of gravity of plane figures in *De Beghinselen der Weeghconst*. Yet, as is clear from the *Weeghconst*, in Stevin's day it was not known that Archimedes had studied

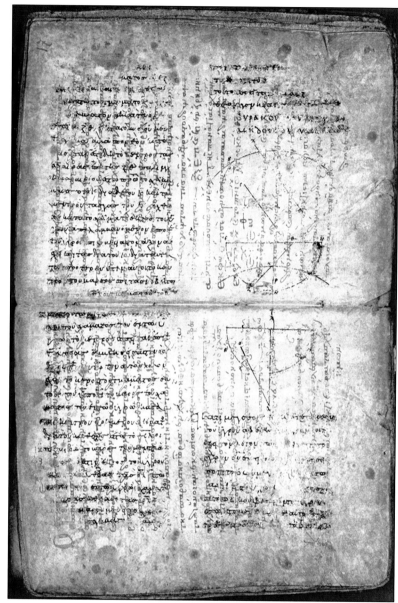

Figure 4b: This is the same page of the palimpsest as that shown in fig. 4a but now turned through 90° and after digital enhancement. The original oldest known text of a work by Archimedes, contained in a manuscript dating from the tenth century, is now visible. Shown here is a passage from *On Floating Bodies*.
New York, Christie's Images Ltd. (ektachrome: JTD).

centres of gravity not only of plane figures but also of solids. In his works on mathematics, Stevin had used printed Latin translations of Greek works. In his *Problematum Geometricorum*, for example, he made use of the 1574 translation of Euclid's *Elements* by the Jesuit Clavius, an astronomer at the Vatican.

Stevin also came to know of a number of texts of the ancient Greeks, thanks to his friend Hugo de Groot, the renowned jurist, as he indicates in his *Wisconstige Gedachtenissen* (1605–1608).

Chapter Two

Simon Stevin, religious exile?

Simon Stevin was born, probably in 1548, as the illegitimate son of well-to-do parents. His mother had Calvinist leanings. This is not unimportant, for in 1584 in Stevin's hometown of Bruges, Catholic order was reimposed after a short Calvinist intermezzo. By that time, however, Stevin was already living in the northern Netherlands. He was enrolled as a student in the recently founded university at Leiden, was the friend and private tutor of Maurice of Nassau, had made a trip to Poland and, in the style of a true homo universalis, *had published on all manner of subjects – mathematics, fortifications, politics and much more.*

1. The Bruges period

1.1. The birth of a natural child in Bruges

Little is known about Simon Stevin's earliest years. The few facts that do appear here and there in historical sources provide a possible but far from conclusive picture of Stevin's youth.

All his biographers agree that Simon Stevin was born in the great Flemish mercantile city of Brugge, or Bruges. Indeed, Stevin refers to this fact himself in nearly all his works. On the title pages of his books he calls himself Simon Stevin *van Brugghe*, or *Brugghelinck* – 'of Bruges' (see fig. 1). In the Bruges city archives, Albert Schouteet (1937) discovered a series of deeds in the municipal registers, dating from 1551 to 1597, connected with a certain 'Symoen Stevin' living in Bruges in the period in which we can situate Simon's youth (see box 1). According to Eduard Dijksterhuis (1955), two arguments support the identification of this Symoen with our famous scientist. First, in the notice of his intended marriage in The Hague in 1616, Simon Stevin is referred to as Simon Anthonis Stevin: Anthonis was the first name of Simon's natural father. Second, in his *Wisconstige Gedachtenissen* (Part v, *Van de Ghemengde Stoffen* (*Miscellaneous Subjects*), v, 2: *Van de Vorstelicke Bouckhouding* (*On Princely Bookkeeping*), Stevin himself mentions that in his early years he worked in the financial administration of the 'Brugse Vrije' or Liberty of Bruges. This is borne out by the deeds of 1577. From the first of these, dated 30 October, it appears that Stevin was the natural child of Cathelyne van der Poort and Anthonis Stevin.

WISCONSTIGE
GEDACHTENISSEN,
Inhoudende t'ghene daer hem
in gheoeffent heeft

DEN DOORLVCHTICHSTEN
Hooghgheboren Vorst ende Heere, MAVRITS Prince van Oraengien, Grave van Nassau, Catzenellenbogen, Vianden, Moers &c. Marckgraef vander Vere, ende Vlissinghen &c. Heere der Stadt Grave ende S'landts van Cuyc, S^t. Vyt, Daesburch &c. Gouverneur van Gelderlant, Hollant, Zeelant, Westvrieslant, Zutphen Vtrecht, Overyssel &c. Opperste Veltheer vande vereenichde Nederlanden, Admirael generael vander Zee &c.

Beschreven deur SIMON STEVIN *van Brugghe.*

TOT LEYDEN,
Inde Druckerye van Ian Bouvvensz.
Int Iaer cIɔ Iɔ CVIII.

Figure 1: Simon Stevin refers to his home town on the title pages of nearly all his books. Here, on the title page of *Wisconstige Gedachtenissen*, his birthplace of *Brugghe* – Bruges – is clearly mentioned.
Bruges, Public Library, B 1919 (ektachrome: JTD).

Box 1: Documents in the Bruges city archives

In 1937, A. Schouteet discovered a number of documents relating to Stevin and his immediate family in the Bruges city archives. They are listed below in chronological order. But for a few small details these documents contain the only historical data pertaining to the first 30 years of Stevin's life. They also contain information about one of the dwellings of Cathelyne van der Poort, Simon Stevin's mother, as documented in Beernaert *et al.* (2000).

1551, FEBRUARY 26: Municipal Archive, Bruges orphans' court, register of oaths taken by guardians, 1545–1558, fol. 73.
Frans van Beane, pinmaker, is appointed in place of Jacop Vaseur as guardian of Hubekin (Hubrecht) and Emerentiana, the natural children of Noël de Caron and Cathelyne van der Poort, daughter of Hubrecht.

1551, MAY 4: Municipal Archive, Bruges orphans' court, register of orphans' properties in the parish of St Donatian, 1536–1551, fol. 176.
Frans van Beane and Jan van der Lutte, guardians of Hubrecht and Emerentiana, bastard children of Noël de Caron and Cathelyne van der Poort, daughter of Hubrecht, give notice to the orphans' court of money intended for the support of the bastards.

1562, MARCH 9: Municipal Archive, Bruges orphans' court, register of oaths taken by the guardians for the years 1558–1576, fol. 185^v.
Joachim de Fournier is appointed guardian of Emerentiana, the natural daughter of Noël de Caron and Cathelyne van der Poort in place of Jan van der Lutte.

1567, MARCH 17: Municipal Archive, Sale of decree, Reg. 1554–1581, f° 28,2.
On 17 March 1567 Be(e)rnaert Winckelman had the house called 'The Golden Shield', Hoogstraat 37 in Bruges, seized. At a public auction he is able to acquire the premises. The last known owner of this house since 23 September 1564 was Joost Sayon, husband of Cathelyne van der Poort.

1577, OCTOBER 30: Municipal Archive, 'feriën' of the Bruges orphans' court for the years 1576–1580, fol. 64.
Joost Sayon and Joachim de Fournier take their oaths as guardians of Simon, the natural son of Anthonis Stevin and Cathelyne van der Poort. At the same time Simon is declared of age by the court (see also the next deed).

Text:
Actum ter camere den 30en Octobre 77.
Joos Sayon ende Joachim de Fournier zwooren voochden van Simon de natuerlicke suene van Anthuenis Stevin by Catelyne vander Poort.
Dezelve Simon wiert aldoe byden collegie zyn zelfs man ghemaect ende uut voochdie ghestelt; clerc: J. Geeraerts.

1577, OCTOBER 30: Municipal Archive, register of Jan Geeraerts, clerk of the Bruges Tribunal, for the years 1577–1581, fol. 4^r-4^v (see also fig. 4).

Simon Stevin, natural son of Anthonis Stevin and Cathelyne van der Poort, is declared of age by the aldermen of Bruges. His friend Jan van der Houve is mentioned as witness.

Text:
Gheconsenteert present scepenen: Despars, Boodts, Cabootere, Volden, Sproncholf, Lernoult, den 30en in Octobre 1577, present burchmeester vander courpse: Heinric Anchemant. - Dat quamen voor ons als voor scepenen ende in 't ghemeene college van scepenen ter camere deser voorn. stede, uppervoochden van alle weesen onder hemlieden resorteerende ende behoorende, Joos Sayon ende Joachim de Fornier, als wettelicke voochden van Symoen, de natuerlicke zone van Anthuenis Stevin, die hy ghehadt heeft by Cathelyne vander Poort, dewelcke vertoochden hoe dezelve huerlieder weese ghecommen was ter oude van acht en twintich jaeren ofte daeromtrent ende, uute goede experiëntie die zy van hem ghehadt ende ghenomen hebben in zyn affairen, hemlieden oorboir ende proffyt dochte, overmidts dat hy oudt ende vroedt ghenough was, den voors. Symoen zyn selfs man te makene ende uut voochdie te doene, omme van nu voortan zyn zelfs affairen ende goet te mueghen regierene ende zyn proffyt daermede te mueghen doene, alzo zy voochden dat presenteren te verclaersene by huerlieder eede, inde presentie vande voors. Cathelyne vander Poort, zyne moedere, ende Jan vander Houve, zynen vriendt ende maech, die in tguent dies voorseyt es, zo verre als 't hemlieden anneghynck, verclaersden te consenteren; ter cause van denwelcken de voors. voochten verzochten ende begheerden van dezelve voochdie ende huerlieder eede ontsleghen te zyne, ende den voors. Symoen, aldaer present zynde ende tzelve instantelick verzouckende, zyn selfs man ghemaect te werdene ende uut voochdie ghedaen te zyne. Al twelcke byden voorn. college ghehoort ende zonderlinghe 't verclaers van eede van beede dezelve voochden, metgaders 't consent van de voorscreven vrienden ende maghen, heeft, interponerende zyn decreet, denzelven Symoen Stevin zyn zelfs ghemaect ende uut voochdie ghedaen omme van nu voortan zyn selfs goedt te ghebruuckene ende administrerene, ende voorts dezelve voochden van huerlieden eedt ende voochdiescepe ontsleghen.

Summary:
A number of aldermen and the burgomaster of Bruges have assembled as head guardians of all orphans; also present are Joost Sayon and Joachim de Fournier, as legal guardians of Simon, the natural son of Anthonis Stevin and Cathelyne van der Poort. The guardians state that Simon is 28 or thereabouts and that they know from experience that he is competent to conduct his own affairs. They ask that Simon be 'made a man' [declared of age]; this they do in the presence of Cathelyne van der Poort, his mother, and Jan van der Houve, his very close friend. They ask to relinquish their guardianship and to be released from it.

1577, OCTOBER 30: Municipal Archive, register of Jan Geeraerts, clerk of the Bruges Tribunal for the years 1577–1581, fol. 4^r-4^v.
Simon Stevin promises in the presence of Joost Sayon and before the aldermen of Bruges to indemnify Lenaert and Adriaan de Cant who stood surety for him for a sum of 75 Flemish pounds loaned to Simon by Jan de Brune, commissioner of taxes for the Four Members of Flanders in the Brugse Vrije, in whose office Stevin was a clerk. As security Simon gave the annuity of four Flemish pounds that he already received from the two Cants.

1577, OCTOBER 30: Municipal Archive, register of Jan Geeraerts, clerk of the Bruges Tribunal for the years 1577–1581, fol. 5^r-5^v.
Simon Stevin promises before the aldermen of Bruges and in the presence of Joost Sayon, to indemnify Pieter Courtewille who stood surety for him for a sum of 50 Flemish pounds loaned to Simon by Jan de Brune, mentioned in the previous deed. This document also mentions an annuity of two Flemish pounds – a fairly small amount – to be paid on every Saint Bavo's mass (1 October) by the aforementioned Pieter Courtewille and his wife Magdalena, daughter of Simon's uncle, Pieter Stevin.

1584, JUNE 28: Municipal Archive, civil sentences, pronounced by aldermen of Bruges, for the years 1580–1590, fol. 207^r-207^v.
Marie van der Poort, widow of Pieter Inghelbrecht, and Cathelyne van der Poort, widow of Joost Sayon, daughters of Huybrecht van der Poort and Clarcken, daughter of Frans van Beane and the late Marie van der Poort, all heirs of François van der Poort, half-brother of Hubrecht van der Poort, authorize Jan van der Lutte, son of Jan van der Lutte and Magriete van der Poort, daughter of Huybrecht, to go and settle their share of the estate of the aforementioned François in Ypres. (Here we become acquainted with a few of Simon's relatives on his mother's side.)

1578, JULY 6: Municipal Archive, civil sentences, pronounced by aldermen of Bruges, for the years 1586–1587 (loose documents), no. 39.
Cathelyne van der Poort, widow of the late Joost Sayon, and other heirs of the late François van der Poort authorize a number of named individuals to demand before the magistrate of Ypres the payment of an annuity amounting to two pounds ten shillings, for a house in the Korte Nekkerstraat in that town, which has been outstanding for six years.

1597, OCTOBER 15: Municipal Archive, procurations drawn up for aldermen of Bruges for the year 1597 (loose documents), no. 197.
Cathelyne van der Poort, widow of Joost Sayon, on her own behalf; Jacques Inghelbrechts, merchant, on his own behalf; Galiaen de Vriese, husband of Tanneken van Pachtenbeke, on behalf of himself and of Jacques Evens; authorize Jan van der Lutte, bailiff of Wackene, to settle their share of the estate of François van der Poort, deceased in Ypres. This document is signed with the name of 'Catelynn vande Poert'.

> Note that no mention has been made here of archive documents recently discovered in Lille, Veurne and elsewhere by Pieter Donche (2002a, b), which contain information on Stevin's ancestors in the Veurne area. Interested readers should consult Donche's publications.

1.2. Burgomasters and aldermen on the spear side

A number of sources provide us with interesting information about Simon's father Anthonis. The structure of the Stevin family in Veurne has been outlined in a very recent publication by Pieter Donche (2002a). He gives details of an Anthonis Stevin, who after 1543 seems to have severed all connection with his family and disappeared from the Veurne area. The association with Veurne had been suggested previously in an article by Van Acker (1999). Here, however, we shall follow Donche's outline, which is based on a number of hitherto unconsulted archive documents. From these it appears that Anthonis was a son of Adriaan Stevin, burgomaster of Veurne, and Francine de Visch (see also the family tree). According to Donche, Adriaan and Francine were well-to-do residents of Veurne and the holders of various properties in fee simple in Bulskamp, Oeren, Veurne (St Nicolas) and Adinkerke. Adriaan's father and grandfather had also been burgomasters of Veurne, and it is possible that his brother Pieter was provost of the Premonstratensian abbey of St Nicholas. Adriaan and Francine are buried in the church of St Nicolas in Veurne. They are portrayed on their tombstone in long robes, with six sons and six daughters depicted below them. Only five of these children are known by name: Jan, who became an alderman of Veurne; Hendrik; Pieter, also an alderman of Veurne; Magdalena, who wed Lenaert de Cant; and Anthonis.

According to Donche, in 1540 or 1541, Francine de Visch bought from Jan Aelvisch for her son Anthonis a fee of four 'gemeten' (an area equivalent to 300 roods or a little over 4,000 square metres), located in Bulskamp. That the purchase was made by his mother suggests that Anthonis was then still a minor. In 1543 or 1544, Anthonis himself bought another fee in Bulskamp, four 'gemeten' and 25 roods in extent. From these data, we can deduce that he must have been born between 1515 and 1518. By 1559, 13 years after Anthonis's last purchase, we read that he had already been gone from the area for quite some time and had not been heard of since. On his mother's death, which occurred in that year, he was due to inherit a fee of eight 'gemeten' in Oostduinkerke. This, however, did not happen; his sister Magdalena's husband, Lenaert de Cant, paid the necessary rights for his wife, 'her brother having been a long time out of the country without having any news from him'. This assignment to Magdalena was on condition 'that if her said brother should return, the said fee should still belong to him' and this arrangement was made 'in order that the said fee should not remain vacant'. Anthonis was heir only to the fees his mother held in her own right; the paternal fees went to his brothers. This indicates that Anthonis was not the eldest son, but that he received special treatment from his mother.

Donche suggests that Anthonis somehow brought shame on the family, and therefore could not or dared not show his face in Veurne ever again. Whatever the case, in 1548 Anthonis enrolled as a member of the archers' guild of St Barbara in Bruges and there he fathered an illegitimate son, Simon. Anthonis never claimed the fees to which he was entitled, and as a child born out of wedlock Simon would not have been able to inherit any tenures. That the family was aware of Simon's existence, however, is borne out by a number of documents in the Bruges Municipal Archive. Schouteet (1937) mentions two deeds in particular (see also box 1). According to one of these, Lenaert and Adriaan de Cant, related by marriage to his aunt Magdalena, agreed to stand surety for 75 Flemish pounds loaned to Simon by Jan de Brune, the tax receiver of the Brugse Vrije, and Simon promised to indemnify his guarantors with all his present and future property, particularly the annuity he had received for years from the Cants. He contracted this loan in order to be able to take up a post as a clerk – *vanden comptoire* – in the tax office of the Brugse Vrije. According to the second deed, Simon made a similar declaration to Pieter de Courtewille, husband of Francisca, a daughter of Simon's uncle, Pieter Stevin, for another loan of 50 Flemish pounds from the same Jan de Brune, again promising to indemnify his guarantor, in particular with the annuity he received from de Courtewille and his wife. These annuities gave Simon a degree of financial independence. Together they represented six Flemish pounds – according to Donche the equivalent of 72 days' wages (or about one quarter of the annual income) for a mason.

On his father's side, Simon was related to prosperous folk, whose civic positions gave them considerable influence in Veurne and the surrounding area.

1.3. Magistrates and merchants on the distaff side

Stevin's mother, Cathelyne, was the daughter of Hubrecht van der Poort of Ieper (Ypres). The deed dated 28 June 1584 (see box 1) reveals that she had at least three sisters: Marie, married to Pieter Inghelbrecht; a second Marie, married to Frans van Beane; and Magriete, married to Jan van der Lutte. The two last-mentioned husbands appear in the Bruges deeds as witnesses. In 1584, all these members of the family received a substantial legacy from their uncle François van der Poort, their father's half-brother. Cathelyne had two other natural children, Hubrecht and Emerentiana. Both were fathered by Noël de Caron, who can be identified as the selfsame de Caron who was interred in St Donatian's church in Bruges on 11 December 1560. Noël de Caron was an important man. From 1530, onwards he was a member of the magistracy of the Brugse Vrije, sometimes serving as burgomaster, sometimes as alderman. He was the father of a politician of the same name who was ambassador of the United Provinces to London and who died there in 1624. The possible connection with Noël de Caron could explain Simon Stevin's employment as a clerk of the Brugse Vrije.

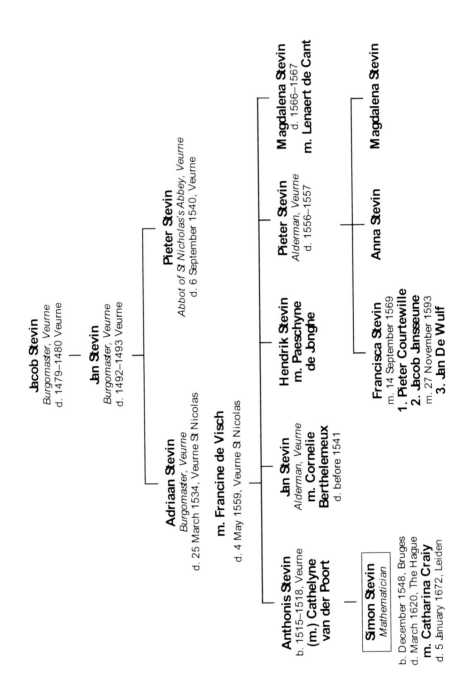

At some point – exactly when is unknown – Cathelyne van der Poort married Joost Sayon. Whether or not there were children of that marriage is likewise unknown. Joost Sayon was a member of a wealthy merchant family with trading contacts in many places, including the Baltic States. He had at least two brothers, Vincent and Jacob. Like Joost, they were employed in the cloth and silk industry. Vincent was also very active in politics and was evidently a confirmed Calvinist (see box 2). Joost owned at least one house in Bruges—'The Golden Shield' on the Hoogstraat (see box 1)—and a silk workshop on the Vlamingdam. Moreover, Ludo Vandamme (1982) suggests that Vincent's son Antoon, who was also active in politics, might have been the husband of the abovementioned Emerentiana. All this goes to show that Stevin's mother had close relations active in political (and Calvinist) circles and that the various men in her life were important figures in Bruges. These families also had relatively large incomes; so there is no reason to doubt that the young Simon Stevin enjoyed a good upbringing and education.

Box 2: Calvinism in Bruges between 1550 and 1584

In comparison with many other towns in the Low Countries, Bruges was late in having adherents to the teachings of Calvin. One of the apostles of Calvinist teachings was Johan Gailliaert. In 1554, he produced the earliest Dutch translation of one of Calvin's works. The breakthrough for Calvinism in Bruges came in 1566. On 3 March a certain Brother Cornelis identified elements of the city's magistracy, especially its legal experts (the lawyers and clerks), as proponents of the new doctrine – 'principally those who have studied in France [at Orleans] and there suckled the venomous liquor from the breast of John Calvin'.

Iconoclasm in Bruges

The bishop and the city council did not react until 15 May 1572, when a complete system of suppression, with eight inquisitors, was brought into operation. The Jesuit college took over the functions of the other Latin schools. Following Alva's departure from the Low Countries in November 1573, the pursuit of Calvinists relaxed.

Between 20 and 26 March 1578, François de la Kethulle (Ryhove) of Ghent seized control of Bruges and established Calvinist rule. This led to the banishing of 77 Catholic notables and their families, as well as the departure of 54 members of the Spanish 'Natie'. Remi Drieux, the bishop of Bruges, had already been imprisoned by Ryhove during the coup in Ghent on 28 October 1577, along with other high-profile captives including the bishop of Ypres and the prince of Chimay, governor of Flanders. Drieux remained in captivity until August 1581.

On 6 August 1578 the Jesuit college was closed, and on 28 August the parish church of St Anne was made available to the Reformed congregation. On 25 September a storm of iconoclasm swept through Bruges. Catholic services were suspended throughout the city. On 31 October the magistracy tried to calm things down. Catholics were permitted to worship in the parish churches again, while the Reformed were allotted the conventual churches of the Carmelites and the Augustinians, and the chapel of St John. On 8 November, Bruges became the first town in Flanders to accept the 'Religious Peace' guaranteeing non-interference to both religious groups.

Vincent Sayon

One remarkable figure in Bruges at this time was Vincent Sayon. He is important to our view of Stevin, because it is very likely that he was the brother of Joost Sayon, who married Stevin's mother acted as Simon's legal guardian in 1577 and owned the house called 'The Golden Shield' (see box 1). Of course, we should be cautious in reconstructing family links: very few documents have been preserved that give definite confirmation of relationships. Vandamme mentions Vincent Sayon as a member of a Calvinist delegation that was admitted to the town hall on 3 November 1566 to petition the magistracy after a disturbance in which some religious innovators had come to blows with the civic militia. There can be no doubt as to Vincent's Calvinist convictions in the years 1578–1584. During the Calvinist Republic he consistently held high office, but his participation in city government dates from as early as 1575. He used his increasing prestige as alderman to get his son, Antoon, into the magistracy and even to usurp profitable posts. Vandamme wonders whether this Antoon Sayon might have been the husband of Emerentiana, the natural daughter of Cathelyne van der Poort and Noël de Caron, and he suggests that the latter may have been the father of a man of the same name who became burgomaster of the Brugse Vrije and later emigrated to the northern provinces (see also box 1 and below).

In 1580, Vincent became a member of the four-man committee charged with regulating the sale of ecclesiastical property. He was one of the heaviest buyers of church goods. He began his last term of office as alderman in March 1584 after the prince of Chimay had reinstituted a more Catholic policy. Farnese remarked in his letters to Philip II that in spite of everything there were still numerous Calvinists on the new bench. Notwithstanding, Vincent played an active part in the civic delegation that negotiated the terms for the restoration of Bruges to royal obedience.

According to Vandamme, it is difficult to define Vincent's professional life because of the wide range of his activities as a merchant. Some refer to him as a *coopman van tapytserie* – a tapestry dealer. He may have taken over this business from his father: Antoon Sayon made a name for himself as a weaver and seller of tapestries. In the years between 1520 and 1530 he provided hangings for the meeting room of the bench of aldermen of the Brugse Vrije. His other son, Jacob, the brother of Vincent and Joost, also dealt in tapestries. Vincent was also active in the silk industry, working out of the house 'The French Arms' on the Vlamingdam, where his brother Joost ran a silk workshop. Vincent gained particular fame as a dealer in textile products. He reactivated Bruges's dormant textile industry, taking on a number of Walloon drapers to do so. He carried on an intensive trade with the Baltic lands.

Vincent was undoubtedly among the wealthiest of the citizens of Bruges. In 1580, he owned at least 15 houses within the city as well as properties outside it. He died before 1598.

The Calvinist brain drain from Bruges

When Alexander Farnese brought the Calvinist Republic to an end on 25 May 1584, Catholic life in Bruges was restored fairly quickly. Ten days before Farnese's triumphal entry, the altars had already been reconsecrated in the churches of St Donatian, St James and St Walburga. The churches of St Gillis and Our Lady followed on 15 July. In May 1585, it was agreed that an anniversary procession would be held on the Sunday closest to 20 May to commemorate the liberation of the city from heretics and rebels.

In the course of 1586 and 1587, the Protestants elected to go or were forced into exile. One hundred and ten Calvinist families left Bruges. After this period a number of scholars from Bruges made their appearance at universities in the northern Netherlands, not only Simon Stevin but also Franciscus Gomarus, Bonaventura Vulcanius and Nicolaus Mulerius.

Franciscus Gomarus, a Reformed theologian, was born in Bruges on 30 January 1563 and died in Groningen on 11 January 1641. In 1587 he was a minister in Frankfurt, and in 1594 he became a professor in Leiden. In 1604 he initiated a theological dispute about predestination with his colleague Arminius (see also box 6). In 1609 he resigned his chair in Leiden and became a preacher and lecturer in Middelburg. In 1615 he was appointed professor in Saumur, and in 1618 in Groningen. He contributed to the *Statenvertaling* (States Translation) of the Bible, which was published in 1631 and holds a position in Dutch language and culture analogous to that of the King James Version in the English-speaking world. We can assume that he left Bruges for religious reasons.

Bonaventura Vulcanius was born in Bruges and died in Leiden in 1614 at the age of 77. Prior to becoming Leiden's professor of Greek he had been secretary to Marnix van St Aldegonde. Vulcanius left a few writings containing comments against the Catholic Church. During his time in Leiden a handwritten work by Cornelius Aurelius (1460–1531) came into his possession. This Dutch historian was a Latin poet and a friend of Erasmus. Vulcanius adapted the text, crossed out those parts of the original manuscript he considered irrelevant, and sent it off to the printer.

We find Nicolaus Mulerius at the Calvinist University of Groningen. He was born in Bruges in 1564 and died in Holland in 1630. In 1617, a year after the work was placed on the Roman Index, he edited the third edition of Copernicus's *De revolutionibus orbium caelestium*. In this edition of the text, explanatory notes were added by Mulerius, who should not be seen as a convinced proponent of Copernicus's theories, though he certainly considered Copernicus to be one of the most important figures in the history of astronomy.

The humanist Franciscus Nansius was also a native of Bruges. He moved to Leiden and later taught at the Latin school in Dordrecht. He had a large private library. When his books were sold after his death, the University of Leiden bought a number of ninth-century manuscripts, copies of works by classical authors of the fourth and fifth centuries, the originals of which had long since vanished.

Martinus Smetius (1525–1578), also from Bruges, had an indirect link with Leiden. This scholar had been to Rome, where he meticulously transcribed thousands of inscriptions from buildings, triumphal arches, sarcophagi and so on. His manuscript was bought in England by Janus Dousa, founder of the University of Leiden (see box 4). Justus Lipsius, who was a professor in Leiden for a while, published Smetius's texts under the title *Inscriptiones Antiquae*; the printer of the work received a subsidy of 500 guilders.

There were others from Bruges in the service of the States General in Holland in the early seventeenth century, among them the aforementioned Noël de Caron. This politician held various positions in the Brugse Vrije that were of some importance to those of Calvinist sympathies. He died in 1624 as ambassador of the United Provinces in London. Another important Brugeois in the service of the States was Jacques Wijts, son of Jean, burgomaster of Bruges in the Calvinist period. The magistrate and man of letters Pieter Cornelisz. Hooft refers to Wijts as Simon Stevin's successor in the States Army. Wijts emigrated to Holland after his father's death, together with his mother and her second husband, Adolf van Meetkerke. In 1606, Wijts was appointed captain of infantry. His marriage to the noblewoman Magdalena van Valckesteyn, daughter of the receiver general of Holland, made him a wealthy and prominent man.

Box 3: Important dates in Simon Stevin's life

(Biographical details in italics)
1545–1563:	Council of Trent
1548:	*Birth of Simon Stevin in Bruges*
1549–1551:	The future Philip II makes a tour of the Low Countries, accompanied by his father, Emperor Charles V
1555:	Abdication of Charles V and investiture of Philip II as the new ruler of the Seventeen Provinces
1566:	Iconoclastic Fury

1567:	The duke of Alva becomes governor-general of the Low Countries
1568:	Execution of counts Egmond and Horne
1572:	William the Silent becomes stadholder of Holland, Zeeland and Utrecht
1575:	Founding of Leiden University
1576:	Pacification of Ghent
1577:	Calvinist seizure of power in Ghent. *At 28 years old, Stevin is declared of age in Bruges. He works as a clerk in the tax office of the Brugse Vrije*
1581:	*Stevin is enrolled in the municipal registers of Leiden. Publication of* Nieuwe Inventie van Rekeninghe van Compaignie ...
1582:	*Publication of the* Tafelen van Interest
1583:	*Publication of* Problematum Geometricorum
1583–1590:	*Stevin is entered on the roll of students attending Leiden University*
1584:	William of Orange is shot in Delft. *Granting of the first patents to Stevin in the Republic*
1585:	Alexander Farnese retakes Antwerp. The Catholic southern provinces and the Protestant northern provinces separate. *Publication of* Dialectike ofte Bewysconst, De Thiende *and* L'Arithmetique (see fig. 2)
Before 1586:	*Stevin and Johan Cornets de Groot carry out an experiment with falling bodies in Delft*
1586:	*Publication of* De Beghinselen der Weeghconst. *Stevin is granted a patent for a 'watermill'*
1589:	*Nine new patents granted to Stevin*
1590:	*Publication of* Vita Politica, Het Burgherlick Leven (see fig. 3)
1591:	*Stevin travels to Gdańsk*
1593:	*Stevin comes into contact with Prince Maurice and enters his personal service*
1594:	*Publication of* Appendice Algebraique *and* De Sterctenbouwing
1596:	Archduke Albert of Austria becomes governor of the Spanish Netherlands
1597:	Marriage of Albert and the Infanta Isabella; new sovereigns of the Spanish Netherlands
1599:	*Publication of* De Havenvinding
1600:	Battle of Nieuwpoort
c. 1601:	*Stevin builds a sailing chariot*
1601:	Founding of the United East India Company (the Verenigde Oost-Indische Compagnie, or VOC)
1604:	*Stevin becomes quartermaster of the States Army.* Destruction of Ostend harbour by Spanish troops

1605–1608:	Publication of *Wisconstige Gedachtenissen*
1609:	Twelve Years' Truce
1612:	Stevin buys a newly built house in The Hague, Raamstraat 42
1612:	Birth of Stevin's children Frederik, Hendrick, Susanna and Levina
1616:	Notice of Stevin's marriage to Catharina Craiy, the mother of his children, is published
1617:	Publication of *Castrametatio* and *Nieuwe Maniere van Stercteboud, door Spilsluysen*
1620:	Stevin dies in The Hague

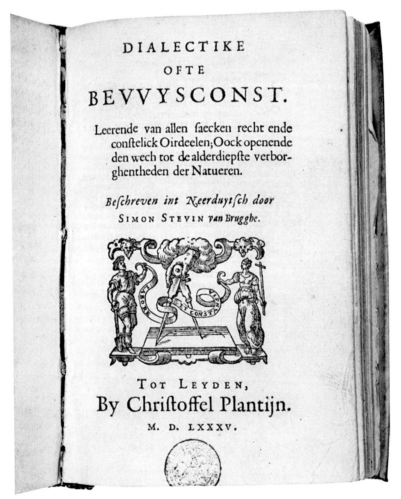

Figure 2: Simon Stevin's *Dialectike ofte Bewysconst* is the oldest treatise on logic in the Netherlands.
Antwerp, Plantin–Moretus Museum (photo: Peter Maes) (ektachrome: JTD).

Figure 3: From 1581 Simon Stevin published a respectable number of scientific works. In his *Vita Politica*, whose title page is shown here, he considers the conduct of the citizen in the state.
Leiden, University Library, 689 F 10:2 (ektachrome: JTD).

1.4. Coming of age at twenty-eight

It was only on 30 October 1577 that guardians were appointed for Stevin. One of the guardians was his stepfather Joost Sayon. The other was Joachim de Fournier, who had already become a guardian of Emerentiana, Simon's half-sister, in 1562. On that same day, Stevin's majority was also officially declared (see fig. 4). This procedure gave him legal status; now, for instance, he could contract loans. This official coming of age occurred rather late: Stevin was already *acht en twintich jaeren ofte daeromtrent* – 28 or thereabouts. We can deduce from this that Stevin must have been born in late 1548 or early 1549. A painting in Leiden University Library gives 1548 as the year of his birth.

We can assume that the appointment of guardians and the formal release from guardianship was arranged with a view to obtaining (a part of) his father's estate. Stevin probably needed the inheritance and legal status in order to lawfully contract loans and securities. As we have seen above, various members of the Veurne branch of the Stevin family stood surety for him with his employer, Jan de Brune, commissioner of taxes for the Four Members of Flanders in the Brugse Vrije.

Figure 4: Several deeds in the Bruges city archives relate to Simon Stevin. This entry in the register of Jan Geeraerts, a clerk of the Bruges Tribunal, declares the 28-year-old Stevin to be of age. Bruges, Municipal Archive, Old Archive r. 198, Clerks of the Tribunal, Register of Jan Geeraerts, October 1577, fol. 4^r-4^v (photo: Hugo Maertens).

1.5. Simon Stevin's travels

Little else is known about Simon Stevin's youth. He must have had an extremely good education. His later publications show that he was familiar with Latin and French, and apparently also had some knowledge of Italian and German. He was probably taught at one of the Latin schools in Bruges. We know from a reference in the *Wisconstige Gedachtenissen* that he worked for a time as a bookkeeper and cashier in Antwerp, although when and for how long he did so is uncertain. In any case, this kind of activity also tallies with the commercial activities of his stepfather Joost and his brothers.

Some authors state that in the period from 1571 to 1577, Stevin travelled to Poland, Prussia and Norway, and indeed, in his writings Stevin does give detailed descriptions of these countries (see for instance *Materiae Politicae*, vol. 1: *onderscheyt van de oirdeningh der steden* (Town planning), p. 57 (Norway) and p. 120 (Cracow, Poland); *Wisconstige Gedachtenissen*, Part. i: *Vant Weereltschrift* (Cosmography); i,2: *Vant Eertclootschrift* (Geography); and *Vant Stofroersel des Eertcloots* (Geomorphology), p. 67 (Prussia)). It is therefore highly probable that he did indeed undertake these journeys, though we have no actual proof. Perhaps he was able to make use of the Sayon family's contacts in the Baltic. However, we do have proof from historical sources of his later journey to Gdańsk in 1591.

1.5.1. ...a certain Simon Stevin, native of Bruges [...], who was a schoolmaster and a mathematician and a very experienced man...

In their endeavour to give more substance to Stevin's biography, a number of authors have elevated their hopes and hypotheses to the level of fact. There is one document we have not as yet discussed – a letter from an otherwise unknown I. H. Verplancken of Bruges in reply to the Leiden town bailiff, Gijsbert Ariens van Rijckhuijsen. Between 1740 and 1760, van Rijckhuijsen responded to numerous requests for genealogical and heraldic research. One of his clients was Master Juliaen van Groenwegen, a grandson of Stevin's elder daughter Susanna. Master Juliaen probably wanted to know more about his illustrious great-grandfather. In February 1753, Verplancken wrote the following (the letter is now in Leiden's municipal archive):

> Op u toegesonde memorie van Simon Stevijn of Stevin, dient gevonden wort in den Boek van de Heijligh Geesters synde Poesie, dat er een sekeren Simon Stevijn geboortigh van Brugghe geworden is confrater data 16 Maart 1571, die schoolmeester was en een wijsconstenaer, en een seer ervaren Man en faute hy niet connen becommen vrydom van de Bier accijns, vertrocken is naar Middelburgh, sonder tot hier te connen aghter haelen syn waepen of wat kinderen hy aghter gelaeten heeft,...

It would appear from this document that in 1753 in Bruges, facts about Stevin might still be looked for. The letter tells us that in 1571, Stevin was a member of the chamber of rhetoric known as 'De Heilige Geest', or 'The Holy Ghost'. Unfortunately this can no longer be verified, as the chamber's membership registers have been lost. Nevertheless, it is clear – given that the fact is correct – that in Bruges Stevin had already developed an interest in the Dutch language, an interest that would come to be ever more explicitly expressed in his works. He is also referred to as a schoolmaster and mathematician. At some unspecified stage he apparently left Bruges for Middelburg. Some biographers have used this statement as a starting point from which to send Stevin on his tour of the abovementioned areas around the Baltic. However, it is worth bearing in mind that the letter's author refers to Middelburg as Stevin's destination. During the religious conflict between the Catholic south and the Protestant north, Middelburg was the southernmost town under Calvinist government. There was a great influx of Protestants from the Antwerp area. The émigré community is traceable as a separate group in Middelburg until late in the seventeenth century. Indeed, it is in Middelburg that the father of the mathematician Isaac Beeckman settled in 1585, and carried on a trade as candle-maker and water-conduit layer.

No official documents dating from after 1577 connected with Simon Stevin are to be found in Bruges. He must indeed have left the town. Perhaps his reasons for leaving were religious, but it could also be that he had simply had enough of the atmosphere in Bruges. The dissension between Catholics and Calvinists, among whom his closest family could be counted, made it impossible for him to devote himself to his scientific work. Even so, by the time he arrived in Leiden in 1581 he must have had a sound scientific education. Perhaps he studied at one of the Protestant universities in France, Germany or Switzerland, or at an Italian university, where he may have had contact with the Italian mathematicians and algebraists of his day.

2. Dedication to the Republic

From 1581 onwards, the course of Stevin's career becomes a little easier to follow. In that year the name of the Bruges scholar was inscribed in the Leiden municipal register as *Symon Stephani van Brueg*. He lodged with Nicholaas Stochius, headmaster of the Latin school on the Pieterskerkgracht. On 16 February 1583, he enrolled at Leiden's recently founded university (see box 4). His name appears in the university registers until 1590 as *Simon Stevinius brugensis, studuit artes apud Stochium*. Most likely he followed courses given by Rudolph Snel (known as Snellius), who taught mathematics, astronomy and Hebrew there from 1581. The decision to study or complete his studies in Leiden was probably prompted by religious considerations. At that time Leiden was the only fully fledged Calvinist university in the Low Countries.

Box 4: The University of Leiden

Sacra Scriptura leads Leiden's inaugral procession

On 28 December 1574 William the Silent, prince of Orange, wrote to the States of Holland and Zeeland proposing that a university should be founded in order to educate the office-holders and professionals who would staff the new state and provide preachers for the public Church. Middelburg, Gouda and Leiden were all considered as locations, and on 3 January 1575 Leiden was chosen. The humanist Jan van der Does or Janus Dousa (1545–1604) was appointed to implement the decision, together with a number of colleagues. Two weeks later, Leiden's town council decided that the inauguration of the new university (see fig. 5) should take place on 8 February 1575. Within the

Figure 5: This anonymous tinted drawing is from Utrecht-born Johannes van Amstel van Mijnden's *Album Amicorum*. On 6 September 1600 he matriculated into Leiden University as a student of law. He was already 22 years old. This picture shows the university's main building on the Rapenburg. To the right of the university gateway was the shop of Lodewijk (Lowys) Elsevier. He had left the southern Netherlands and settled in Leiden in 1580. Here he published his first books under his own name. Six years later he became the university registrar. In 1620 his grandson Isaac became the university printer.

The Hague, National Library of the Netherlands, 74 J 37, fol. 158[r] (ektachrome: JTD).

space of six weeks everything was arranged. However, there was no money, no students, no buildings and no professors. The last problem had to be solved first – a professor-less inauguration would make a distinctly odd impression. Time was too short for permanent appointments to be made, but Dousa and his colleagues sought out scholars willing to do the honours.

The inauguration itself was a great celebration. Dousa had conceived a great procession embodying his vision of the essence of humanism and the practices of scholarship. Figures representing the various disciplines rode in the parade: Justitia for Law, Medicina for Medicine and Minerva for the Humanities. They were all accompanied by their sources, the famous Greek and Roman authors who had written on the subject in question. Heading the procession was a triumphal float bearing the figure of *Sacra Scriptura* (the Bible). It was no accident that the Bible led the parade. It was clearly William of Orange's intention to advance the study of theology at the university, as is likewise suggested by his gift to the university of a Plantin *Polyglot*. This bible contained the official Latin translation (the Vulgate) and the four original languages: Hebrew, Greek, Chaldean (Aramaic) and Syriac. It was also an indication of how William thought the most important subject should be studied, by returning to the sources, the original text. On William's advice the official charter of foundation, the *Octrooi*, was drafted as though issued on behalf of Philip II, at that time still rightful ruler of the Netherlands.

Leiden had become Protestant, and at first the University of Leiden was housed in disused churches and convents. The first convent to be put to academic use was that of St Barbara, but the continuing growth of the university led to the allocation of more Catholic ecclesiastical buildings. The main building was on the Rapenburg. University funding came from confiscated ecclesiastical possessions and control of the new foundation was in the hands of seven curators, three appointees of the States of Holland and four 'burgomasters' (here meaning an elected civic officeholder) appointed by the city. They oversaw the finances and the maintenance of order, and also had some say in the content of teaching.

The university had its own tribunal, the *Vierschaar*, which was made up of six civic representatives and five professors. Judgement was given concerning killings, assaults and affrays. The *Vierschaar* could pronounce the death penalty, but in general the punishments it handed down were considerably milder than those of ordinary courts, usually being suspension or expulsion from the university.

The university's recruitment policy was boldly enterprising. Temporary professors had been appointed for the inauguration, but afterwards permanent appointments had to be made, and several famous professors were poached from other universities (see box 2). Most of them received an annual salary of around 300 guilders. This amount was not considered exceptionally high, and many professors supplemented it by taking in students and charging them for board and lodging. Private tuition could also be a remunerative source of additional income.

It is not out of place to mention a few of the famous professors of the university's formative years. Justus Lipsius (Joost Lips, 1547–1606) was perhaps the most renowned scholar of his day. He had become a Protestant and, at Dousa's request, took up the professorships of history and law in 1578. This great expert on Tacitus returned to Leuven, and to the Catholic faith, in 1591. After his departure, the curators invited Josephus Justus Scaliger (Joseph Juste della Scalla, 1540–1609) to replace him. Between 1572 and 1574 this French classicist and convert to Protestantism had taught philosophy at the University of Geneva. In Leiden he was not obliged to teach, his mere presence being deemed to add lustre to the university. He did give private tuition to a select group of gifted students. One of these was Grotius (Hugo de Groot).

In 1592 Pieter Pauw (1564–1617) was appointed professor of medicine, with the special duty of demonstrating anatomy. He had gained his skills as an anatomist at the University of Padua. Anatomy demonstrations took place in the renowned *theatrum anatomicum Lugdunai Batavorum* – the Theatre of Anatomy at Leiden (a replica of which has been built in the Boerhavemuseum in Leiden). From 1598 Pauw also lectured on botany. In 1592 he had been asked to lay out the *Hortus botanicus*, or botanical garden – a replica of which can still be seen. Pauw himself did little of the actual work, leaving the task to the Delft apothecary Dirck Cluyt (1546–1598), a man without an academic degree. Cluyt became the first director under Carolus Clusius (1526–1609), a renowned botanist from Lille. It was here in Leiden, at Clusius's instigation, that the first tulips in the Netherlands were grown. Clusius was the first botanist to study plants systematically not only for their medicinal value, but also for their botanical significance and beauty.

Tax-free beer, wine and books for Leiden's students

Students had to register in the *Album Studiosorum*. They followed a year or two of literary studies in the arts faculty before they were allowed to specialize. The professors gave one-hour lectures, between 8 a.m. and 6 p.m., every day except on Wednesday and Saturday. These days were reserved for written and oral examinations. Sundays were free. The language of instruction was Latin. Students wanting to become ministers did not have to sit exams at all. Students in other disciplines, who did sit exams, could gain a bachelor's or a licence degree. The highest degree in the humanities was *Magister Artium*, but the student of law, medicine or theology was only truly learned when he obtained a doctorate.

Students were the sons of noblemen or wealthy burghers, because studying was an expense that few could afford. A room with Professor Bronchorst cost 100 guilders a year; meals every day for a year came to the same again. On the other hand, students did not have to pay sales tax on beer, wine or books. First-year students were usually younger than they are now. They were boys only, of around 15, who could speak, read and write Latin. A few were a little older, particularly those who had already spent a year or two studying elsewhere in the Netherlands or abroad.

Military and economic imperatives determine the job market

The young Republic needed many mathematicians, people who could calculate, in the first place for military purposes – for measuring, estimating and designing fortifications; for working out the numbers and transport of soldiers and supplies; and for solving ballistic problems in the artillery. There were also economic motives for studying mathematics: the calculation of interest, the reckoning of time, the determining of longitude and latitude in navigation. Mathematics was taught at the university, but when it came to practical application one had to go elsewhere.

Besides lectures in Latin, from 1600 on lessons were also given in *Nederduyts* (Dutch) for those of humbler origins. They took place in the school of engineering, which was founded at the instigation of Prince Maurice. According to David Bierens de Haan (1878), it was Prince Maurice who asked Simon Stevin to set up a school in which 'would be taught, in good Dutch, arithmetic and surveying, principally for the advancement of those who would dedicate themselves to engineering'. One of the first professors was Ludolph van Ceulen. He taught mathematics, 'surveying on paper, surveying in the field' and fortification-building. He is remembered particularly for his study of π, which he calculated correctly to the 35th decimal place. The result was engraved on his tomb in St Peter's Church in Leiden.

One of the young university's most important resources was its library (see fig. 6). The first volumes were installed in the church known as the *Faliede Bagijnenkerk* in 1591; by 1595 the library already filled an entire floor. The *Polyglot* was the first book in the rapidly growing and constantly improving collection. The original intention was to create a library that would make basic materials available to students who could not afford to buy their own. But within a few years a visit to the library became a question of status. Anybody wanting to consult a book had to apply for the key, as the library was usually locked. New books were bought from publishers and at fairs. The main source of foreign books was the Frankfurt Book Fair, still a leading event in the publishing calendar today. Older editions were also purchased, and some private collections came to the library by donation or bequest.

The university set Leiden's printers to work

The university printers played an important role in the library's acquisition of new books. Although printing had been invented in the mid-fifteenth century,

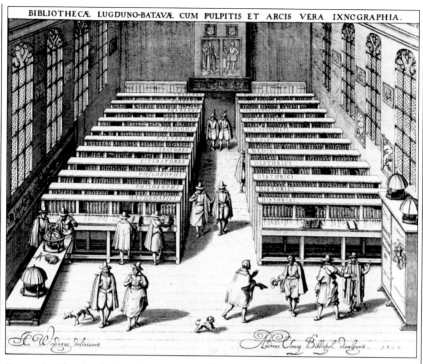

Figure 6: The books in the university library were chained. This engraving by Woudanus dates from 1610.
Leiden, University Library, Bodel Nijenhuis Collection.

most scientific publications were so complicated that only highly specialized houses could take them on. Type for many different languages was required as were compositors able to set it. The correction of proofs was also beyond the capacity of most printers. In 1583 the curators fetched the famous Antwerp printer Christopher Plantin (c. 1520–1589) to Leiden. Antwerp was then a major printing centre, and Plantin brought with him not only three presses but also, and more importantly, tremendous experience and expertise. After two years he returned to Antwerp – the Spanish were again in control of the city and it was peaceful enough to do business there once more – and sent his highly erudite son-in-law Frans van Ravelingen (Franciscus Raphelengius, 1539–1597) to the Leiden printing office. Raphelengius had studied at various universities and had even been professor of Greek at Cambridge. He was the ideal person to take over the running of the printing house. He could select important manuscripts and was able to correct proofs in foreign languages. Besides Latin and Greek he was also very competent in Hebrew. Furthermore, he had taught himself Arabic.

Isaac Elsevier, whose grandfather had been apprenticed to Plantin, started up a printing house in Leiden in 1616. In 1620 he was officially appointed as the university printer. His printing office developed by leaps and bounds, especially under his sons Bonaventure and Abraham: in 25 years they brought out no fewer than 450 editions of high quality. (The current Elsevier publishing house, founded at the beginning of the twentieth century, has no direct relation to its seventeenth-century predecessor, only the name was adopted.)

2.1. The young Stevin is enormously productive. He also applies his theoretical research in practice

Even before he enrolled at the university Stevin had published a number of works, which makes it fairly certain that he had already received an excellent education. Recently a short text has been discovered in the National Library of the Netherlands in The Hague that may well be Stevin's earliest published work. *Nieuwe Inventie van Rekeninghe van Compaignie...gheinuenteert ende nu eerst int licht gheghegeven door Simonem Stephanum* (*New Invention for Company Accounting ... invented and now published for the first time by Simon Stevin*) was published in Delft in 1581 by Albert Hendricz. In 1582, in Antwerp, Christopher Plantin published Stevin's *Tafelen van Interest*. Both short works are to do with bookkeeping and working with money; their contents are thus closely related to Stevin's activities in Antwerp and Bruges. They were followed in 1583 by *Problematum Geometricorum*, a work on geometry printed by Joannes Bellerus in Antwerp. All the later works were published in the northern Netherlands, for the most part by Plantin and his son-in-law Frans van Ravelingen (see box 4). Among the important scientific works Stevin produced in these years are *Dialectike ofte Bewysconst* (1585), the oldest treatise on logic in the Netherlands; *De Thiende* (1585), an argument for the introduction of decimal fractions; *L'Arithmetique* (1585), a summary of algebra of great educational value; *De Beghinselen der Weeghconst* and *De Weeghdaet* (1586), an original treatise on mechanics and in particular statics; and *De Beghinselen des Waterwichts* (1586), an original treatise on hydrostatics. These works followed one another rapidly. Stevin's first years in Leiden and perhaps also in Delft were evidently filled with rigorous scientific study. He achieved the same edifying level in his *Appendice Algebraique* (1595, though probably written earlier). This brief tract is a mere six pages and in later editions it would be added to *L'Arithmetique*. It contained a method for approximating a root of an algebraic equation. The only known original copy of this work was lost in 1914, when the Leuven University Library was burned to the ground. Fortunately, we know its contents via a 1625 reprint of *L'Arithmetique*.

Stevin certainly did not confine himself to theoretical considerations in these years; he also applied his knowledge. In 1584, via his friend Johan Cornets de Groot, he began negotiations with Delft's town council for permission to test one of his inventions to do with drainage. In addition, and usually in consultation with the same friend, he improved the working of the mills used for pumping water from marshlands. Johan Cornets de Groot was burgomaster of Delft and father of the jurist Hugo Grotius. In 1588 Stevin came to an agreement with him to put his hydraulic inventions into practice. With this end in mind he acquired a number of patents. Again we see that, just as in Bruges, Stevin sought the support of the leading figures in his circle. These relationships led to his contacts with Poland, in particular with the towns of Gdańsk and Toruń.

We have already noted that in some of his works Stevin refers to detailed descriptions and maps of these areas. In 1979, Bernard Woelderink found confirmation of the journey that Stevin made to Poland in 1591 in the national archives in Gdańsk. He came upon a letter dated 22 June 1591 (see box 5) from the Rotterdam town councillor Fop Pietersz. van der Meijde to the town council of Gdańsk. Van der Meijde was a cheese merchant and herring shipper. He became a member of the Rotterdam town council in 1580 and remained so until his death in 1616. In this period he served as burgomaster 11 times. From 1590 his attendance at the assemblies of the States of Holland as a member of Rotterdam's delegation was more or less uninterrupted. He was one of the most enterprising and important representatives of the merchant class in Rotterdam, and played a major role in the Baltic trade. From the letter it appears that he had received several communications from Gdańsk's town council in the previous two years, asking him to recommend a suitable individual or individuals to draw up a plan for the deepening of Gdańsk's harbour and probably also to supervise the execution of the project. Van der Meijde replied that he had discussed the matter with Johan de Groot of Delft and Simon Stevin, and that they were prepared to take on the task. But it was of course necessary to study the situation *in situ*. Stevin, who had been too busy for a visit the previous summer (of 1590), was on his way to Gdańsk as the 'bringer' of the letter. The description of Stevin as the letter's 'bringer', combined with the presence of the letter in the archives at Gdańsk, makes it virtually certain that Stevin was indeed in the Polish port in the summer of 1591.

Box 5: A letter to the town council of Gdańsk

Letter written by Fop Pietersz. van der Meijde. The (Dutch) original is preserved in the national archives in Gdańsk, rubric 300,53.
Correspondence of Gdańsk town council
(Inventory no. 796: Correspondence with the Netherlands, 1580–1600).

Noble wise and worthy Sirs,

As your Worthies wished to have your harbour deepened and wrote to us here about talking to those who would be able and willing to do this, I discussed this (as also mentioned in my previous letter) with Mr Jan de Groot, at present burgomaster of Delft, and with Mr Simon Stevin, who are ready to deal with the matter in every detail.

> But since nothing of this can be done without inspecting the places and indeed the whole situation, as your Worthies also said was necessary in your letter, I arranged – both to satisfy your desires as expressed in several letters to me, and to please numerous merchants and mariners with whom I also spoke – that the aforementioned Mr Stevin, bringer of this letter, should come there to see the situation himself, which he was not able to do last summer due to other engagements.
>
> God grant that the matter may be undertaken and brought to a successful end to the prosperity of the royal trading town of Danssick [Gdańsk] and to the advantage of mariners and other merchants – amen.
>
> Noble wise learned and very worthy Sirs, may the Almighty Lord grant you all a long and happy life – amen.
>
> Written in haste in Rotterdam this 22nd of June 1591.
>
> Your faithful servant and friend Fop Pietrsz. van der Meijde
>
> Written on the address side of the folded letter:
>
> Noble wise learned and very worthy Sirs, my Lords Burgomasters of the royal trading town of Danssick.
>
> Furthermore, the secretary to the town council has noted here:
>
> > R. augustij anno 1591
>
> That is to say, received in August 1591.

2.2. Friendship with Maurice of Nassau ensures Stevin's future in the northern Netherlands

Stevin's friendship with Maurice of Nassau – later prince of Orange, stadholder and captain-general of Holland and Zeeland – dates virtually from the moment the Bruges scholar moved to the north (see box 6). Maurice was already resident in Leiden in July 1582. He was matriculated into the university on 19 April 1583, a few months after Stevin. He left suddenly on 10 July 1584, following the assassination of his father William the Silent, prince of Orange. Thus Maurice and Simon were in Leiden at the same time. The student body was still small at that time (see box 4) and it is more than likely that the two men met. Some biographers even assert that Stevin acted as the prince's *praeceptor* in his study of geometry as early as 1584.

What is certain is that Stevin entered the prince's service around 1593. In letters to both the States General and the Council of State, the prince indicated that Simon Stevin would be entitled to a salary of at least 50 pounds per month 'as designer of the army's quarters, which he need no longer do for nothing as he has done for ten years now'. By then Stevin had already been working in the army for quite some time without ever having been appointed to an official position by the States. He was probably in the prince's personal service. In any case it is striking that from this time onwards all Stevin's works contributed to the construction and fortification of the northern Republic, or were written especially for the prince.

2.3. The scholar from Bruges becomes involved with the politics of the northern Netherlands

A noteworthy work dating from this period is the *Vita Politica, Het Burgherlick Leven* (*Civic Life*) (1590), in which Stevin considers the subject of good and bad citizenship. At a time when the Dutch Republic was being wrought into shape, Stevin sought to impart to its inhabitants a sense of public spiritedness and responsibility. This 56-page booklet was printed by Frans van Ravelingen at the Plantin press in Leiden, and there were many reprints, including an edition brought out in 1611 by Jan Andriesz. at Delft. It was also included in the *Materiae Politicae*, published by Stevin's son Hendrick in 1649. The *Vita Politica* was dedicated to Govert Brasser, burgomaster of Delft, where Stevin was probably staying when he wrote the book. It is written in 'good Dutch words', no doubt to make it accessible to the ordinary citizen to whom it was addressed.

According to Stevin, it is easy to practice good citizenship as long as the civil, natural and divine rules from which the organization of the citizens receives its form are neither in contradiction with each other nor in conflict with the views of the citizen. However, when conflict does exist or when the citizen finds the obligations laid upon him to be unjust, being a good citizen becomes more difficult. In *Het Burgherlick Leven*, Stevin sought to provide a guideline that could help to determine the right course of conduct in such cases. His premise rests on the dictum that 'everyone has to be a faithful and loyal subject of his authorities, as being obliged by the civil laws'. Who is that authority? Here too, Stevin provides an answer: 'Everyone must always consider as his rightful authority those who at the present are actually governing the place where he chooses his dwelling, without concerning himself with the question of whether they or their predecessors have reached their position justly or unjustly.' This definition must be set in its historical context. The Republic had sundered itself from the Spanish Netherlands and repudiated the Spanish king as head of state. Many of those living in the Republic were, like Stevin, immigrants. Stevin's definition is an appeal for respect for the present form of government. He also demands that wherever one settles, one should respect and conform to the authority in place. But keeping to the rules can be difficult when there is internal discord between two parties, each of which has a lawful ruler at its head. Here Stevin has in mind the Spanish king on the one hand and the stadholder on the other. Stevin believes that one behaves as a good citizen when one takes account of the nature of the form of government. As an example he cites the case of a *Staetvorst*, or constitutional monarch, who rules with a parliament at his side and may not disregard parliament because he has sworn to uphold this system and must therefore respect it. If he seeks to override the parliament it is the good citizen's duty to oppose him. As this is how the Republic came into being, we may be fairly confident that Stevin is describing the formation of the United Provinces and the split from the Spanish-ruled south.

The citizen's behaviour in respect of the laws of God and religion is a particularly important aspect of *Het Burgherlick Leven*. Stevin takes it for granted that all parents like to see their children grow up 'in virtue and righteousness'. To this end a religion is necessary. Stevin's conclusion is clear: 'There must then of necessity be a religion, and without religion everything is lost.' Likewise Stevin deals with the conduct of a citizen in a country whose religion he cannot profess. This was a very real situation in the Republic, where Calvinism was the official religion and yet where many Catholics still lived. Stevin outlines two possible cases. The first is that of the individual who, 'without any belief in God or the Devil, or doubting and having made no certain decision about it, considers religion only as an invented instrument suitable for keeping a tight rein over the communities.' The second is that of those who 'with full confidence do hold a religion as certain and sacred, but one which is not allowed in the country they live in'. Of those who fall into the first category, 'reason demands that they conform to the present ordinances of the place where they chose their dwelling' and they should not 'combat and fiercely ridicule' its religion. Those who find themselves in the second category can 'serve their God in secret according to their wits'; however, they may not 'propagate their views secretly, secretly doing as much as they can against the country's religion'. It is worth noting that in this work Stevin proposes very clear and generally applicable rules while giving no hint at all as to his own opinions or ideas on religion. In conclusion, he states that it is best to find a place to live in which one can comfortably conform to the customs that prevail. If we apply this view to Stevin's emigration to the north, we can conclude that he chose for himself a dwelling place where he could conform to Calvinism, which was the official religion of Holland and Zeeland.

2.4. Stevin's mathematical approach to fortification-building is translated into German, French and English

With a view to serving both the Republic and the interests of Prince Maurice, shortly after *Het Burgherlick Leven* appeared, Stevin published two books that were of practical use for the defence of the country and the development of its nascent fleet. *De Sterctenbouwing* (*The Art of Fortification*) was printed by Frans van Ravelingen in 1594; as the title suggests, it deals with the theory of fortification-building. It could be that it was intended as course material for the engineering school that Stevin had helped to set up in Leiden (see box 4).

Since the start of the sixteenth century, fortification construction had been constantly evolving. The development of heavy artillery made the need for new defensive systems imperative. Stevin had experience of the practice; now, in

De Sterctenbouwing, he provided a theoretical mathematical foundation for it. He also introduced Dutch words in place of the French and Italian terms then in common use. A German translation appeared under the title of *Festung-Bawing* in 1608 (and was reprinted in 1623). In that same year a French version came out, entitled *La Fortification*. Remarkably, in the library of Trinity College in Cambridge is a handwritten English translation dated 1604 and entitled *The Building of Fortes*. Its author is referred to as 'Symon Stephen of Bridge'.

In 1599, the second of these useful books was published, again by Frans van Ravelingen. *De Havenvinding* (*The Haven-Finding Art*) describes a method by which a ship can determine its bearings at sea. It is only to be expected that the ever-inventive Stevin, who was now an inhabitant of a burgeoning sea power, would put his mind to the problems of navigation.

2.5. Stevin's sailing chariot

As far as the ordinary citizen was concerned, Stevin's most popular invention was his sailing chariot (see fig. 7), which he is known to have designed between 1600 and 1602. Sailing chariots were not new: the Chinese had built them and Dutch seafarers described them on their return from the Orient. Stevin probably

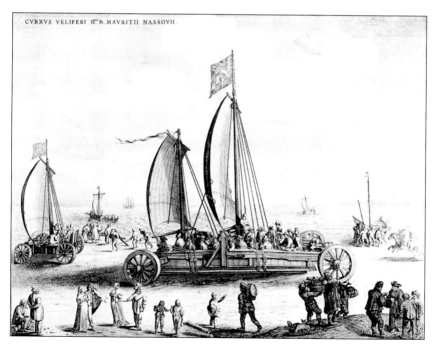

Figure 7: Simon Stevin's sailing chariot gained him his greatest popularity with ordinary folk. Many artists depicted the chariot. Shown here is an engraving from Joannes Blaeu's *Tonneel der Steden* (1652).
Leiden, University Library, Bodel Nijenhuis Collection, Atlas 49-1.

constructed several models and 'sailed' them as well. The fame of his chariot spread well beyond the borders of the Dutch Republic. Indeed, in Laurence Sterne's *The Life and Opinions of Tristram Shandy*, Uncle Toby dilates on 'the celebrated sailing chariot, which belonged to Prince Maurice, and was of such wonderful contrivance and velocity, as to carry half a dozen people thirty German miles, in I don't know how few minutes [...] invented by Stevinus, that great mathematician and engineer.'

Hugo Grotius, who was a boy when Stevin's chariot made its first runs, recalled the experience in several poems. Various drawings were made of the sailing chariot. One of the best known is by De Gheyn, and the text accompanying the first edition of the drawing gives a lively description of one of the chariot's outings. We are told that 28 people had joined Prince Maurice aboard the chariot. The prince himself was at the helm. A strong southeasterly wind was blowing. The chariot gathered such speed that it seemed rather to fly than to roll and was no sooner seen than gone. The story continues: 'At one moment, for fun and in order to play a trick on the gentlemen, his Excellency steered the chariot into the sea, which movement struck many with great fear; but as he moved the helm in good time the chariot struck the beach again and sped along its former course.'

The sailing chariot inspired other artists too. The Royal Danish Collection (Rosenborg Castle) in Copenhagen contains a tablecloth made in the royal silk workshops of Christian IV of Denmark. Depicted in the border is Simon Stevin's sailing chariot. Karl Thiessen, chief supervisor of the silk workshops, was most likely the tablecloth's maker, but the design seems to have been by Passchier Lammertijn and was based on De Gheyn's picture. Lammertijn was known in Holland for his application to the States General in 1601 for a patent for an invention intended to simplify the weaving of patterns into damask. Christian IV must also have been interested in the sailing chariot, seeing that one of its passengers – at least according to the story accompanying De Gheyn's drawing – was apparently his brother, Ulric of Holstein.

Maurice was extremely interested in applied mathematics. In the voluminous *Wisconstige Gedachtenissen* (published in two parts in 1605 and 1608), we find the contents of the lessons Stevin devised when he was appointed as the prince's private tutor. Stevin published little more after this magnum opus. In 1617 he dealt with subjects derived from his practical experience with the army in *Castrametatio* or *Legermeting* (*The Marking Out of Army Camps*).

In his *Nieuwe Maniere van Sterctebou, door Spilsluysen* (*New Manner of Fortification by means of Pivoted Sluice-Locks*), published together with *Castrametatio*, Stevin worked out plans to use water for the defence of fortified places. Both works were printed by Jan van Waesberghe in Rotterdam. In the dedication of *Castrametatio* to the States General, Stevin styled himself '*Legermeter*' or 'army measurer' – a function that will not have differed much from that of a quartermaster general. However, there is not a single document to indicate that he was appointed as such.

From the discussions above, we can see that in the northern Netherlands Simon Stevin did indeed contribute to the building of the new Republic.

Box 6: Maurice of Nassau

Maurice was born in Dillenburg in Westphalia on 13 November 1567, the son of William I of Orange and Anne of Saxony. His mother's loose living and wayward conduct (which included adultery with Jan Rubens, father of the painter Peter Paul Rubens) made his parents' marriage a hell. In 1571 the marriage was annulled and Anne was imprisoned in Beilstein Castle, near Herborn, where she died insane on 18 December 1577. Maurice was brought up and educated as a German nobleman at the court of his grandmother, Juliana, at Dillenburg. From April 1583 to September 1584 he studied at Leiden University. In August 1584, following the assassination of his father the previous month, the 16-year-old Maurice was nominated head of the new Council of State by the States of Holland, Zeeland, Utrecht and Friesland, and nominally also those of Flanders, Brabant and Mechelen. This council had no real power, however. On 1 November 1585, the States of Holland and Zeeland appointed him as stadholder of those provinces. He was strongly supported by Johan van Oldenbarnevelt, Grand Pensionary of Holland. In 1590 and 1591 the States of Utrecht, Guelders and Overijssel also named Maurice stadholder and captain-general of their provinces. On 8 March he became commander-in-chief of the Republic's army (see fig. 8).

Figure 8: Maurice, prince of Orange (1567–1625), was stadholder and captain-general of the States of the Republic. He was considered to be one of the most important military leaders of his day. His friend Simon Stevin instructed him in all kinds of scientific subjects. Here Michiel Jansz. van Mierevelt has portrayed Maurice as a military commander. The painting is now in the Rijksmuseum, Amsterdam (ektachrome: JTD).

Maurice defeats the Spanish in the Nieuwpoort dunes

In the meantime, Maurice's confrontation with Alexander Farnese, duke of Parma and Spain's governor-general of the Netherlands, had begun. Like Parma, Maurice was a master of the new, scientifically based warfare and after Parma's death was indisputably the greatest strategist of his day. His capture of Breda in 1590 signalled the start of a series of military successes. Towns such as Hulst, Nijmegen, Zutphen, Deventer and Groningen were taken in rapid succession. The battle of Nieuwpoort in 1600 was a remarkable military achievement. The States ordered Maurice to strike along the Flemish coast and wipe out the pirates who were launching increasing attacks on the merchant fleet from their base in Dunkirk. At this time the Spanish army was paralysed by mutiny and Maurice advanced as far as Nieuwpoort with little opposition. On 2 July, however, the Spanish unexpectedly moved into action once more and Maurice suddenly found himself cornered by crack Spanish infantry, with pitched battle his only option. Fighting began on the beach around midday and spread into the dunes. The outcome remained in doubt until four o'clock. Then Maurice sent in the cavalry he had held in reserve and the Spanish were routed. Maurice had won the battle of Nieuwpoort, but he failed to reach Dunkirk. His army was dangerously exposed and under-provisioned, and he was apprehensive of a renewed Spanish offensive. Deeming discretion the better part of valour, he took ship and withdrew from the Flemish coast.

Maurice's military successes greatly increased his standing and wealth. In the years that followed, the Spanish general, Ambrosio Spinola, won the odd engagement here and there, but for both the North and South, the war reached a stalemate.

In 1607 peace negations between Spain and the Republic began in The Hague. Spain was prepared to recognize the Republic as an independent power on the condition that Catholics were granted open and public toleration of their faith, and that the Republic withdrew from the Indies and dissolved the East India Company. In the face of these unacceptable demands the talks broke down. Then the possibility of an armistice was raised. Maurice was vehemently opposed to this, but Oldenbarnevelt, championing peace, prevailed, and in 1609 the Twelve Years' Truce was signed. As armed conflict abated, however, domestic differences provided a new battlefield. In dispute were foreign policy, autonomy of the provinces and the relationship between Church and State. Discord polarized into two parties, the Arminians or 'Remonstrants', religious moderates and republicans, and the Gomarists or 'Counter-Remonstrants', who advocated doctrinal orthodoxy and supported the stadholder (see also box 2). The Remonstrants were led by Oldenbarnevelt; Maurice himself headed the Counter-Remonstrants. Oldenbarnevelt lost not only the argument but also his head; he was executed in 1619. This failed to heal the opposition between the parties, however. It persisted throughout the history of the Republic, until the Republic itself became a casualty of the French Revolution. In 1620 Maurice also became stadholder of Groningen and Drenthe, succeeding Willem Lodewijk.

Maurice surrounds himself with specialists

Maurice's private life was not always exemplary. He lived an active soldier's existence and was usually surrounded by German nobles who were known for their rough lifestyle. When not campaigning he resided at the Binnenhof in The Hague. As a result, the town was increasingly frequented by foreign statesmen and dignitaries and it grew into a royal power centre. Maurice had many problems to contend with, and therefore he needed specialist knowledge. He was particularly interested in mathematics and related subjects. He surrounded himself with intellectuals and erudite officers. In the winter months the court at The Hague must have resembled a university, where Maurice, together with Willem Lodewijk and his staff, pored over such subjects as fortification-building, ballistics and maritime navigation. They also made a thorough study of the history of classical warfare. Drill models were developed and then tried out in the Buitenhof. Maurice was a consummate horseman, and had his own stud at Rijswijk near The Hague. His equestrian interest explains his quest for a good bridle (Stevin gives a description of this in his *Wisconstige Gedachtenissen*). Between 1596 and 1624 his stud was run by Jan Evertsz., a German from Mülheim. A recently discovered register of 1604 tells us that this important courtier received a salary of 600 guilders per annum (see also Chapter Five, section 4).

Maurice never married. From around 1600 to 1610 he lived with Margaret of Mechelen. This Southern Netherlandish noblewoman was the daughter of Cornelis, alderman of Lier, and Barbara of Nassau Conroy, a descendent of a bastard branch of the Breda Nassaus. Margaret was brought up by her aunts in Leiden and she probably remained a devout Catholic to her death. Her relationship with Maurice produced three illegitimate sons: Maurice (1601–1617), William of Nassau-de-Leck (1602–1627) and Louis of Nassau-Beverweerd (1604–1665). In Maurice's will these sons were favoured significantly over his other bastards. To Margaret he bequeathed an annuity of 6,000 guilders, a very considerable sum. Besides Margaret, Maurice had at least five other mistresses, who between them gave him at least five more illegitimate children. The court at The Hague also served as a training school for Maurice's pages and natural sons. Jacques Wijts (see box 2) most likely played a leading role in their education. Maurice died in The Hague on 23 April 1625.

3. Married life and descendants

Simon Stevin married Catharina Craiy (also written as Craey, Krai, Kraai, Carels or Caerls). The date of the actual wedding ceremony is unknown. Her brother, Jehan Carels, was an engineer in the army of the Swedish king, Gustaaf Adolf, and was described by Stevin in a letter to the king as *frere de ma femme*. G. Van Rijckhuijsen tells us that she was *een zeer mooy vrouwspersoon* – a very beautiful female.

3.1. Three natural children

Before his marriage, probably in anticipation of it, Stevin bought a newly built house (see fig. 9) in The Hague – number 42 in the Raamstraat – on 24 March 1612, paying 3,800 guilders for it. The house was not yet finished for the deeds stipulated that the house should be ready to be handed over on 1 April 1612. Thanks to the investigative persistence of Theo Morren (1899) we know something of the house's history. Its builder, carpenter Abraham Jansz., had bought the ground from the apothecary Jan Claesz. Splinter, who originally had a herb garden on the site. After Stevin's death in 1620, his widow sold the house to

Figure 9: In the autumn of his life Stevin married Catharina Craiy. The couple had two sons and two daughters. The family occupied this handsome town house on the Raamstraat in The Hague. The medallion, added to the facade in 1897, proclaims that Stevin lived here from 1612 to 1620 (photo: G. Vanden Berghe).

Dr Clement Overschie on 6 May 1623 for 4,800 guilders. In 1897 the facade was improved with a bust of Stevin set in a roundel.

It was probably in this house that Stevin's four children were born: Frederik in 1612, Hendrick in 1613 and then the two girls, Susanna on 29 April 1615 and Levina, whose date of birth is unknown. The birth dates we do have are not derived from baptismal registers, for these were not introduced in The Hague until 1630, but have been inferred from references to the children's ages in various other official documents. The Hague's register of intended marriages records that the notice of Simon and Catharina's marriage was given on 10 April 1616. If these data are correct, it follows that certainly three of the children were born out of wedlock. This is the deed, incidentally, which mentions Stevin's middle name as Anthonis.

On 2 February 1619, according to a deed drawn up by notary Johan Adriaensz. van Warmenhuysen, Stevin bought from Didolff Duyren 'a certain house and grounds located in the Nieuwe Houtstraat on the corner of the Doelstraat in this town'. The purchase price was 6,300 Carolus guilders in cash money. The family probably continued to live in the Raamstraat, but it may be observed that at this point Stevin was clearly doing very nicely indeed – the sums he paid for the two houses were quite considerable.

3.2. Stevin's death

Simon Stevin died in 1620. Definite data about his death are as scarce as those that relate to his birth. The painting in Leiden University Library mentions 1620. On 20 February of that year Simon Stevin was still alive; he signed a certificate in which he declared that Dirck Gerritsz. Langedijck, having taken an examination, was competent to act as a surveyor. On 10 April 1620, the Rotterdam mathematician Jan Jansz. Stampioen submitted a request to the Council of State to be appointed to the position in the army left vacant by Stevin's death. Moreover, on 8 April 1620 the Court of Holland refused a petition by Stevin's widow Catharina, in which she requested the Court, in her late husband's name, to charge Bartholomeus Panhuysen, councillor and chief treasurer to the prince, and Willebrord Snellius, professor of mathematics at Leiden, with the guardianship of the children – a responsibility which, for reasons unknown, they declined. On 1 May 1620, Andries de Hubert, secretary-in-ordinary to the Court of Holland, and Anthonis Beyts were appointed as guardians. The States General granted Catharina's request for the sum of 50 pounds for services rendered by her husband. On 5 August 1621, the States granted to her and the children's guardians the exclusive right to print or publish the *Wisconstige Gedachtenissen* or translations thereof in the United Provinces for the next seven years. Until 1626, Stevin's children received from Prince Maurice an annual allowance of 400 pounds. It is extraordinary that so little is known about the death and burial place of Simon Stevin, who in the last 20 or 30 years of his life had been one of the Republic's most notable figures.

On 14 March 1621, Stevin's widow wed Maurice de Viry (or Virieu), steward of Hazerswoude and son of François de Viry, lord of St Raphiau, a native of

Orange in southern France. In the notification of the nuptials, Catharina Craiy was referred to as the widow of *Symon van Stevyn in sijn leven Raet ende Super-Intendent van de finantiën van Z. Furstelycke Genade den Prince van Oraengien* – 'Simon Stevin, in his life Councillor and Superintendent of finances of His Royal Grace the Prince of Orange'. On 6 May, Catharina and Maurice sold the Stevin home on the Raamstraat and on 4 October 1631 they sold the house on the Houtstraat. The second house had been let for 540 guilders per year. In 1627 it was home to Margaret of Mechelen, the mother of Prince Maurice's three children (see box 6). From The Hague the family moved to Hazerswoude. There, on 15 June 1624, Isaac Beeckman borrowed from Stevin's widow a number of papers the Bruges mathematician had left. In 1639 the family was once again living in Leiden, on the Houtmarkt. Here Maurice de Viry died in 1650 and Catharina on 5 January 1673; both are buried in St Peter's Church in Leiden. Their marriage produced six children.

3.3. Stevin's descendants

What of Simon Stevin's children? Frederik, Simon and Catharina's eldest son, first studied mathematics under Jacob Beeckman in Rotterdam but on 5 December 1629 he enrolled at Leiden University as a student of theology. He enrolled once more on 11 February 1639. On that occasion he was referred to as a doctor of law. It is not known where he had acquired this degree. He died unmarried on 15 December 1639 in Leiden.

Hendrick's career resembled his father's, though he lacked his father's brilliance. When he matriculated into Leiden University on 14 February 1639 as a student of mathematics and stated that he and his brother Frederik lived with their mother and stepfather on the Houtmarkt, he already had a military career behind him. On 13 January 1633 he was appointed quartermaster of the lord of Brederode's regiment, where he succeeded his uncle Pieter Craiy. Injury obliged him to quit his military service. He devoted himself to the study of physics and technical matters. In addition, he evidently considered that he owed it to his father's memory to trace and publish the manuscripts Simon Stevin had left. His mother had treated the papers very carelessly, and Hendrick complained about it more than once in no uncertain terms. He retrieved the papers from Isaac Beeckman. Thanks to Hendrick, today we have the *Materiae Politicae*, in which several of Stevin's unpublished works finally appeared, 30 years after his death. Other works by Stevin appear in Hendrick's own *Wisconstich Filosofisch Bedryf* (*Mathematical Philosophical Transactions*).

On 20 June 1642 Hendrick married Joanna van Leeuwen, lady of Alphen, widow of Gerrit Beunen van Wender, lord of Alphen-on-Rhine. After the death of his wife in 1655, Hendrick himself officially became lord of Alphen, though in fact he had assumed the manorial title even earlier. In January 1670 he died, leaving no children. It is worth mentioning that Hendrick Stevin derived his coat of arms from the famous figure of the *clootcrans* or 'wreath of spheres', to which his father was so attached. Until 1795 his arms hung in the church at

Alphen. G. Van Rijckhuijsen describes them as follows: 14 red spheres on a white field, in the left quarter a black carpenter's square. Hendrick also took his motto from that of his father, albeit with an addition. *Wonder en is gheen wonder* – 'Magic is no magic' – Simon had inscribed around the wreath of spheres, to which Hendrick added *En gheen wonder is wonder* – 'And no magic is magic'.

We know that Stevin's elder daughter Susanna married Petrus Vlietentoorn (or Vliedthoorn), a Protestant minister in Katwijk, on 29 April 1635. The couple had three children: Simon, Cornelis and Catharina. One of Catharina's sons was the aforementioned Master Juliaen van Groenwegen, who contacted the Leiden town bailiff to see what he could discover about his great-grandfather. Simon and Catharina's younger daughter, Levina, married Johan Roosenboom, a Court of Justice prosecutor. The baptismal register of the Groote Kerk at The Hague records the christening of their daughters Maria (8 December 1652) and Alida (20 February 1656); in the baptismal register of the Kloosterkerk the christening of their sons Simon (19 June 1654) and Karel (1 April 1661) is recorded. From other sources it appears that two more daughters had been born, Geertruyt (15 March 1647) and Catharina (13 February 1650); Catharina later married Joannes Vollenhove.

Chapter Three

The man who 'invented' the decimal system

Simon Stevin's De Thiende (The Dime), *published in 1585, was one of his most influential works. Aided by its rapid translation, first into French and then other languages including English, this short essay on the decimal system came to be very widely read. In a time of increasing trade contacts, there was a great need for uniform standards of measures and money. Even so, it would be the end of the eighteenth century before the political climate was right for the sort of rationalization that Stevin proposed.*

1. Decimal fractions for elementary calculations

Simon Stevin's most radical mathematical invention was his elegant way of representing decimal fractions and using them to carry out elementary calculations. It was on this subject that he published his essay *De Thiende* (fig. 1) in 1585. Stevin had then been in Holland for at least four years, probably spending most of his time in Leiden. The slender 37-page volume contains the description of decimal fractions, their application in addition, subtraction, multiplication, division and, briefly, root extraction. The pamphlet was published in Leiden at the printing office of Christopher Plantin, at the time when the northern provinces had broken away from the Spanish Netherlands and were confronted with the task of forming a state. It is a constant in history: when states are on the point of forming or consolidating, a standardization of measuring systems is invariably sought. Charlemagne attempted it in his empire. In France, in 1558, Henry II required that weights and measures throughout the realm should conform to those used in Paris. In the Appendix to *De Thiende*, Steven writes in the same spirit of the practical application of decimal fractions for dividing money and measurements of length, area and volume into tenths.

2. Translations of Stevin's *De Thiende*

Stevin wanted his work to be read as widely as possible. At virtually the same time as the Dutch version appeared, he published his own French translation of *De Thiende* in his work *L'Arithmetique* (1585) (fig. 2), sonorously entitled *La Disme, enseignant facilement expédier par nombres entiers sans rompuz, tous*

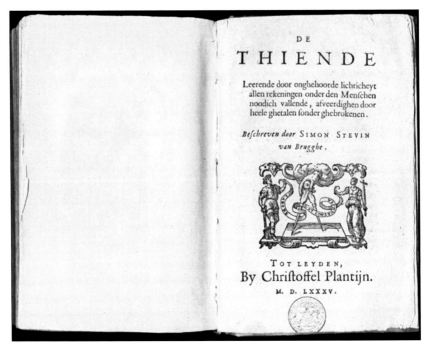

Figure 1: In 1585, Simon Stevin published the short essay *De Thiende*, a proposal for the introduction of the decimal system. He gave the arithmetical rules for the addition, subtraction, multiplication, division and root extraction of decimal numbers. He also described the practical application of decimal numbers in everyday life.
Antwerp, Plantin–Moretus Museum (ektachrome: JTD).

comptes se recontrans aux affaires des Hommes. Premierement descripte en Flameng, et maintenant convertie en François, par Simon Stevin de Bruges.

This title tells us that the Dutch version was indeed the first and that Stevin employed the term *flameng* for the language he himself used. In 1602, a Danish translation of *De Thiende* appeared, produced by Christoffer Dybvad, under the title of *Decarithmia, det er Thinde-Regnskab*. An English translation by Robert Norton was published in 1608 as *Disme or The Art of Tenths*. Norton, the son of an English lawyer and poet, was an engineer and gunner in the royal service and the author of several works on mathematics and artillery. The English edition of *De Thiende* is a literal translation of Stevin's French text. However, Norton made a number of additions: a short foreword 'to the courteous reader', a table for the conversion of sexagesimal fractions (based on sixtieths) into decimal fractions (based on tenths) and a short exposition on integers and the use of the rule of three. In Stevin's day, numbers were still very often recorded in sexagesimal notation: the number was written as a sum of fractions with denominators of 60, the second power of 60, the third power and so on. Nowadays, we are familiar only with the decimal system, in which division takes place with fractions whose denominators consist of whole powers of 10.

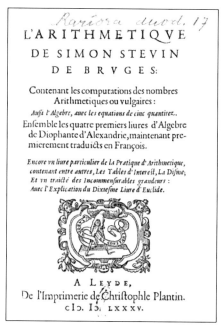

Figure 2: To ensure that his *De Thiende* was widely read, Stevin published his own French translation in *L'Arithmetique*. A few years later English and Danish editions of the work appeared. *Utrecht, University Library, Rariora duod. 17.*

The English and Danish editions appeared more than 15 years after the original Dutch publication. The reason for this was a privilege granted to the printer, Christopher Plantin, by Maurice of Nassau and the Council of State of the United Provinces. An extract from it appears on the last page of the original pamphlet.

3. Uniform ways of counting and measuring

In the growing towns and cities, with their rapidly expanding commerce, there was a great need for a definition of elementary rules of reckoning. Administration, bookkeeping and surveying demanded some skill in calculating and measuring. The expansion of the wine trade between the Flemish towns and France and Germany made the conversion of liquid measures necessary. The collection of levies on beer and wine or other merchandise presupposed the power to control the volume of casks and barrels. Since the fourteenth century numerous departments and posts had been created to deal with this: the gauger of weights and measures, the wine gauger, corn gauger, land gauger, bookkeeper and so on. These officials, master masons and merchants, possessed no great mathematical knowledge. Often their only education was experience. As apprentices they absorbed the expertise and traditions of their masters. But as the fifteenth century progressed, the growing importance of these offices stimulated a demand for better schooling and teaching devices. For the nascent printing trade, this increasing interest in immediately available and applicable knowledge opened up a very promising field of activity at the start of the sixteenth century. In her 1992 study Marjolein Kool mentions 22 arithmetical works that had appeared in the Dutch language before 1585. She examines the technical mathematical vocabulary that originated in this period and is still in use today.

Simon Stevin's *De Thiende* belongs to this context of applied mathematics for the specialist craftsman. Indeed, he begins his work with the following dedication:

> *To Astronomers, Land-meters, Measurers of Tapestry, Gaugers, Stereometers in general, Money-Masters, and to all Merchants, SIMON STEVIN wishes health.*

Stevin hoped that the 'land-meters' (surveyors) would apply his decimal division to the rod, the tapestry-measurers (who included clothiers) to the ell, the wine gaugers to the aum, the astronomers to the degrees of a circle and the 'money-masters' to their ducats and pounds. From his Antwerp period Stevin already knew of one example of decimal division: the Antwerp aum was divided into 100 'pots'.

It is clear that Stevin regarded the content of *De Thiende* as his own invention and that he was very proud of it. In his introduction he expresses it as follows:

> *Therefore, if any will think that I vaunt myself of my knowledge, because of the explication of these utilities, out of doubt he shows himself to have neither judgment, understanding, nor knowledge, to discern simple things from ingenious inventions, but he (rather) seems envious of the common benefit; yet howsoever, it were not fit to omit the benefit hereof for the inconvenience of such calumny. But as the mariner, having by hap found a certain unknown island, spares not to declare to his Prince the riches and profits thereof, as the fair fruits, precious minerals, pleasant champions, etc., and that without imputation of self-glorification, even so shall we speak freely of the great use of this invention; I call it great, being greater than any of you expect to come from me.*

> [If anyone thinks that, in explaining the usefulness of decimal numbers, I am boasting of my cleverness in devising them, he shows beyond doubt that he has neither the judgement nor the intelligence to distinguish simple things from difficult, or else he is envious of something that is for everyone's benefit. However this may be, I shall not fail to mention the usefulness of these numbers because of such a man's calumny. Just as the mariner who has by chance found

an unknown isle may declare all its riches to the king, such as its having beautiful fruits, precious minerals, pleasant plains, etc. without it being imputed to him as conceit, so may I speak freely of the great usefulness of this invention, a usefulness greater than I think any of you anticipates, without constantly priding myself on my achievements.]

Although in Stevin's day it was not the custom to refer to the work of predecessors, he himself took great pains in some of his works to mention every source, giving credit where it was due. However, following contemporary practice, nowhere in *De Thiende* does he mention a source except for a reference on page 20 to the *Boogpees Tafelen* – the goniometric tables of Ptolemy and Jan van Kuenincxberghe (Johann Müller of Königsberg, who signed himself Regiomontanus) – because in these tables decimal fractions were multiplied with an appropriate multiple of 10 (see also box 1). Stevin makes a similar reference in his *Wisconstige Gedachtenissen*, an immense volume of 'mathematical memoirs'.

Box 1: The decimal system before and after Stevin

In the course of his history, man has devised and used various kinds of numerical notation. The oldest known and sometimes still practiced method is that of cutting notches in a stick, with each notch representing one unit. A slightly more advanced approach, still in use today, involves making a vertical line for each of the first four units and representing the fifth unit by an oblique line struck through the first four.

The Romans gave us Roman numerals. The letters I, V, X, L, C, D and M stand for 1, 5, 10, 50, 100, 500 and 1,000, respectively. By the cumbersome combination of these letters all the smaller numbers could be represented. This system is still used in certain contexts – for indicating dates on buildings, for example. Systems that use letters for numbers were widespread in antiquity – the Phoenicians, Hebrews and Greeks all employed them – but they had two significant disadvantages. First, calculating with these numbers was not simple; and secondly, still other letters had to be introduced to represent greater numbers.

The Sumerians were the first to use a form of decimalization

In the fourth millennium BCE, the first system to use decimal subdivisions was introduced in Mesopotamia by the Sumerians. They based their numerical system on powers of 60, which were subdivided in multiples of 10. The organization of time that we still use today derives from this system: every hour is subdivided into 60 minutes, which are in turn subdivided into 60 seconds. The Sumerians had no symbol for 0. This appears to have been introduced in the first millennium BCE. It is thought that 0 was proposed by Hindu mathematicians because the concept of nothingness was central to their religion and philosophy. In addition to 0, the Hindus introduced the idea of the place value system, which was based on a pure decimal concept. Thus, each decimal number could be written without any ambiguity with the help of 10 different symbols – the numerals 1 to 9, plus 0. Although these numerals were introduced in India, they are referred to worldwide as 'Arabic' numerals.

The Maya of Central America had a place value system with a base number of 20. It was claimed that they had adopted the system of the Olmecs, who were believed to have used it around 1500 BCE. They represented every numeral between 0 and 19 by just three symbols. These numerals formed the basis for a vigesimal system. Although this is in all probability the most ancient place value system, it has had no influence on European science – and mathematics in particular – for the Hindu–Arabic system, with its 10 symbols, was already in widespread use by the time the New World was discovered.

Despite the simplicity and utility of the decimal numerical system, centuries would pass before it was accepted in Europe. It was not until the time of the crusades that works by Arab writers such as Al-Khwarizmi were heard of. Books in Arabic were translated into Latin, the lingua franca of European learning, by scholars such as Adelard of Bath. In 1202, Leonardo of Pisa, better known as Fibonacci, explained the decimal system to his European colleagues in his *Liber Abaci*. Fibonacci's father was a customs official in the North African city of Bugia, so the young Fibonacci had grown up with the Moorish tradition of the decimal system. Many translators of Arabic works considered the Arabic numerals to be a part of the Arabic text, and when transcribing them into Latin replaced them with their Roman equivalents. In fact, these translators were somewhat shortsighted, for they failed to perceive that the new way of writing had not only a symbolic value but also brought with it a change in content.

The Hindu–Arabic notation became increasingly known in Europe from the late thirteenth to the fifteenth century. In 1253, in England, for instance, John Halifax attempted to promote the decimal system among his countrymen, albeit without much success. According to Sarton (1935) the use of the new way of representing numbers and calculation with decimal whole numbers was relatively common in Stevin's day; Roman numerals were only employed for ceremonial and decorative purposes.

In the West the earliest acquaintance with decimal fractions dates from around 1343. In his *Quadripartitum numerorum* John of Meurs, an astronomer in Paris, describes a calculation of the root of 2:

$$\sqrt{2} = 1/1000\sqrt{2000000} = 1/1000 \cdot 1414$$

In the *Tafel der houck-maten*, a table for the measurement of angles devised by Regiomontanus (1436–1476), we also find the decimal concept recurring.

He expressed goniometric line segments by the unit $R/10^a$. The given values are whole numbers and the value of each numeral is known only if the magnitude of 'a' is known. Thus, sine 15°21' is given as 2647147; this only acquires meaning when we know that $a=7$, so that the value of the sine is in fact 0.2647147. Stevin was familiar with this work, referring to its author as Iehan de Montroial in *L'Arithmetique* and as Jan van Kueninexberghe in *De Thiende*. He even acknowledges that Regiomontanus had a very good idea of what could be done with decimal fractions: 'He understood in its entirety the tenth progression that I sought'. In other words, Stevin confirms that Regiomontanus was on the right track as regards the representation of decimal numbers but something was still lacking in the notation.

The notion of a decimal system was already in the air

It can be seen from the foregoing that the use of a decimal place value system was not a novelty in Stevin's day. It was understood that numerals could be written down in a certain order and that the place each numeral occupied was important. If one begins with the units, the numeral to the right of this 'unit numeral' occupies the place of the tenths, the numeral to the right of that the place of the hundredths, and so on. Many mathematicians had already applied the idea, up to a certain point.

Stevin's *De Thiende* presented an orderly and well-thought-out description of decimal fractions and simultaneously introduced a sound and viable notation. In scientific circles, Stevin's work was certainly appreciated, employed and further elaborated, albeit very slowly. At first there were attempts to simplify the notation. Giovanni Antonio Magini used a comma as a symbol with which to separate whole and decimal parts in his *De planis triangulis* (Venice, 1592). He even spoke of a decimal notation (*denaria ratio*). In the same year, Thomas Masterson published his *First booke of arithmeticke* (London, 1592). Masterson used a vertical line as a separator. The Jesuit Christopher Clavius, or Christoph Klau, made use of a period after the units in the sine tables that appeared in his treatise *Astrolabium* (Rome, 1593). The German physicist Johann Hartmann Beyer (1563–1625), who worked in Frankfurt am Main, was another important figure. In 1619, he published the *Logistica Decimalis: Das ist: Kunstrechnung der Zehentheyligen Brüchen*. This work is often cited as a source of decimal ideas, though it was published long after Stevin's *De Thiende*. Presumably, Beyer was unacquainted with Stevin's work; in any case, he made no mention of it (but as we know, it was not generally the custom to refer to other works). Beyer's first chapter is entitled: '*Was die Mechanische Kunstrechnung sey: Und wie sie erfunden*'. *Mechanische Kunstrechnung* is Beyer's German term for decimal calculation. It bears a significant resemblance to Stevin's. Beyer uses *Primen, Secunden, Terzen* for the names of the decimal numbers – the equivalents of Stevin's primes, seconds and thirds, which lead us to suspect that he might, after all, have known Stevin's work.

> Johannes Kepler had a fine understanding of decimal calculation. In his *Wein-Visier-Buchlein* (1616), he used the simple notation 3(65×6 = 21(90, but he attributed this technique to the Swiss Joost Bürgi. In his *Arithmetica*, completed after August 1592, Bürgi placed a zero beneath the last integer to signal the decimal part. In fact, this would normally go unmentioned, for Bürgi's work was never published – but as Kepler ascribes the new type of decimal calculation to Bürgi rather than Stevin we can hardly ignore it.
>
> It was the Scot, John Napier (1550–1617), who brought decimal numbers into general usage. In chapter five of his posthumously published *Mirifici logarithmorum canonis constructio* he wrote, 'In numbers distinguished by a period in their midst, whatever is written after the period is a fraction'. This is now recognized and applied everywhere: Napier was responsible for the period or point that separates fraction from integer in our present notation. In his work *Rabdologia* of 1617, he adopted Stevin's notation for decimal fractions, simultaneously proposing a notation using a decimal point.
>
> The early seventeenth century witnessed the publication of lengthy logarithmic tables. In Henry Briggs's *Arithmetica Logarithmica* (1624), which lists logarithms to base 10, we again encounter decimal notation, albeit Briggs used the comma as a decimal separator.
>
> In the Dutch-speaking world the works of Ezechiel De Decker, *Eerste Deel van de Nieuwe Telkonst* (1626) and *Tweede Deel van de Nieuwe Telkonst* (1627), were eloquent witnesses to the triumph of the decimal system in all its aspects. This two-volume publication brings together various works, including those of Simon Stevin, John Napier and Henry Briggs: here are Stevin's *De Thiende* and his translation of *Rabdologia*, as well as Briggs's logarithms of every number between 1 and 1,000,000. These works emphasized three fundamental mathematical principles: Hindu-Arabic notation with modern numerals, decimal fractions and logarithms to base 10.

Great discoveries are often slow to develop and it was no different with the introduction of decimal fractions. The important factor is the individual who fully comprehends the value of the new concept and grasps its implications. This Stevin did to a great degree; he is no less than the author of the first textbook about decimal fractions.

In his *cortbegryp* – his argument – Stevin states that *De Thiende* consists of two parts: *Bepalinghen ende Werckinghe* – Definitions and Operations. We encounter this typical division in most of his works. Today we would speak of theory and practice. In the Appendix, he says, we shall put into practice 'the use of the Dime in many things by certain examples'.

3.1. Stevin begins with a definition of the concept of the *Thiende* – the Tenth, or Dime

> *Dime is a kind of arithmetic, invented by the tenth progression, consisting in characters of ciphers, whereby a certain number is*

described and by which also all accounts which happen in human affairs are dispatched by whole numbers, without fractions or broken numbers.

[Decimal numbers are a kind of arithmetic based on the idea of progression by tens, using ordinary Arabic numerals, in which any number may be written and by which all calculations required in business can be performed by integers alone, without the aid of fractions.]

This sounds rather odd, when working with decimal fractions is precisely what he is about to do. Nevertheless, Stevin's work, and especially the notation he introduced, resulted in a calculation model using decimal fractions, not with denominators but 'decimal point numbers', which is as simple as working with integers. Stevin explains this way of thinking as follows:

Let the certain number be one thousand one hundred and eleven, described by the characters of ciphers thus 1111, in which it appears that each 1 is the 10th part of his precedent character 1; likewise in 2378 each unity of 8 is the tenth of each unity of 7, and so of all the others. But because it is convenient that the things whereof we would speak have names, and that this manner of computation is found by the consideration of such tenth or dime progression, that is that it consists therein entirely, as shall hereafter appear, we call this treatise fitly by the name of Dime, *whereby all accounts happening in the affairs of man may be wrought and effected without fractions or broken numbers, as hereafter appears.*

[Let the number one thousand one hundred and eleven be written in Arabic numerals thus, 1111, in which form it appears that each 1 is the 10th part of the next higher figure. Similarly, in the number 2378, each unit of the 8 is the tenth part of each unit of the 7, and so on. But because it is convenient that the things we study have names, and because this kind of calculation is based solely on the idea of progression by tens, as we shall see below, we may appropriately call this treatise *Dime*, and we shall see by it that all the calculations we meet with in business can be performed without fractions.]

Stevin introduces a number of definitions, demonstrating his complete understanding of the structure of a decimal system (see box 1).

Every number propounded is called COMMENCEMENT, whose sign is thus ⓪.

[All whole numbers propounded are called the units and are indicated by the sign ⓪.]

And each tenth part of the unity of the COMMENCEMENT we call the PRIME, whose sign is thus ①, and each tenth part of the unity of the prime we call the SECOND, whose sign is ②, and so of the other: each tenth part of the unity of the precedent sign, always in order one further.

[The tenth part of a unit is called a prime and is indicated by the sign ①, and the tenth of a prime is called second, and is indicated by the sign ②, and so on for each tenth part of the unit of the next higher figure.]

He explains this definition as follows:

As 3① 7② 5③ 9④, that is to say 3 primes, 7 seconds, 5 thirds, 9 fourths, and so proceeding infinitely, but to speak of their value, you may note that according to this definition the said numbers are 3/10, 7/100, 5/1000, 9/10000, together 3759/10000, and likewise 8⓪ 9① 3② 7③ are worth 8, 9/10, 3/100, 7/1000, together $8\frac{937}{1000}$, and so of other like. Also you may understand that in this dime we use no fractions, and that the multitude of signs, except ⓪, never exceed 9, as for example not 7① 12②, but in their place 8① 2②, for they value as much.

[Thus 3① 7② 5③ 9④ means 3 primes, 7 seconds, 5 thirds, 9 fourths, and this might be continued infinitely. It can be seen from the definition that the numbers are 3/10, 7/100, 5/1000, 9/10000, and that this number is 3759/10000. Likewise 8⓪ 9① 3② 7③ has the value 8, 9/10, 3/100, 7/1000, making this number $8\frac{937}{1000}$, and so on for other numbers. Also, you should realize that in these numbers we use no fractions, and that the number under each sign, except ⓪, never exceeds 9. For instance, we do not write 7① 12②, but instead 8① 2②, for it has the same value, that is, 7/10 1 12/100 5 8/10 1 2/100.]

Stevin then moves on to 'Operations' and sets out the basis for elementary calculations using decimal fractions: he deals with addition, subtraction, multiplication, division and 'the extraction of all kinds of roots'. In boxes 2 and 3 are the texts (as translated by Norton) for addition and root extraction, with a modern English 'translation' and explanation where necessary.

Box 2: An addition of decimal fractions

Here is Stevin's text concerning the addition of decimal fractions (cf. p. 13 in *De Thiende*), followed by a translation into modern English with the decimal fractions written as modern decimal point numbers.

THE FIRST PROPOSITION: OF ADDITION
Dime numbers being given, how to add them to find their sum.
THE EXPLICATION PROPOUNDED: There are 3 orders of dime numbers given, of which the first 27⓪ 8① 4② 7③, the second, 37⓪ 6① 7② 5③, the third 875⓪ 7① 8② 2③.
THE EXPLICATION REQUIRED: We must find their total sum.
CONSTRUCTION: The numbers given must be placed in order as here adjoining, adding them in the vulgar manner of adding of whole numbers in this manner.

[Proposition 1
To add decimal fractions.
Given: three decimal numbers: 27.847, 37.675 and 875.782.
Required: to find their sum.
Construction: arrange the numbers as shown in the figure and add them up as you would usually add integers.]

$$\begin{array}{cccccc} & ⓪ & ① & ② & ③ \\ 2 & 7 & 8 & 4 & 7 \\ 3 & 7 & 6 & 7 & 5 \\ 8 & 7 & 5 & 7 & 8 & 2 \\ \hline 9 & 4 & 1 & 3 & 0 & 4 \end{array}$$

The sum [...] is 941304, which are (that which the signs above the numbers do show) 941⓪ 3① 0② 4③. I say they are the sum required. Demonstration: The 27⓪ 8① 4② 7③ given make by the 3rd definition before 27, 8/10, 4/100, 7/1000, together $27\frac{847}{1000}$ *and by the same reason the 37⓪ 6① 7② 5③ shall make* $37\frac{675}{1000}$ *and the 875⓪ 7① 8② 2③ will make* $875\frac{782}{1000}$*, which three numbers make by common addition of vulgar arithmetic* $941\frac{304}{1000}$*. But so much is the sum 941⓪ 3① 0② 4③; therefore it is the true sum to be demonstrated. Conclusion: Then dime numbers being given to be added, we have found their sum, which is the thing required.*

[This gives the sum 941304, which as the symbols above the numbers show, is 941.304. This is the sum required.

Proof: By the third definition, the given number 27.847 is 27, 8/10, 4/100, 7/1000, or $27\frac{847}{1000}$. Similarly, 37.675 is $37\frac{675}{1000}$ and

875.782 is $875\frac{782}{1000}$. Added, these numbers make $941\frac{304}{1000}$, but 941.304 has this same value and is therefore the true sum that was to be shown. Conclusion: having been given decimal numbers to add, we have found their sum, which was what we set out to do.]

NOTE that if in the number given there want some signs of their natural order, the place of the defectant shall be filled. As, for example, let the numbers given be 8⓪ 5① 6② and 5⓪ 7②, in which the latter wanted the sign of ①; in the place thereof shall 0① be put. Take then for that latter number given 5⓪ 0① 7②, adding them in this sort.

[If, in the numbers in question, a figure in the natural order is missing, put a zero in its place. For example, in the numbers 8⓪ 5① 6② and 5⓪ 7②, where the second number is lacking a prime, insert 0① and take 5⓪ 0① 7② as the given number, and add as before.]

```
         ⓪  ①  ②
         8   5   6
         5   0   7
        ─────────
      1  3   6   3
```

Box 3: Extracting roots

At the end of *De Thiende* Stevin gives a short explanation of how the roots of decimal fractions can be found. He is extremely concise, but with a little extra explanation his text is perfectly comprehensible. He starts from the assumption that the interested reader is already able to extract the root from an integer. Now he extends this technique to decimal numbers. The reader who can no longer recall how to find a square root by hand – pocket calculators having made this ability virtually superfluous – can safely skip this passage.

The extraction of all kinds of roots may also be made by these dime numbers; as, for example, to extract the square root of 5② 2③ 9④ which is performed in the vulgar manner of extraction in this sort...

[Decimal numbers can be used in the extraction of roots. For example, to find the square root of 0.0529 work according to the ordinary method of extracting the square root...]

The man who 'invented' the decimal system

$$\frac{\cancel{5}\ \cancel{2}\ \cancel{9}}{2\ \ \ \ \ \ 3}$$
$$\overline{\ \ \ \ \ \cancel{4}\ \ \ \ }$$

We can interpret the above calculation as follows:

Let us consider the integer 529; the highest square smaller than 5 is 4; thus 2 is the root; because five is in the position of the hundredths, the square is 400, thus the 2 stands for tenths. So we know that $(20+x)(20+x)$ must give the result 529, thus $400+40x+x^2 = 529$ and consequently $40x+x^2 = 129$; well then, 3 is the required x because $40*3 = 120$ and $3^2 = 9$. Now all we have to do to find the root of 0.0529 is to put the decimal sign in the right place. Having achieved this result Stevin continues:

> *Likewise in the extraction of the cubic root, the third part of the latter sign given shall be always the sign of the root; and so of all other kinds of roots.*
>
> [...and the root will be 0.23. The last sign of the root is always one half the last sign of the given number. If, however, the last sign is odd, add a zero in place of the next sign and extract the root of the resulting number as above. By a similar method with the cube root, the third of the last sign of the given number will be the sign of the root – and likewise for all other roots.]

In the Appendix, Stevin comes to the application of his theory. In his preface he states what he intends to show:

> *Seeing that we have already described the Dime, we will now come to the use thereof, showing by 6 articles how all computations which can happen in any man's business may be easily performed thereby; beginning first to show how they are to be put in practice in the casting up of the content or quantity of land, measured as follows.*

The six articles, or applications, he describes are 'land-meting', the 'measures of tapestry or cloth', the 'measures of all liquor vessels', 'stereometry (the 'art of measuring bodies') in general', 'astronomical calculations' and 'the computations of money-masters, merchants and of all estates in general'. Boxes 4 and 5 contain background information on weights and measures in Stevin's day and their evolution from that time to the present, as influenced by Stevin's writings.

Box 4: Of weights and measures

Weights and measures were and indeed still are derived from nature in general and from the human body in particular. Terms such as 'thumb', 'foot' and 'ell' – the last of these stems from 'ulna'; the measure was originally linked to the length of the human forearm – are the offspring of this linguistic ancestry. Commerce made the standardization of measures necessary. As a rule it was left to the authorities to keep the standard measure. Standards already existed at the time of the Egyptians; some have been discovered in the ruins of Memphis. In Athens, a few of the most important citizens were charged with the custody of the standards, while in Rome the standard *libra* (pound) and the *pes* (foot) were kept in the temple of Jupiter. In the Middle Ages the Roman-style regulation of weights and measures was lost. Official involvement was limited to keeping a standard, usually in the outside wall of a public building, as an aid to the public.

The flourishing of towns and cities and their function as independent economic entities in the later Middle Ages, and particularly in the sixteenth and seventeenth centuries, once again gave rise to the need for properly regulated weights and measures. Municipal autonomy made for differences between towns and this became an ever-increasing hindrance to their rapidly expanding trade. Here are a couple of examples of local premetric weights and measures.

Length:
  Amsterdam ell: 68.78 cm   Brabant ell: 69.23 cm
  Amsterdam foot: 26.31 cm   Guelder foot: 27.19 cm
  Amsterdam rod: 4.53 m   Antwerp rod: 5.74 m

Area:
  Amsterdam *morgen*: 0.8129 ha   Den Briel *morgen*: 0.9183 ha

Weight:
  Amsterdam pound: 494.09 g   Nijmegen pound: 476.56 g

In the Appendix of *De Thiende* (see above), Stevin reacted against this diversity of measures, proposing instead a universal system of measurements with a decimal structure. Figure 3 shows what he writes concerning *tapytmeterie* – the computing of cloth measurements. This is truly revolutionary; its historical significance is enormous. Even so, not until the French Revolution would the climate be right for the creation of a new system of units. In 1790, the National Assembly instructed the French Academy of Sciences to prepare a new system that could be used by all people for all time. It had to meet the following criteria:

a. The system must be made up of standard measures based on unvarying quantities in nature.
b. All units must be derivable from a restricted number of base units.
c. Multiples and subdivisions of a unit must be decimal.

The commission that assembled to study the problem consisted of highly renowned scientists, including mathematicians such as Pierre-Simon Laplace and Joseph-Louis Lagrange. They chose the 'metre' as the unit of length and the 'kilogram' as the unit of weight. On 22 June 1799, the platinum prototype of the metre was deposited in the Archives of the French Republic. The standard of length had been determined by measuring the distance from the Equator to the North Pole along the Paris meridian. The ten-millionth part of that distance was accepted as one metre. To give the system an international character, the multiples and subdivisions received Greek and Latin names, such as centi, deca, milli and so on.

The French revolutionary government also attempted to reorganize the calendar on a decimal basis but after a few years the project was abandoned, as the proposed scheme did not correspond with astronomical observations and thus failed to meet the first criterion.

A crash on Mars

From 1820 the metric system also came into use in The Netherlands, Belgium and Luxembourg, and was subsequently introduced into the greater part of the civilized world. But the Anglo-Saxon countries, with Great Britain and the United States in the forefront, clung to the British Imperial System (which is certainly not decimal) for decades – and still do to a certain extent. In many parts of the former British Empire, metric measurement has replaced Imperial. However, the United States, which was so progressive with its monetary units, still adheres to this rather inefficient unit system. The tenacious use of inches, feet and pounds can be something of a hindrance to scientific and technical cooperation. On 23 September 1999, NASA lost the Mars Climate Orbiter because a segment of ground-based, navigation-related mission software was not transferred from Imperial units into metric units. As a result the Orbiter approached too closely to Mars and burned up in the planet's atmosphere.

In 1862, Carl Friedrich Gauss drew attention to the use of the metric system in physics. He was the first to carry out absolute measurements of the earth's magnetic force, using a decimal system based on the three mechanical units: the millimetre, the gram and the second. Together with Wilhelm Eduard Weber, Gauss extended the principle to the measurement of electrical phenomena. The application in the area of electricity and magnetism was further developed between 1860 and 1870 by James Clerk Maxwell and Joseph John Thomson, who formulated the need for a coherent system of units. From this grew the CGS System, based on the centimetre, the gram and the second. These units proved to be insufficiently flexible, and a number of practical units were added, such as the ohm for electrical resistance, the volt for electrical potential and the ampere for electrical current. In 1901, Giovanni Giorgi, an Italian electrotechnician and professor of mathematical physics at

Palermo (1926) and in telephony and telegraphy at Rome (1934) and Milan, added to the three-dimensional CGS system the ampere as the fourth basic unit. This four-dimensional system was named after him. Over the years new units were added to it, including units of magnetic flux, temperature and so on. In 1960, the International System of Units, or SI, was established by the 11th General Conference on Weights and Measures. The logical evolution of the metric system, the SI is continuously supplemented to keep pace with the development of science and technology.

The definition for the three most important standards, based on physically measurable units, thus runs as follows:

The *metre* is the length of the path travelled by light in a vacuum during a time interval of 1/299,792,458 of a second.

The *kilogram* is the unit of mass. Today the standard for the unit is the international prototype, a platinum–iridium cylinder, kept at the BIPM (International Office of Weights and Measures) at Sèvres in France. Throughout the world there are eight copies of this prototype; six are official. Scientists continue to search for newer definitions of the kilogram, based on atomic units.

The *second* is defined as the duration of 9,192,631,770 periods of the radiation associated with the transition between two hyperfine levels of the ground state of the caesium-133 atom.

Box 5: A chaotic monetary situation in the Low Countries

Before the Burgundian period (until around 1400), each county and duchy in the Low Countries had its own coinage. The uniting of the principal regions under the dukes of Burgundy was accompanied by more unified and stable currency. In 1433, Philip the Good introduced a uniform monetary system for Brabant, Flanders, Hainaut and Holland, which for decades remained essentially unchanged. In each region the coins had the same weight, shape and value, though the image on the obverse might be different.

In the second quarter of the sixteenth century this monetary system was extended with heavy silver coins – first the silver guilder and later the larger *daalder* issued by Philip II. Copper coins of smaller value were also introduced. The Habsburg emperor, Charles V, unified the monetary system throughout the whole of the Low Countries. In 1521, he brought the gold Carolus guilder into circulation. The use of foreign currency was strongly discouraged; only foreign gold pieces and the large silver pieces such as the *rijksdaalder* and Spanish *piastre*, which at that time were regarded as international currencies, remained in use. In general, the guilder became the common unit of reckoning. The gold Carolus guilder was divided into 20 stuivers, and each stuiver into eight *duiten* or 16 *penningen*.

Once the Revolt had begun, the northern provinces organized their currencies fairly independently of each other, so that many coins of different appearance and varying quality came into circulation. It was not until 1606 that the States General issued decrees that brought some order into the affair: the provinces of the Republic would mint their own coins but would do so according to the regulations laid down by the States General for the monetary standard and the image to appear on the obverse. Thenceforth, the gold ducat and the silver *rijksdaalder* that had been adopted from the Holy Roman Empire in 1583 and the silver *rijksdaalder* introduced by Holland in 1575 were minted. The small coins – the silver shilling, *dubbeltje*, stuiver and copper *duit* – had various designs on the obverse. However, the continuous rise in the exchange rate of the large coins could not be stopped. The chief cause of this was the continuing circulation of countless old and foreign coins, which though of the same nominal value contained less gold or silver.

After Spain's reconquest of the southern Netherlands, the old system of Philip II was initially restored. However, in 1612 the *patagon* was introduced (a silver coin worth 48 stuivers and also known as the Albertus *daalder*) and in 1618 the *ducaton* (a silver coin worth three guilders). Both were proportionally lighter than the specie of the Dutch Republic. The coins of the northern provinces were completely supplanted in the south.

It was against this chaotic monetary background that Simon Stevin proposed a simple decimal-based monetary system that could be accepted and used throughout a large area. We find the first application of this idea at the end of the eighteenth century in the 13 states of the newly formed United States. With the introduction of the euro in 2002 in most of the countries of the European Union, one of Stevin's visionary ideas has become reality.

In each of the applications, Stevin proposes as the 'commencement', or unit, a measure that was already familiar to ordinary folk. For 'land-meters' (surveyors) this was the perch or rod. Stevin divides the rod into primes, seconds and thirds. He notes that as a rule, 'in land-meting, divisions of seconds will be small enough'; thirds will be useful only for things requiring more exactness, such as the thickness of lead roofs. He links the terms used for existing linear measures to the concept of primes and seconds, associating primes with feet and seconds with the familiar unit of measurement, the 'thumb'. The most important thing in Stevin's argument is that he proposes a division based on tenths, though this was certainly not common practice in his day.

For tapestry makers and clothiers (fig. 3), Stevin suggests the ell as a unit. This too he divides into tenths and hundredths. For wine gaugers he falls back on an existing Antwerp measure, the aum. The aum, marked on a 'wine rod', should be 'divided into length and deepness into 10 equal parts (namely equal in respect of the wine, not of the rod, of which the parts of the depth shall be unequal)', by which parenthesis Stevin indicates that the shape of a cask or bottle must be taken into account. Each tenth part thus contained 10 'pots', the Antwerp

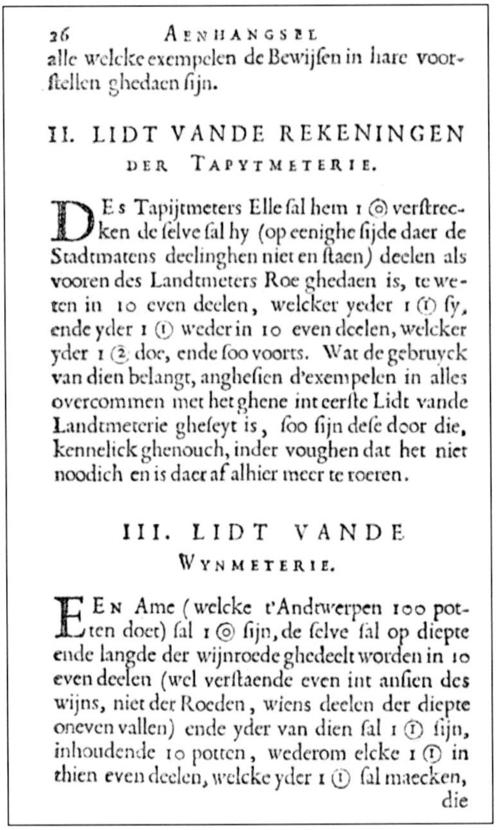

> 26   AENHANGSEL
> alle welcke exempelen de Bewijſen in hare voor-
> ſtellen ghedaen ſijn.
>
> ## II. LIDT VANDE REKENINGEN
> ### DER TAPYTMETERIE.
>
> DEs Tapijtmeters Elle ſal hem 1 ⓪ verſtrec-
> ken de ſelve ſal hy (op eenighe ſijde daer de
> Stadtmatens deelinghen niet en ſtaen) deelen als
> vooren des Landtmeters Roe ghedaen is, te we-
> ten in 10 even deelen, welcker yeder 1 ① ſy,
> ende yder 1 ① weder in 10 even deelen, welcker
> yder 1 ② doe, ende ſoo voorts. Wat de gebruyck
> van dien belangt, angheſien d'exempelen in alles
> overcommen met het ghene int eerſte Lidt vande
> Landtmeterie gheſeyt is, ſoo ſijn deſe door die,
> kennelick ghenouch, inder voughen dat het niet
> noodich en is daer af alhier meer te roeren.
>
> ## III. LIDT VANDE
> ### WYNMETERIE.
>
> EEn Ame (welcke t'Andtwerpen 100 pot-
> ten doet) ſal 1 ⓪ ſijn, de ſelve ſal op diepte
> ende langde der wijnroede ghedeelt worden in 10
> even deelen (wel verſtaende even int anſien des
> wijns, niet der Roeden, wiens deelen der diepte
> oneven vallen) ende yder van dien ſal 1 ① ſijn,
> inhoudende 10 potten, wederom elcke 1 ① in
> thien even deelen, welcke yder 1 ① ſal maecken,
>   die

Figure 3: On page 26 of *De Thiende*, Stevin describes the practical use of decimal fractions in *tapytmeterie* – the measures of tapestry or cloth – taking the ell as the basic measurement. *Utrecht, University Library, Dijns 21-727, vol. 2.*

subdivision of the aum, so that the hundredth part or 'second' corresponds exactly to one pot.

With 'stereometry in general' Stevin demonstrates the measurement of volume in decimal terms, using the example of a column with a length, width and height. He reminds us that the volume of three-dimensional objects is to be calculated and that therefore it is necessary to work with cubic measurements. In the fifth article he goes on to deal with astronomical computations, concluding with a vow to the reader that he will use decimal division in every astronomical table in future. And he declares, in lyrical terms, that he intends to use the *Duytsche Tale* – the Dutch language – the most richly adorned and perfect tongue of all. In Chapter Eight, we shall look more closely at Stevin's contribution to the Dutch language.

When Stevin describes the last area of application, that of the money-masters and merchants, he shows himself to be very well aware of the various currencies used in international commerce. He also proves to have a good idea of the exchange value of these currencies, as appears from the beginning of the paragraph:

> *To the end we speak in general and briefly of the sum and contents of this article, it must always be understood that all measures (be they of length, liquors, of money, etc.) be parted by the tenth progression, and each notable species of them shall be called* commencement: *as a mark,* commencement *of weight, by the which silver and gold are weighed, pound of other common weights, livres de gros in Flanders, pound sterling in England, ducat in Spain, etc.* commencement *of money.*

> [In summary we might say that all measures (linear, liquid, dry and monetary) may be divided decimally. And each best-known type may be called the unit. Thus the mark is the unit of weight for gold and silver, and the pound for other common weights. The livre gros in Flanders, the pound sterling in England and the ducat in Spain are the units of money in those countries.]

Stevin also remarks that a legal definition of currencies should be introduced so that the same norms are used throughout a region or state. It would be a commendable thing, he says, if 'the same tenth progression might be lawfully ordained by the superiors for everyone that would use the same; it might also do well, if the values of moneys, principally the new coins, might be valued and reckoned upon certain *primes, seconds, thirds*, etc.'

We might ask ourselves whether Stevin really hoped that the decimal system would be applied in all these areas. He gives us an answer himself in the last paragraph of the Appendix. 'But if all this be not put in practice so soon as we could wish, yet it will first content us that it will be beneficial to our successors, if future men shall hereafter be of such nature as our predecessors, who were never negligent of so great advantage.'

Stevin's *De Thiende* was read. This is clear from the reactions of Napier and Briggs, who referred to the work and actually applied it (see box 1). And in the literature of the time, particularly the plays of William Shakespeare, we find references to Stevin's work (see box 6). But the real breakthrough of the decimal system came with the French Revolution. At the end of the eighteenth century (see box 4), it came to be implemented in France and not long afterwards in other European countries as well. The United States followed with the introduction of the dollar as a monetary unit (see box 7). The introduction of a decimal numerical system, together with a method of carrying out every important arithmetical calculation with decimal numbers, has been seen by many as an important step in the development of calculating machines. Thus, Stevin takes his place among the scholars and scientists whose work paved the way for the development of the modern day computer. In this context his name is usually included in the company of Pascal, Babbage, Lady Lovelace, Hollerith and Von Neumann.

Box 6: Shakespeare and De Thiende

William Shakespeare (1564–1616) was a contemporary of Simon Stevin. But what did he know about *De Thiende* and Stevin's contribution to mathematics and technology? Some lines from *Troilus and Cressida* suggest that Shakespeare had come into contact with the English translation of Stevin's work: *DISME: the Art of Tenths, OR; Decimall Arithmetike, Teaching how to performe all Computations whatsoever, by whole Numbers without Fractions, by the foure Principles of Common Arithmeticke: namely, Addition, Subtraction, Multiplication and Division. Invented by the excellent Mathematician, Simon Stevin. Published in English with some additions by Robert Norton, Gent. Imprinted at London by S.S. for Hugh Astley, and are to be sold at his shop at Saint Magnus' corner. 1608.*

Troilus and Cressida is a remarkable play. A manuscript copy of the text was entered in the register of the Stationers Company in 1603. Six years later, in 1609, it was registered for a second time and in that same year a printed quarto edition appeared. Willy Courteaux, an eminent Flemish translator of Shakespeare, suspects that the text registered in 1609 was a reworked version that had been adapted for performance at the Globe on Bankside.

In Act II, scene ii, which is set in a room in Priam's palace in Troy at the time of the Trojan wars, we find the following dialogue (lines 8–26) between Hector and Troilus:

Hector

Though no man lesser fears the Greeks than I
As far as toucheth my particular,
Yet, dread Priam,
There is no lady of more softer bowels,
More spongy to suck in the sense of fear,
More ready to cry out 'Who knows what follows?'
Than Hector is: the wound of peace is surety,
Surety secure; but modest doubt is call'd
The beacon of the wise, the tent that searches
To the bottom of the worst. Let Helen go:
Since the first sword was drawn about this question,
*Every **tithe** soul, 'mongst many thousand **dismes**,*
Hath been as dear as Helen; I mean, of ours:
*If we have lost so many **tenths** of ours,*
To guard a thing not ours nor worth to us,
*Had it our name, the value of **one ten**,*
What merit's in that reason which denies
The yielding of her up?

Troilus

Fie, fie, my brother!
Weigh you the worth and honour of a king
So great as our dread father in a scale
*Of common **ounces**? Will you with counters sum*
The past proportion of his infinite?
And buckle in a waist most fathomless
*With **spans** and **inches** so diminutive*
As fears and reasons? Fie, for godly shame!

In this passage Shakespeare uses the word 'dismes', which Norton employed to translate Stevin's *De Thiende*. In making several references to multiples of 10, he shows that he well understands the intent of Stevin's work. Moreover, he has Troilus speak of ounces and inches, to draw attention to the complicated division of measures of weight and length in the English system – while it is precisely for a 10-part division that Stevin argues in his Appendix.

Box 7: The dollar and the disme

Thomas Jefferson (Shadwell 1743 – Monticello 1826), American statesman, philosopher and artist, is often regarded as the spiritual father of the American nation. He was deeply involved in the growing conflict with Great Britain in the early1770s. In 1774, he published a pamphlet, *A Summary View of the Rights of British America*, reminding George III that 'our ancestors, before their emigration to America, were the free inhabitants of the British dominions in Europe, and possessed a right which nature has given to all men, of departing from the country in which chance, not choice, has placed them, of going in quest of new habitations ...' Having found these new habitations, 'the emigrants thought proper to adopt that system of laws under which they had hitherto lived in the mother country, and to continue their union with her by submitting themselves to the same common sovereign, who was thereby made the central link connecting the several parts of the empire thus newly multiplied.' Which did not mean, he pointed out, that Great Britain had rights of sovereignty over the new nation.

In 1776, at the request of Congress, Jefferson drafted the Declaration of Independence. In resounding phrases he summarized the ideals of individual liberty in the self-evident truths 'that all men are created equal, that they are endowed by their Creator with certain unalienable Rights, that among these are Life, Liberty and the pursuit of Happiness' and he set forth a list of grievances against George III – 'The history of the present King of Great

Britain is a history of repeated injuries and usurpations, all having in direct object the establishment of an absolute Tyranny over these States' – in order to justify to the world the breaking of ties between the colonies and the mother country. The declaration's adoption on 4 July 1776 officially marked the birth of the American nation.

One of the newly independent nation's first decisions concerned the introduction of a monetary system for use in commerce. For the first 13 united states this was doubly important. Although Spanish silver pieces were the generally accepted standard currency, English pounds, shillings and pence were also in circulation in the young nation. The conduct of business and trade between the various states was hampered because each state used a different rate of exchange between the two monetary units making intricate conversion tables necessary. There was much confusion, and by the end of the 1780s the need for a reliable, non-fluctuating system of coinage was being hotly discussed.

European and colonial monetary systems had originated at various points in history and were non-decimal. The British system, for instance, had its roots in ancient Rome, and some Saxon and Norman additions had been grafted onto it. In the eighteenth century the unit, the pound, was divided into 20 shillings and each shilling into 12 pence, with innumerable additional, sometimes temporary, silver and copper subdivisions. Spanish *reales* were 'pieces of eight'. Together with Alexander Hamilton and the financier Robert Morris, Thomas Jefferson strongly argued for the adoption of a decimal currency. This not only represented a clean break with the past, but also was truly revolutionary and entirely consistent with the new philosophy of the Age of Enlightenment. The champions of this currency referred to Stevin's *De Thiende* and more specifically to Robert Norton's English translation, *Disme: The Art of Tenths* (fig. 4). In 1785, the dollar was adopted as the monetary unit of the United States. The weight and fineness were specified and the decimal system for the relationship of each of the coins was authorized. The Mint Act of 1792 specified 'that the money of account of the United States shall be expressed in dollars or units, dismes or tenths, cents or hundredths and that all accounts in the public offices and all proceedings in the courts of the United States shall be kept and had in conformity to this regulation.'

The first silver coins struck by the Federal Mint in 1792 were half-dismes or twentieth-of-a-dollar pieces (the spelling 'dime' as a reference to a tenth of a dollar has been in use only since 1837). Only 1,500 half-dismes were struck; Jefferson gave many away as gifts. In November 1792, in his annual speech to Congress, President George Washington spoke of 'a small beginning in the coinage of half-dismes'. This was not only the beginning of a decimal monetary system in the United States but also had great political significance, for the silver coin was generally regarded as an expression of national sovereignty.

DISME:
The Art of Tenths,
OR,
Decimall Arithmetike,

Teaching how to performe all Computations whatsoever, by whole Numbers without Fractions, by the foure Principles of Common Arithmeticke: namely, Addition, Subſtraction, Multiplication, and Division.

Invented by the excellent Mathematician,
Simon Stevin.

Publiſhed in Engliſh with ſome additions by *Robert Norton,* Gent.

Imprinted at London by *S. S.* for *Hugh Aſtley,* and are to be ſold at his ſhop at Saint Magnus corner. 1 6 0 8.

Figure 4: Like Shakespeare, Jefferson probably came to know Stevin's work via Norton's English translation of *De Thiende*, entitled *Disme, The Art of Tenths, or Decimall Arithmetike*. © The British Library Board. All Rights Reserved (C. 112 c. 8).

Specifications of the half-disme:

Diameter: 16.5 mm
Weight: 1.35 g
Composition: 0.8924 silver, 0.1076 copper
Edge: diagonally reeded
Netto weight: 0.0387 ounce pure silver

Chapter Four

Engineer and Inventor

Simon Stevin the scholar was also an eminently practical man. That he lived in the Low Countries, a land of water and wind, is easily deduced from the patents he applied for: mills, sluices, dredging machines. He made a lasting mark, too, on navigation (Holland was a maritime power) and warfare (Stevin was quartermaster of the States Army), and as an architect and town planner.

1. Patents

Simon Stevin's few texts on engineering are contained in works on the fortification of towns (*Castrametatio* and *Nieuwe Maniere van Sterctebou, door Spilsluysen*) published by himself, his son Hendrick (*Materiae Politicae*), or others. To gain an idea of his technological designs, we generally have to rely on his various patent applications or on documents detailing disputes in which he became involved as a result of one or other of his inventions. Then again, some of his engineering achievements were so highly thought of that contemporaries penned accounts of them, or even dedicated odes to them. The design and construction of a sailing chariot gained Stevin particular fame in the eyes of the general public (see Chapter Two, section 2).

1.1. Raising water

The earliest documents that provide clues about Stevin's work as an engineer date from 1584. On 17 February of that year the States of Holland granted him a patent for three inventions (see box 1):

1. To bring all sorts of ships across shallow waters.
2. To bring ships over dams.
3. To raise water by other means than those used so far.

On 22 February 1584, the States General also granted him a patent to allow him 'to have an instrument installed to withdraw water from land'.

Box 1: Patent granted by the States of Holland on 17 February 1584

From the Minutes-Resolutions of the States of Holland (State Archives, The Hague, no. 336)

> *Opt versouck van Simon Stevin van Brugge versouckende Octroy voor twintich jaeren omme te mogen practiseren drie inventien, d'eerste omme over d'ondiepe waeteren allerhande schepen te brengen, de tweede omme de selve schepen over dammen te voeren ende de derde omme het water te verheffen ende op te trecken deur andere middelen als nu ter tijt gebruijckt worden, dienende niet alleene omme de wateren uut den platten Lande te trecken, maer oock als het water buyten dijcx hoger is dan binnen 't selffde niettemin in grooter quantiteijt te doen loosen, jae oock omme in corter tijt de geheele Haven drooge te maecken ende anders, is geappostilleert de Staten van Hollant hebben omme redenen in desen geroert gegunt ende geoctroyeert, gunnen enden octroyeren bij deze dat de suppl[ian]t alhier binnen den Lande van Hollant sal mogen practiseren ende gebruijcken sijne inventie, cunste ende instrumente tot de welcken in dese geroert voor dengeenen die met den suppl[ian]t diesaengaende sullen begeeren te handelen ende 't accorderen tot dienste vanden Lande, sonder dat ijemant binnen voorsz. Lande alsulcke instrumenten ofte middelen als hij supp[ian]t daer toe gebruyckende is, sal moghen te wercke stellen, buijten des supp[ian]ts wille ende consent, binnen de tijt van vijfftien jaeren eerstcomende, opte verbeurte vande selve instrumenten ende de peyne van twee duysent guldens, 't appliceren d'eene helft tenbehouve vanden suppl[ian]t, ende de ander helft voorden Officier van der plaetse voor d'eene helft ende d'ander helft tot behouff vande gemeene saecke, mits dat den suppl[ian]t sijne voorn. inventie ende cunste binnen s jaers int werck stellen sall.*

[Simon Stevin of Bruges applies for a 20-year patent in order to put into practice three inventions: the first to bring all sorts of ships across shallow waters, the second to bring ships over dams and the third to raise water by other means than those used so far. This last serves not only to raise water from low-lying lands but also to drain water in greater quantities if the water is higher outside the dike than inside. Harbours will also be drained and filled again in a shorter time. By the authority vested in them, the States of Holland grant the applicant permission, by means of a patent, to practice and employ his invention, art and instruments in Holland. This also applies to those who shall, together with the applicant, desire and employ this practice in the service of the country, it being forbidden to any person in this country to use such instruments as are utilized by the applicant without the latter's wish or consent within the next 15 years, under pain of seizure of the instruments and a fine of

> 2,000 guilders, one half for the applicant, one quarter for the town officer and one quarter for the commonweal, providing that the applicant puts his invention into practice within one year.]
>
> In tribute to Simon Stevin, 450 years after his birth, the Netherlands Industrial Property Office placed the text of this patent on the cover of its journal next to a photograph of the modern dredging vessel belonging to the dredging company Ballast Nedam Baggeren.

It is difficult to determine precisely what was meant by the inventions referred to under items 1 and 2, unless it may be supposed that Stevin intended to employ his 'Almighty' (*Almachtich*), which he describes in *De Weeghdaet* (see Chapter Six). Item 3 perhaps refers to a piston pump, which Hendrick later described as one of his father's inventions. Or could it allude to a water-drainage mill, with which Stevin would later be much occupied?

Whatever the case, in that same year Stevin tried to exploit these things: the Delft comptroller's accounts for 1585 contain an entry of five guilders and twelve stuivers for expenses connected with 'a meeting with one Mr Simon of Leiden, inventor of certain arts in the field of drainage'.

On 24 November 1586, the States General granted Stevin a patent for a type of water-drainage mill that was intended to have a greater pumping capacity and would drain the land more quickly than those in common use at the time. This patent was registered at the Rekenkamer (Audit Office) of Holland on 15 September 1588, at the same time as another patent that had been granted on 23 February 1588 by the earl of Leicester and was connected with the same subject. This latter document specifies the capacity of the mill, which was to pump five times the amount of an ordinary mill.

To put his new invention into practice, Stevin entered into an agreement with his friend Johan Cornets de Groot of Delft, which gave the partners equal shares in the rights and proceeds of the two patents. On 22 August 1588 the Delft authorities agreed to pay De Groot and Stevin a fee of 100 crowns for 'the new work on the water mill at the Duyvelsgat' on condition that they should also, if required, 'install a new invention of the said Mr Simon Stevin in the water mill on the town wall at the end of the Langendijck', for which they would receive the same sum. On 29 August 1590 the Delft magistrates gave Stevin a testimonial, at his own request, confirming that the two mills 'rebuilt according to the art of the said Stevin have scoured at least thrice as much water as the two former mills usually did'. Stevin and De Groot also incorporated their inventions in the mills at Stolwijk and Cralingen.

1.2. A roasting spit

On 28 November 1589, the States General granted Stevin patents for no fewer than nine new inventions:

1. Water mills working in tandem to raise the water to a higher level in two, three or four stages as required.

2. A water mill that can not only drain a polder but in dry periods can be made to pump water back again if the outer water level is lower.
3. A method for draining water from ditches and canals by manpower.
4. A method for dredging clay, sand and similar solid materials from harbours and the bottoms of watercourses.
5. A method for discharging these materials into the water elsewhere.
6. A method for dredging mud, peat and similar fluid material 'with great profit'.
7. A method for discharging this fluid sludge into the water or on to the land.
8. A roasting spit which, when not used for roasting, 'can drive clockwork for twelve hours or rock a cradle for half an hour or more'.
9. A method for operating the pumps described in invention 3 by wind power. (Stevin mentions in the accompanying text that this process was used to drain the Rapenburg, a canal that still exists in Leiden.)

Patent 1 will be examined in more detail in the section on mills below. It comes as something of a surprise to find patent number 8 among the other useful inventions. Nonetheless, it contains a very significant clause, for it stipulates that the inventor 'will mark the iron parts of this apparatus with his common trademark, called the *clootcrans*, which is the first figure accompanying the nineteenth proposition of the first book of his *Weeghconst*' (see Chapter Six, section 2.3 and box 1). Thus we see that Stevin used the *clootcrans*, or 'wreath of spheres', which he had also chosen as the vignette for the *De Beghinselen der Weeghconst* and with which he was to decorate many of his later works, as his inventor's trademark – a necessary precaution at a time when the construction of mechanically driven spits and other automata was very popular.

With each of these applications for patents Stevin supplied a complete description of the apparatus or method, usually accompanied by an explanatory drawing. Below is a typical example: the original text and drawing that were appended to his application for a patent for invention number 4 (see fig. 1). They were entered in a file entitled *Verscheyden Inventien van Simon Stevin* (State Archives).

> *Verclaring van een manier om cley, sant of dierghelycke vaste stof, met overvloet ende voordeel uyt havens ende gronden van wateren te trecken.*
>
> *A beteeckent een groot net, meer cleys vervatende dan 25 ander ghemeen baggaertnetten, tselve heeft inden bodem twee halve yser rynghen, die open en toegaen op twee carnieren an B en C, ende sluyten int middel met een slot. Dit net wort vol cleys ghetrocken duer den as D, met een tau vanden selven as commende over tcaterol E, ende van daer over tcaterol F, van daer ant net. Tcaterol F staet in een stock, welckmen hoogher en leegher steken can, om het tou FA te wyle men treckt, altyt langs de gront te doen strecken, soo naer alsmen wil. Tnet alsoo vol cleys getrocken wesende, soo treckmen de kruck vanden as D, ende men steeckse anden as G, van*

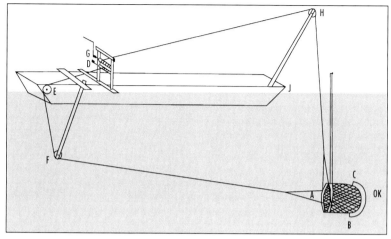

Figure 1: Stevin submitted several patents to the States of Holland to safeguard his own inventions. This diagram was added to a patent of 1589. It shows how fine materials such as clay and sand can be dredged from harbours and the bottom of watercourses.

welcken as een ander tou comt over tcaterol H, tot ant net, ende tnet wort daer mede opghewonden soo hoogh tot dattet boven de cant van tschip is datmen vullen wil, welck schip gheleyt wort nevens de schauwe, alsoo dat syn middelt comt omtrent teinde der schauwe an J. Tnet alsoo boven dat schip hanghende, men steeckt met een stock in een ront oogh K daer mede tslot open springt, ende duer tghewicht des cleys, soo wycken die twee halve ringen B, C van malcander, ende de cley valt daer duer int schip.

[Explanation of a method of drawing clay, sand and similar solid materials in great quantities and economically from harbours and the bottom of watercourses.

A represents a large net, holding more clay than 25 ordinary dredging nets, and having two semicircular iron rings at its bottom, which open and close on two hinges at *B* and *C* and close in the middle by means of a lock. The net is drawn full of clay by means of the winch *D* and a rope running from this winch via the pulley *E* and thence to the pulley *F* and the net. Pulley *F* is fixed on a stick, which can be raised and lowered in order to let the rope *FA* come as near to the ground as desired when dredging. When the net is full of clay the crank is pulled from winch *D* and put on winch *G*, from which a rope leads via pulley *H* and the net, and the net is raised by this means until it is over the rim of the barge to be filled, which is anchored alongside the scow in such a way that its centre is at the end of the scow at *J*. The net being thus suspended above the barge, a stick is pushed into a round eye *K*, thus opening the lock, and by the weight of the clay the two semicircles *B* and *C* separate and the clay falls into the barge.]

2. Mills

When Stevin speaks of water mills he does not mean the sort of water-driven mills common in other parts of Europe but windmills that were used to drain water – a constant preoccupation in the Low Countries. We are very well informed about Stevin's work on water mills, for in addition to the descriptions in his patents, there are also references in Isaac Beeckman's *Journal* and in Hendrick Stevin's publications, such as his *Wisconstich Filosofisch Bedryf*, which has a section on his father's treatise on water mills. And Bierens de Haan's work includes an essay *Vande Molens* (*On Mills*) with a series of Stevin's drawings and calculations. Stevin himself never completed his manuscript on mill construction, though he did leave draft notes that were found here and there. It should be mentioned that the data given in the *Journal*, in Hendrick's work and in Bierens de Haan's publication are not always consistent.

2.1. Stevin turned his mills by theory

A diagram derived from Dijksterhuis (1943) shows the type of mill in general use in Stevin's day (see fig. 2). The purpose of the mechanism is of course to transfer the motion of the sails to the scoop wheel. The machinery consists of the following parts: B is the wind shaft, which is turned by the revolving of the sails. Mounted on the wind shaft is brake wheel C, whose cogs engage the staves of the wallower S, which is mounted on vertical shaft K. Vertical shaft K turns on the top journal I in centre beam Y and thrust journal or pintle L in bottom M. Thus wallower S, driven by brake wheel C, turns vertical shaft K, which turns attached crown wheel N and thus pit wheel O. Both pit wheel O and scoop wheel R are mounted on horizontal scoop wheel shaft W. Thus scoop wheel R turns on scoop wheel shaft W. To keep it dry, the pit wheel is enclosed in a timber-lined pit at A. The scoop wheel turns between the two vertical walls of the wheel pit, also called the scoop wheel race. When the mill turns, the inner water level is raised and the water gate, which gives into the higher water, is opened. As the supply of water stops, this gate is closed by the pressure of water on the other side.

Stevin's innovations and improvements were chiefly connected with two parts of the mill machinery:

a. He tried to improve the contact between the cogs of the brake wheel and the staves of the wallower.
b. Basing his ideas on detailed calculations he proposed that the scoop wheel should revolve at a much slower rate and should have a smaller number of much wider floats. To stop the water flowing back between the wheel and the vertical walls of the wheel pit as a result of the slower rotation of the scoop wheel, Stevin fitted the outer edges of the floats with leather flaps that would brush betweem the wheel pit walls and the water guide. As we shall also see in Chapter Six, the innovation here lies in the fact that Stevin used a quantitative mathematical calculation as the basis for the building of a mechanical instrument.

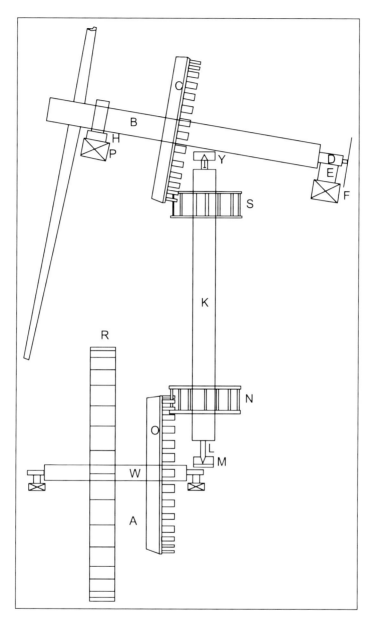

Figure 2: Stevin designed many mills for draining the land. The sails were turned by the wind and the water was 'milled' out of the polders by the scoop wheel.

The changes Stevin proposed were based on theory. In other words, he described the operation of the mills by drawing on static and hydrostatic ideas. Stevin's approach is important; the way in which he combined theory with its actual application to a mechanism as complex as a mill is extraordinary. Years later, even Galileo had still not ventured so far.

From the disagreements between Stevin and his patrons, however, it would seem that not everything worked quite as effectively as Stevin had predicted. The lengthiest dispute in which he was involved was connected with the drainage mill at IJsselstein. We can follow the course of the argument in two files containing some 70 documents still in the IJsselstein archives, among them two letters in Stevin's own hand to the bailiff of IJsselstein (see box 2). Dijksterhuis (1943) estimates that throughout Holland and Zeeland Stevin built some 20 mills of this kind, but none have withstood the ravages of time, so the author tells us.

Box 2: The IJsselstein mill

Much of the fertile land in the Netherlands is polder land (low-lying land reclaimed from the sea or a river and protected from flooding by dikes) and it is vital to the country's food production and economy. On 8 April 1589, Johan de Groot signed a contract, in his and Stevin's name, with the representatives of Leege Biesen, Achtersloot, Meerloo and Broek polders in the land of IJsselstein. He had committed himself and Stevin to deliver 'a good new water mill made of timber and iron ... for the sum of 630 Carolus guilders and an extra gift at discretion' and to guaranteeing that the new mill would drain 'as much water as two of the best mills of thereabouts could do'.

According to the contract, payment was to be made in four instalments; the first within eight days of the polder representatives signing the contract, the second on delivery, the third after a year had elapsed and the fourth a year after that. The contract stipulated that the mill should be in operation before St Mathew's Day (21 September), but it appears that this was not the case, and in the autumn of 1589 the polder remained undrained, to the farmers' great distress. The receipt for the third instalment of the fee is dated 24 June 1591, so the mill must have been completed by 24 June 1590. However, it seems not to have lived up to expectations; the land was inundated and the inhabitants of the polder infuriated, and when the final instalment was due the polder authorities flatly refused to pay it. De Groot protested to Princess Mary of Nassau who, since the death of her father William the Silent, had managed the barony of IJsselstein on behalf of her brother Prince Philip William, at that time in Spanish captivity. Thus began a protracted correspondence between the princess and her Council on the one hand, and the bailiff of IJsselstein on the other.

Stevin, wishing to safeguard his reputation as an inventor, provided himself with a series of testimonials as to the effectiveness of his constructions in other parts of the country. He repeatedly visited the IJsselstein mill, finally concluding that its poor performance was due to neglect, bad management and sabotage. He even had this assessment recorded in a document drawn up by the notary Paulus Viruly on 26 August 1593. The document still exists.

> The actual problem seems to have been that the thrust journal had penetrated the wood of the vertical shaft, causing it to split, so that it no longer turned smoothly in the bottom. Eventually a commission of four impartial lawyers was appointed, and the warring parties agreed to abide by their decision. The commission's conclusions have not come down to us, but their decision must have been announced at the end of 1594 or the beginning of 1595. The polder administration apparently did pay De Groot, having deducted the cost of the innumerable repairs to the mill that had meanwhile been carried out.
>
> Among the documents in the IJsselstein archives, two letters in Stevin's own hand were found, both addressed to the bailiff of IJsselstein.

2.2. A long tradition of impoldering

The Low Countries have a long tradition of draining lakes and reclaiming land from coastal waters and estuaries. Three technical factors play important roles in this:

1. The holding back of higher outer water by dikes of sufficient height and strength.
2. The discharging of excess water from the polder into the outer water.
3. The draining and transformation of the former lake or seabed into good agricultural land.

As we have seen above, in 1589 Stevin took out a patent on a way of discharging water from polder to washland. His method, known as the *molengang* – mills working in series – was already in use in Stevin's day in land reclamation, where the outer water level was consistently higher than that of the water in the polder. Mills with a scoop wheel were used; these raised the water between two inner walls to the outer water. *Wachtdeuren* or 'guard doors' prevented the water from pouring back. These scoop wheels could raise the water to a limited height – 1.5 m at most. So only shallow stretches of water could be drained and reclaimed in this way. However, by 'milling' the water stepwise with the *molengang* system, the water could be raised to a higher level, though of course it took a little longer. As already mentioned, the mills were powered by the wind.

In his explanation of the first of the nine patents listed above, Stevin outlined the operation of his *molengang*. He confined himself to describing a simple tandem system, but it could obviously be extended by using more mills.

Stevin was evidently a pioneer of this sort of drainage installation. The celebrated hydraulic engineer Jan Adriaensz. Leeghwater (1575–1650) used the tandem system for the first time in 1609 – the year Stevin's patent expired – to drain the Noord-Schermer, a marshy area to the north of Amsterdam. It is likely that he was inspired by the text of Stevin's patent and made use of it for his own project.

This kind of *molengang* is still in operation in the same region today, draining the area known as 'De Schermeer' (see fig. 3).

Figure 3: Stevin built mills in series, as this allowed water to be raised two, three or even four times. This type of *molengang* is still in use today in the polder known as 'De Schermeer'
(photo: AeroCamera – Michel Hofmeester, Rotterdam).

3. Sluices

We have two texts relating to Stevin's work on sluices, his *Nieuwe Maniere van Sterctebou, door Spilsluysen* and *Van den handel der Waterschuyring onses Vaders Simon Stevin (Essay on Waterscouring by Our Father Simon Stevin)*, posthumously published by his son Hendrick as part of his *Wisconstich Filosofisch Bedryf*. Simon Stevin was greatly interested in hydraulic engineering, a typical preoccupation in the Low Countries and a field in which monastic orders like the Cistercians and Benedictines had played a significant part in previous centuries. But all the work had been purely practical, and no theories had been propounded. Modern hydraulic engineering begins with Bernard Forest de Bélidor's work on hydraulics, *Architecture Hydraulique ou L'Art de Conduire, d'Elever et Demenager les Eaux pour les Differens Besoins de la Vie*, published in Paris between 1738 and 1742. Bélidor (1693/1698–1761) spoke highly of Stevin's work, calling him 'the famous engineer of the United Provinces and the first to have written on sluices in 1618...'

In his book on pivoted sluice locks, Stevin first explained what sluices are used for: 'To scour harbours, to drain low areas, and to allow vessels to pass through with upright masts'.

3.1. Of sluices and locks

The type of sluice used for scouring a harbour was vertical with a door that was raised and lowered by hoisting. The door shut off a reservoir or chamber from the water of the sea or estuary. The door would be raised at high tide so that the reservoir was filled, then lowered again. At low tide the door was raised again and the body of water trapped behind it surged through, washing away the deposited sand and thus deepening the harbour. But no large masted ship could pass through such a sluice, for as Stevin pointed out, 'the doors and the (windlass) shaft with which they are raised are in the way'.

'For the second type of lock, serving to drain low areas,' Stevin tells us, 'the most suitable are sluices with swivel-doors (which some called swaying or mitred doors), which are built into the dikes...' The edge of the obtuse angle formed by the doors was directed towards the outer water. At low tide the doors were opened automatically by the pressure of the higher water behind them and closed again as the level of the outer water rose. This type of sluice did not allow the passage of masted vessels either, because of the section of dike above it.

The third type, 'serving for the passage of large vessels with upright masts', which has scarcely changed since Stevin's day, was the familiar canal lock consisting of two pairs of gates 'which are not built under the dike, like the second type, but which extend in the dike up to its top so that they reach from the bottom to the top of the dike in order to dam up the highest water'.

The possibility of combining the different types of sluices and locks into one had long been considered in the Low Countries. As Stevin tells us, 'several people seriously studied this problem particularly in Holland, where in towns, in villages and in the country such large numbers of locks have been built and are still being built... that I believe that now no country on earth has more experience in these matters'. In his work Stevin discourses on various inventions, finally giving his preference to the swivel gate lock.

The new invention consisted of the addition of sluice doors to the mitred gates of a lock. Figure 4 shows the lock gates, each having a square frame *ABCD* in which there is sluice door *EFGH* that pivots on vertical shaft *IK* set into the frame in such a way that part *IFGK* is slightly wider than part *IEHK*. The three sides *IE*, *EH* and *HK* rest in rebates on the inside of the frame, the three sides *IF*, *FG* and *GK* do not. The sluice door is kept closed by means of an upright iron spindle *LM*, which is turned with rod *MN* that is secured at the top of the frame with a catch. The moment this iron rod is released, the difference in the water pressure on the two unequal parts *EIKH* and *IFGK* causes the door to pivot through 90°, allowing scouring to take place. When necessary the entire gate can be opened for ships to pass through. This is Stevin's pivoted sluice lock, which he also used in his fortification work.

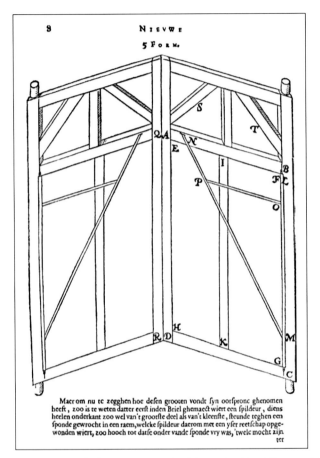

Figure 4: Stevin improved the structure of sluices. In his *Nieuwe Maniere van Stercebou, door Spilsluysen* he includes a survey of existing sluices and the way in which they were used. Shown here is the pivoted sluice lock that was also incorporated into the fortifications protecting towns. From: *Nieuwe Maniere van Stercebou, door Spilsluysen*.
© *Royal Library of Belgium (Brussels). Precious Works, VH 8489 C 2.*

Before turning to the practical application of his pivoted sluice locks in fortifications, Stevin deals with the way in which the foundations of dams, locks and sluices must be constructed if they are not to be undermined and their masonry ruined by the effects of scouring and seepage.

Stevin reflects that 'though the foundations of locks and dams are made in these countries with great forethought and expense, still it has not been possible to attain such certainty that they will not often cause considerable difficulties ... the foundations are sometimes washed away so that locks are useless, the dams collapse and sink deep into the water, drowning the land'. He therefore recommends the use of piles dovetailed together over their full length to form retaining walls – in other words, sheet piling. The walls thus built enclose a rectangle, from which the sand or peat is excavated with dredge-nets. Regular piles are driven into this rectangular pit and the crevices between them are filled with clay, thus providing a solid foundation for the masonry dam. Simon Stevin is regarded as the inventor of sheet piling, which is still used in many hydraulic works.

4. Hydraulic designs

Simon Stevin's hydraulic studies are extremely extensive. What distinguished him from other hydraulic engineers of his time was his consistent concern to test the physical theories he developed in *De Beghinselen der Weeghconst* and *De Beghinselen des Waterwichts* (see also Chapter Six) in practice. This was in contrast to his contemporaries, who put forward many impracticable hydraulic designs. Stevin worked out his ideas and tried to give them a theoretical basis by using models.

Hendrick Stevin collected the various opinions and numerous drawings his father left. They can be found in *Van den handel der Waterschuyring onses Vaders Simon Stevin*, in Hendrick's *Wisconstich Filosofisch Bedryf* and *Plaetboec, Vervangende de figuren of formen gehorig tottet Wisconstich Filosofisch Bedryf*. The latter was published in Leiden in 1668 by Philippe de Croy and contains improved versions of the drawings and figures accompanying the treatise in the former book. There is a copy in Ghent University Library.

Stevin's thinking about water scouring with the help of sluices, which he connected with the invention of the pivoted sluice lock in his *Nieuwe Maniere van Stertebou, door Spilsluysen*, had already in earlier years – presumably on his own advice – been put into practice in plans for the improvement of harbours and waterways. Some of the plans were very general; others were worked out for specific locations. In Hendrick's work the following general cases are dealt with:

1. A town lies at some distance from the sea, but near enough for the rise and fall of the tide to be perceptible. From the harbour two canals are dug, which both lead into a moat or basin that surrounds the town. They are closed off from the sea by sluices, which are opened alternately at low water. Hendrick mentions Bruges, Middelburg and Leiden as examples.
2. A town lies by the sea. In this case the moat or basin is built with an extension running inland. The two canals either lead separately into the sea or flow together into a harbour. Hendrick gives Ostend and Calais as examples of such an arrangement. (Box 3 gives Hendrick Stevin's description of each of these two cases)

There are specific descriptions of Gdańsk, Elbing, Deventer, Zutphen, Schiedam and Calais, among other places. Gdańsk and Elbing are dealt with in great detail. We know that Stevin stayed in Gdańsk in 1591 (see Chapter Two, section 1). Presumably he put forward suggestions to the Gdańsk town council and set them down on paper. We find the text in Hendrick's *Van den handel der Waterschuyring onses Vaders Simon Stevin*. The scheme Stevin proposed was very extensive. In Gdańsk two rivers, the Montlau and the Radaune, flow into the Weichsel at virtually the same point (see fig. 5), creating a sandbank in the process. To scour away this hindrance to shipping, Stevin proposed building sluices on the Radaune and even diverting the mouth of the river. In this same part of the Weichsel he proposed digging connecting channels between waterways, adding several pivoted sluice locks and providing a basin for the harbour at Weichselmünde on the Baltic by enlarging the Sasper Sea. None of these works were carried out in Stevin's day, however. Stevin also proposed the annual removal of the sandbank in the harbour at Weichselmünde. He was prepared to undertake this himself, writing to the council that 'I offer to contract everything at my own risk, both as to the machinery and as to the labour, at one and a half *groschen* per *last*, for which sum I will constantly maintain the depth at 7 ells over a width of 6 roods.' Whether or not this was actually done, we cannot be sure. However, we do know that between 1619 and 1623 Dutch hydraulic engineers carried out works extremely similar to those Stevin had proposed.

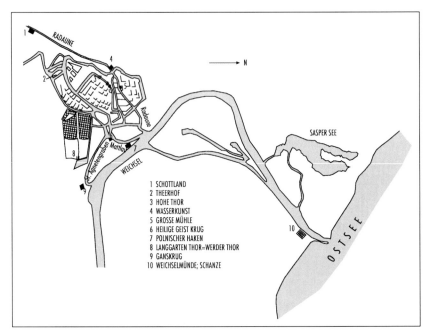

Figure 5: Stevin's hydraulic advice was called on far beyond the borders of the Netherlands. In 1591 he was in Gdańsk, for whose town council he drew up several suggestions for the better control of waterways in the area between the Baltic and the town.

Box 3: Extract from *Van den handel der Waterschuyring onses Vaders Simon Stevin*

(fol. 46 ff. Vande Schuyring met sluysen (On scouring with locks))
Scouring by combining the ebb and flow of the tide with sluices, as commonly used until now, can be described as follows. In fig. 6 (12 form) *AB* is the sea. In the saltings or embanked land is a harbour *CD* and its basin *DE*, closed off by a sluice at *D*. This basin having been filled at high tide, the sluice-door is lowered and left thus till low tide when the sluice-door is raised and the pent-up water falls to a lower level, scouring away the silt from the harbour and piers, if necessary. But the basin *DE* is bound to silt up, and at the rear end most, in the same way that creeks do, as explained above. It is true that usually

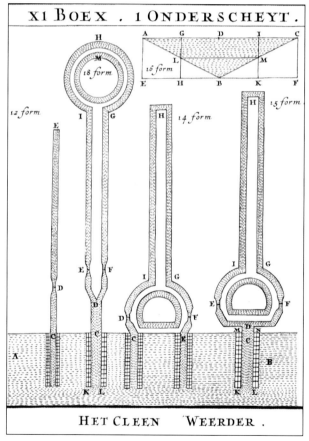

Figure 6: This diagram from *Van den handel der Waterschuyring onses Vaders Simon Stevin*, included in Hendrick Stevin's *Wisconstich Filosofisch Bedryf*, indicates how inland and coastal harbours can be scoured by variously adapting a sluice system.
Ghent, University Library (photo: University of Ghent).

the largest possible amount of rainwater from inland is introduced into it, but this does not help much, as appears from the many places in Holland and elsewhere where dredging is carried out manually, which involves great cost and discomfort, because as long as the work lasts, ships cannot berth there. But even if one cannot obtain sufficient depth for large loaded ships to berth there, it may serve for the wintering of empty ones.

To avoid these difficulties and obtain depths in which the largest loaded vessels can sail, one should proceed thus. In fig. 6 (13 form) *AB* is the sea. In the saltings or embanked land harbour *CD* is built. At point *D* the harbour splits into two parts, one part going towards *E*, the other part going towards *F* and continuing to *G*, *H* and *I*. Two sluices have been built, one at *E*, the other at *F*. At high tide the basin is filled through these two sluices; then the sluice doors are closed until low tide. Then sluice-door *E* is opened and sluice-door *F* is left closed. As a result the entire basin will be scoured from *F* to *G*, *H* and *I* through the sluice *E* to the sea. For the next scouring, sluice *E* should be kept closed and sluice *F* opened (because if *E* was constantly opened and *F* kept closed, there would be silting-up at *F* and subsequently in the whole basin). Thus the whole basin will be scoured from *E* to *I*, *H* and *G* through sluice *F* to the sea. If scouring was always carried out with the two sluices open simultaneously, silting-up may be anticipated, starting at *H*.

If we assume that area *GHI* represents a town, then it has been shown how towns built between embanked lands, such as Bruges, Leiden, Middelburg and the like, can make harbours and basins with a constant depth. This method can also be used to fortify the towns, because if a relatively shallow cut is made from *I* to *G,* the town will be completely surrounded by water. Moreover, inside the basin around the town, a ditch for standing water can be dug, such as *M*, and the earth between the two waters can be used as a covered way. In this manner, in addition to greater strength, the scouring of the bulwarks would also be prevented. At Bruges for this purpose use might be made of the two canals (the salt water canal and the freshwater canal dug from there to Sluys), and the two moats around the town. At Leiden one might use the Rhine to Katwijk for one canal; for the second waterway a new canal would have to be dug alongside the Rhine and properly embanked. At Middelburg two new canals would have to be made from the town northwards through the dunes, for in the direction of Flanders no good outlet could be expected because of the great silting up that takes place there. Other places can be dealt with in a similar way, according to the situation. It would also be possible to place the two sluices *E* and *F* together inside a strong redoubt, in order to keep them under constant control.

Now follows an explanation of scouring not only in front of towns in flat countries not far from the sea, such as those described above, but also for those that actually lie on it, such as Ostend, Calais and the like. In fig. 6 (14 form) *AB* is the sea. In the saltings or embanked lands or dunes two harbours *CD* and *EF* have been dug. There is also a basin *FGHID* with two sluices, one at *D* and the other at *F,* where one sees that good scouring can be effected both in the basin and in the harbour by using sometimes sluice *D* and sometimes using sluice *F*, and the rest as above.

But two sluices scouring a harbour together as in fig. 6 (13 form) produce better depth than two each in its own harbour, as in fig. 6 (14 form) But if so wished, for towns lying on the sea this might be done as in fig. (15 form), where *A, B, C, D, E, F, G, H, I, K, L* have the same meaning as in fig. (14 form). The corners *M* and *N* would not then be so suitable as fortification for the town, because the enemy could easily use them to protect their approach from the beach.

Note the following too: towns that lie on the sea and project beyond the general line of the dunes, such as Ostend, are very costly to maintain against the powerful scouring of the sea. As a solution to these difficulties, the outer parts of the town walls can be constructed in front of the line of the dunes. Building them further inland is pointless because the sea would fill the bay with sand, as it tends to form a straight continuous beach. If the town walls reach just as far as the dune line, they will suffer the same damage as the dunes. But if the town projects further and its walls are properly maintained, no expense being spared, it would be stronger, owing to greater depth and a narrower beach at low tide. We may also assume that Ostend and similar towns were not originally built to project this much, but that afterwards the sea encroached on the dunes. Similar erosion might be expected within a few years for those towns whose outer fortifications have been built on the outside of dunes. All these things should be considered when such works are undertaken or altered.

In September 1604, after a three-year siege, Ostend finally fell to the Spanish. In the course of the siege the harbour was destroyed. The rebuilding of the harbour and its defences was directed from 1605 by Wenceslas Cobergher, the archducal architect and engineer. In construction and design, and in the use of sluices, Simon Stevin's theoretical ideas can clearly be perceived. Piet Lombaerde (1983) has suggested that here Stevin's concepts of urban planning, fortification and hydraulic engineering were united.

4.1. Stevin's son Hendrick and the impoldering of the Zuiderzee

Hendrick may have been less gifted than his father but he elaborated the latter's hydraulic ideas nonetheless. His *Wisconstich Filosofisch Bedryf* contains not only his father's posthumously published work but his own ideas as well – for instance, his proposal to tame the Zuiderzee, which threatened the northern Netherlands from within, and to transform part of it into fertile agricultural land. His plan was to 'expel the violence and venom of the North Sea' by joining the Wadden Islands to each other and the mainland, which would 'produce enough fresh water and much land that can be easily embanked'. The plan also involved cutting a channel through the narrowest part of Holland to provide Amsterdam with access to the sea. But it would be 1920, 250 years after Hendrick's work appeared, before work actually began, with a dam joining the North Holland mainland to the island of Wieringen.

5. Navigation

By the end of the sixteenth century the Dutch Republic had become a major sea power, and Amsterdam had replaced Antwerp as the greatest centre of trade. The city was very favourably situated at a meeting of waterways and was increasingly used as an entrepôt. In 1602, as the fruit of consultations on the federation of Dutch maritime trade between merchants and the States, in which Grand Pensionary Oldenbarnevelt played a major role, the United East India Company (the Vereinigde Oost-Indische Compagnie or VOC) was established and received a monopoly on trade with the Indies. The Company was empowered to maintain troops and garrisons, to conduct warfare, to sign treaties and alliances with foreign states and to appoint Dutch governors within the colonies. Indeed, the Company provided the foundation of Dutch colonial power in East Asia.

It was thus understandable that the authorities in the Dutch Republic – Prince Maurice and the States General – were greatly interested in a safe and speedy method of maritime travel. Maurice showed considerable enthusiasm for nautical affairs, and it is likely that he asked Stevin to prepare a study of the subject. Stevin's works on navigation must therefore be seen as a service to the nation. They cover two different subjects: in *Vande Zeylstreken* (*The Sailings*) (a subsection of *Vant Eertclootschrift,* which is the second section of the first part of the *Wisconstige Gedachtenissen*), he deals with courses and distances (navigation, in other words); while in the following subsection, *Van de Havenvinding* (an earlier-published work which is also included in a revised version in the *Wisconstige Gedachtenissen*), he describes a method for finding one's intended landfall without knowing the latitude.

5.1. The straight and curved tracks...

In *Vande Zeylstreken* the 'sailing tracks' are defined as 'the lines which ships describe when they are sailing'. They fall into two categories, straight and curved tracks, or, in modern terminology, great-circle sailing, in which a ship follows the shortest line connecting the places of departure and destination, and loxodromic sailing, in which a ship sails a constant course. (A loxodrome is another term for a rhumb, an imaginary line on the earth's surface cutting each meridian at the same angle, still used as the standard method of plotting a ship's course on a chart.). Stevin proposes a mathematical theory based on spherical trigonometry. In the fifth chapter of the *Appendix of the Loxodromes* he describes 'How one might steer more exactly by the mariner's compass than is usually done.' *Vande Zeylstreken* can be regarded as a form of highly theoretical navigation and was not of immediate use to seamen, who would have been unable to follow Stevin's mathematics.

De Havenvinding (1599), printed at Leiden on the Plantin presses by Christopher van Ravelingen, official printer to the University of Leiden, is Stevin's work on finding one's position at sea. Unlike the highly theoretical *Zeylstreken*, it was aimed at the ordinary mariner and is set out in a very practical style. In that same year, 1599, Hugo de Groot brought out a Latin translation, *LIMENEVPETIKH, sive portuum investigandorum ratio*, and this was swiftly followed by a French edition, *Le Trouve-Port*, which is included in *Les Œuvres Mathematiques*, published by Girard. Yet another translation came out in 1599, in London, an English version prepared by the mathematician and nautical expert Edward Wright and entitled *The Haven-finding Art or the way to find any Haven or place at sea, by the Latitude and Variation*. This is further discussed in Chapter Twelve. The book's rapid translation into several languages indicates just how great was the interest it aroused in seafaring nations. *De Havenvinding* is the only work by Stevin not to include his name as author on the title page, not even in the original Dutch publication. However, his authorship is mentioned in Grotius's Latin translation and in the English version by Wright.

Stevin explains what he intends right at the start of his book. 'It is known,' he says,

> *that for a long time past, principally since the great voyages to the Indies and America began, a means has been sought by which the navigator might know at sea the longitude of the place where his ship is at the moment in order thus to get to the harbours to which he wishes to go, but that hitherto it has not been possible to arrive at such accurate knowledge of the longitude. For some people, hoping to find it through the variation of the compass, ascribed a pole to the said variation, calling it magnetic pole, but it is found upon further experience that these variations do not obey a pole. Nevertheless the search for this has furnished a means for reaching a desired harbour, even though the true longitudes of both the harbour and the ship are unknown.*

In navigation, the 'variation of the compass' is understood as the angle between the geographic and magnetic meridian. Gerardus Mercator and Pieter Plaetevoet, known as Plancius, a Dranouter-born pastor of the Reformed Church in Amsterdam, had written on the phenomenon of magnetic variation before Stevin, and had tried to make use of it in practical navigation and in attempts at determining longitude. Stevin's aim was clearly much more modest: he sought to enable the seafarer to reach a given harbour without having recourse to longitude.

Moreover, Stevin was not convinced of the existence of a magnetic pole, conceived as a rock located somewhere in the Arctic. In *De Havenvinding* he makes no attempt to explain the variation of the compass needle or terrestrial magnetism in general, as his predecessors Plancius and Mercator had done. But he did make a thorough study of the observational data Plancius had collected and expanded on them.

5.2. Stevin replaces Plancius's four agonic lines with six

Plancius assumed that there were four meridians on earth where the variation was zero: the prime meridian, which at that time passed over the island of Corvo in the Azores, and the meridians of 60°, 160° and 260° east longitude. In each of the four lunes into which these meridians divide the earth's surface, the needle is supposed to deviate from magnetic north in the same way, that is northeasterly in the lunes I (0–60°) and III (160–260°), and northwesterly in the lunes II (60–160°) and IV (260–360°). The northeasterly variation would increase in lune I from 0° to 30° eastern longitude and decrease from 30° to 60° and so on. Stevin concurred with Plancius as regards lunes I and II, for which there existed sufficient observational data, but in place of four agonic lines he introduced six, at 0°, 60°, 160°, 180°, 240° and 340°, and cautiously presented his system as conjecture or supposition. Although Stevin criticized certain aspects of Plancius's work and used his method in a more restrained form, he made no attempt to hide his admiration for his predecessor's data gathering, 'listing in a table the variations that have already been observed, which the learned geographer Mr Petrus Plancius has collected by protracted labour and not without great expense from different corners of the earth, both far and near, so that, if navigators shall find land and harbours generally in this way, as some in particular have already found them, the said Plancius may be considered one of the principal causes of this.'

Stevin advocated the use of the tables of variation (see fig. 7) to find harbours or even to enable ships belonging to the same fleet to regroup at a specific point. He interpreted the proposition in an appendix to *De Havenvinding* thus:

> *Since the given variation and latitude in combination indicate a definite point, both at sea and on the land, it follows from this that it is possible for ships to find each other at a given point at sea, far from the land. This is useful, among other things, to help the ships of a fleet to reassemble after a storm. By this means it is also possible to fix a rendezvous where ships coming from different directions may meet at a predetermined time.*

If this technique was to be truly reliable, the collection of compass variations at as many positions in the world as possible was essential. Prince Maurice ordered that navigators should thenceforth 'find out actually and very carefully the variations of the needle from the north' and faithfully report the results of their observations to the Admiralty so practical tests could be made. Stevin studied the various methods of observing variations, finally recommending a way of measuring them in *De Havenvinding*. In a section entitled 'How the True North and the Variation are Found' he explains how observations can best be taken. The navigator should use 'an azimuthal quadrant, the horizontal plane of which, notwithstanding the movement of the ship, always remains level.' In the margin of the page, Stevin provides a Latin translation for the description of this

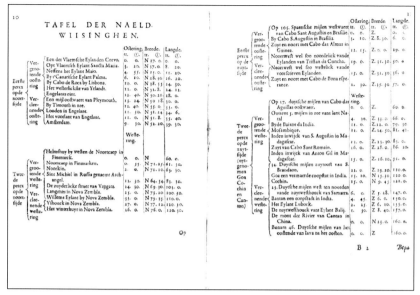

Figure 7: The method used for navigating a route to a harbour was based on the magnetic variation of the compass. Tables of variations were compiled on the basis of many hundreds of measurements.
The Hague, Royal Library (ektachrome: JTD).

instrument: *Quadrantem Azimuthalium seu verticulú cuius planú horizontale* – an azimuthal quadrant that turns about a vertical axis over a horizontal graduated circle. This instrument was built by Reynier Pietersz., also known as Reynier Pieter van Twisch. An inhabitant of Hoorn (in the province of North Holland), Reynier Pietersz. worked for the Hoorn ship owners. In 1598 he applied to the States of Holland and Westfresia for a subsidy for the building of two instruments. One of these was no doubt his azimuthal quadrant. On 13 March 1598, the States appointed a committee consisting of Scaliger, Snellius, Van Ceulen and Stevin, together with the deputies of Amsterdam, Rotterdam, Hoorn and Enkhuizen, to examine and test the instruments and report on them. The committee's conclusions are not known, but the fact that Stevin recommended the instrument and explained how it should be used suggests that the committee deemed Reynier Pietersz.'s device to be useful and usable. Certain sources even describe it as the 'Golden Compass'.

6. The Art of War (military science)

Stevin's contribution to the art of warfare was twofold. In *De Sterctenbouwing* and *Nieuwe Maniere van Stercteboû, door Spilsluysen* he describes the building of fortifications. In *Castrametatio, Dat is Legermeting*, he deals with the construction and layout of military camps in Prince Maurice's day. Some of

Stevin's other works on military science were posthumously published in *Materiae Politicae, Burgherlicke Stoffen* (*Civic Matters*), compiled by his son Hendrick. In Section VII is a plan for the organization of regular changes of garrisons and in Section VIII we find Stevin's non-mathematical writings on warfare under the title *Van de Crijchspiegeling* (*On the Theory of War*). Here too it can be seen that Stevin was putting his technological ingenuity to use in the service of his adopted country.

Castrametatio, which was not published until 1617, when Stevin's contribution to the structuring of the army was already much less than it had been, is not primarily a theoretical treatise but should rather be regarded as an account of the way in which army camps were traditionally laid out by the princes of Orange and the counts of Nassau. Stevin was aware that other armies and army staffs admired the discipline and symmetry with which the States Army on the march laid out their camp on arrival at a new place. As quartermaster (in Stevin's day an officer responsible for the layout and care of an army's quarters) he would have played a significant part in bringing this about, and a note of satisfaction can be detected in the dedication given at the beginning of the book, where Stevin comments that 'in previous years the ground plans of the encampments of the States Armies have been in great demand not only with private persons but also with great Princes in distant countries'.

6.1. Stevin's presence in military camps

Stevin's *Castrametatio* was not based merely on theory. He actually spent time in the kind of entrenched quarters he described. A ground plan of an army camp in the Army Museum in Delft (see fig. 8) shows Stevin's tent pitched very close to that of Prince Maurice.

The almost immediate translation of *Castrametatio* into French and the publication, just a few years later, of two German editions, indicate the depth of interest the work aroused. The book is in two parts: the first three chapters deal with the practice of the laying-out or construction of encampments, while the fourth is part history of the subject and part manual for the improvement of the army's organization. Stevin's erudition is again apparent in his easy references to Greek and Roman strategists such as Xenophon and Polybius. In the section *Vande legering int ghemeen, mette form des Romeynschen Leghers* (*On encampment in general, including the figure of a Roman Army camp*), he takes as his starting point the layout for an army camp Polybius devised during the Second Punic War and reproduces his ground plan (taken from Justus Lipsius's *De Militia Romana*, first published in 1585). In the course of his historical survey of army camp planning, he proposes a division of troops on a decimal basis, reflecting the organization of the Hebrews, Greeks, Romans and Tartars, namely a *rije* ('file') of 10 men, a *vendel* ('ensign') of 100 men, a *wimpel* ('pennon') of 1,000 men and a *standaert* ('standard') of 10,000 men. In this, however, Prince Maurice did not follow his quartermaster's advice.

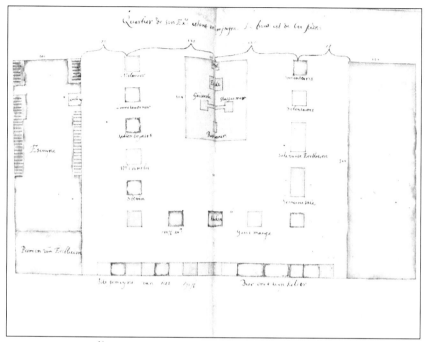

Figure 8: Fols. 68^V–69 of the 'Stevin manuscript': a tinted pen drawing of the *Quartier de son Ex/celence/ estant en Campaigne, le front est de 600 pieds*. As quartermaster, Stevin was responsible for the construction of army camps. We can see from this ground plan that his tent was in the immediate vicinity of that of Prince Maurice.
Delft, Royal Army & Arms Museum 'General Hoefer' (photo: Hans de Lijser, echtachrome: JTD).

6.2. A proposal for army administration

Stevin wrote extensively on the marking-out of army camps and this is not the place to go into every detail. At a time in which most other army units were made up of mercenaries, the States Army was a professional fighting force, active the whole year through. Unlike mercenary troops, they were also paid for the whole year. In winter the garrisons remained in camps and were kept in shape by rigorous drilling. In *Van de geduerighe verlegghing des crijchsvolcx (On the Necessity of Moving Garrisons Frequently)*, Stevin tackles the problems that arise when a garrison remains in one place for any length of time, observing that the chance of disloyalty increases and that differences with the local inhabitants can arise, as there are soldiers who will tie themselves to the place by marriage or the purchase of real estate. Here, Stevin demonstrates his practical sense yet again. He proposes an administration that will give the commander a constant and accurate overview of his army when it is continuously on the march. Stevin proposes that three registers are kept. The first, the *Jornael* (Journal), consists of a list showing the division of the various ensigns of each regiment at the start of a particular period. Every movement with its date, the reason for the move, the name of the captain, and of the old and the new place are also noted here. The second register, the *Plaetsregister* (Register of Place), lists every place and records which garrisons camp there. In the *Capiteinsregister* (Register of Captains), finally, the movements of each captain are logged. This form of administration is still in use in modern armies.

It is noteworthy that Stevin also devoted a great deal of attention to tactics and drill. The *slachoirden* (battle array) consists of a 'rectangular body of men' with 1½, 3 or 6 ft between each man. This formation and marching order had already been in use for some time. It can be seen in Hogenberg's engraving of the taking of Den Briel by the Geuze on 1 April 1577. To change direction the commands *rechtsom* ('right turn'), *slijnxom* ('left turn'), *omkeert* ('about turn') were given; to change the distance between the men the orders *opent* ('open ranks'; to 6 ft), *sluyt* ('close ranks'; back to 3 ft) and so forth, were used. These commands were new for the time and were learnt and practised during periods of inaction, especially in the winter. (For further details see Puype, 1998, 2000.)

The besieging of a town invariably involved a great deal of digging – of ditches, tunnels, redoubts and so on – and for this the troops needed the right equipment. In Section VIII of *Materiae Politicae, Burgherlicke Stoffen*, Stevin writes that the use of digging 'is held to be one of the principal causes of his [Maurice's] famous victories in the besieging of cities'. However, it was impossible for a soldier to carry all the equipment required, such as spade, pickaxe, saw and so forth, so Stevin invented an implement that was a combination of all these things and called it the *spabijhou* – literally the 'spade-axe-pick' (see fig. 9). In this ingenious piece of materiel the forerunner of the Belgian Army's *legerspaatje* can be recognized, an entrenching tool that can be transformed into a spade or pickaxe and so on, as the occasion demands.

Simon Stevin's place in the history of fortification corresponds in many respects to the place he occupies in the evolution of mathematics. He had a thorough knowledge of the literature on the subject, particularly the works of Italian authors (including Tartaglia), and also Daniel Speckle's *Architectura von Vestungen*, published in 1589. He gave a very clear exposition of the earlier material, but he introduced new elements as well. His books were quoted everywhere. He dealt with the subject in exactly the same way as he did the rest of his topics, starting from clear definitions and principles:

> *First the proper words and names of this treatise will be explained in 21 definitions.*

The literature on fortification chiefly employed Italian or French terms. Stevin, however, was a master in the use of his own Dutch language (his contribution to Dutch will be examined in Chapter Eight). Whether or not the words he introduced are neologisms is difficult to verify. It may be assumed that by and large they were words already used sporadically, which Stevin brought into general use. Below are some of the terms that apply to fig. 10.

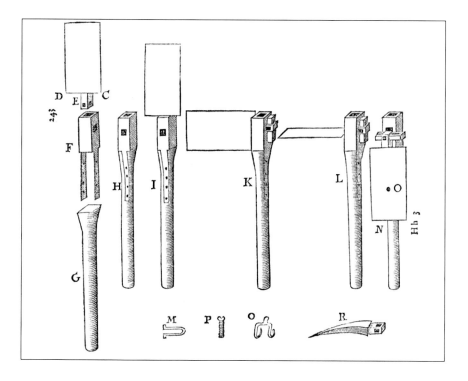

Figure 9: The besieging of a city invariably required a good deal of digging. Stevin invented a very handy piece of equipment and called it the *spabijhou* – literally the 'spade-axe-pick'.
© Royal Library of Belgium (Brussels). Rare Books Department, VH 28529 A.

Bolwercken sijn de uijtstekende hoofden der sterckten, als ... inde 8ᵉ form de twee bolwercken A, B.

[Bastions or bulwarks are the projecting parts of fortresses as in...the two bastions A, B in 8ᵉ form.]

Wallen sijn de eerde dammen tusschen twee bolwercken ligghende..., C inde 8ᵉ form.

[Walls or ramparts are the dams or banks of earth that stretch between two bastions, as...C in 8ᵉ form.]

Caden sijn de buytecanten des grachts...inde 8ᵉ form D.

[Banks are the outsides of the ditch, as...D in 8ᵉ form.]

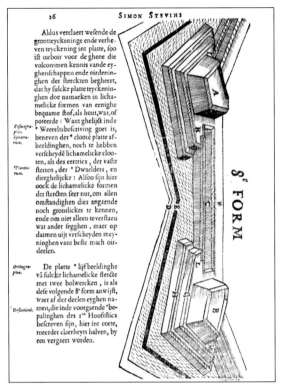

Figure 10: In *De Sterctenbouwing*, Stevin illustrates the new way of fortifying towns. This figure shows the various parts of a rampart.
© *Royal Library of Belgium (Brussels). Rare Books Department, II 16691 A.*

Groote gracht noemtmen t'gene tusschen bolwercken en wallen te eender, ende caden ter ander sijde begrepen is.

[Large ditch we call that which is shut in between the bastions and the walls on the one hand and the banks or outsides of the ditch on the other.]

Middelgracht is die omtrent het middel der groote gracht light..., Inde 8^e form E.

[Middle ditch (cunette) is that which lies about the centre of the large ditch, as...E in 8^e form.]

Borstweer is het ghene daermen achter beschut can staen tot de borst toe.

[Parapet is that behind which men may stand protected breast-high...]

From these definitions Stevin moves on to the necessary adaptations to fortifications. At Prince Maurice's prompting, the States Army asked Stevin's advice several times on alterations to defensive works. One of the most important locations in Holland during the Eighty Years' War was the fortified town of Hardewijk, which barred the entry into Holland of enemy troops who had crossed the Veluwe and the Zuiderzee. Hence the town's fortifications were systematically improved. There are many archive documents that refer to Stevin as an advisor, though it should be mentioned that when it came to the practical working out of urban fortification Stevin was not the first the Republic might call on: the States General consulted Adriaan Anthonisz., David van Orliens and other engineers far more frequently. Nonetheless, as has been mentioned earlier, Stevin was responsible for the text used by the engineering school at Leiden. Originally, 'engineer' referred to those who designed and built fortifications, and who could join the army. We know that Stevin advised on the construction of the fortifications at Flushing, Batavia (Jakarta) and The Hague.

Stevin made all kinds of calculations to introduce changes to existing defences. As the author of the first work in Dutch on military architecture, who took the first steps in adapting the Italian system of fortress building to the requirements of the north, Stevin is an important figure in the history of the science of fortification. Even though the young Republic's financial resources did not permit Stevin's recommendations to be carried out without modification, many (if not all) of the suggestions he made to Maurice of Nassau did have their effect in practice: Stevin's ideas were used in an adapted form in fortifications that were built under the supervision of the French engineer Vauban 100 years later.

7. Stevin as an architect

We know of several manuscripts by Stevin on architectural topics, but he never published a printed version of his ideas, despite more than one announcement of a treatise on architecture in his *Wisconstige Gedachtenissen*. In the fifth chapter, *Van de Ghemengde Stoffen* (*Miscellaneous Subjects*), he explains in a note that he had been unable to finish *Huysbou* (*On the Building of Houses*) in time for the printer and would therefore publish the work later. This, however, he never did. In his son Hendrick's *Materiae Politicae*, we do find writings (depending on the edition) on some of these topics:

- *Van de oirdening der steden. Bijvough: Van de oirdening der deelen eens huys met 't gheene daer ancleeft. (Town Planning. Supplement: On the planning of the parts of a house with that which relates to it.)*
- *Van den Huysbau (On the Building of Houses).*

In 1624, Isaac Beeckman visited Simon Stevin's widow, now remarried and living at Hazerswoude, and made a list of the manuscripts that Stevin had left. He compiled indices and copied extracts into his journal. According to Charles van den Heuvel (1998), Constantijn Huygens also took extracts and discussed them in his correspondence with researchers in other countries. In this way, even Stevin's unpublished texts on architecture became internationally known to some extent. Beeckman proposed the following division of Huysbou:

- *On the symmetry of the building*
- *On the underground structures*
- *On facades*
- *On stairs*
- *On ceilings and vaults*
- *On roofs*
- *On the layout of the parts of a house, such as the rooms, courtyards, galleries, gardens and finally of the parts of a room*

Van den Heuvel (2005) has attempted to reconstruct Stevin's *Huysbou* by means of often contradictory indices and surviving fragments of published and unpublished material. He maintains that water played an important part in Stevin's architectural theory. The Bruges scholar was well aware of the problems arising when building on boggy ground or below the surface of the water. He was also concerned about water supply, damp, water purification and sewage disposal, be it in an individual house or a whole town (see fig. 11).

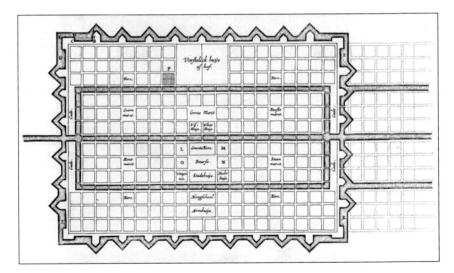

Figure 11: Stevin authored various manuscripts on architecture and town planning. His ideal town was laid out on a strictly geometrical plan. From: *Van de Huysoirdening* in *Burgherlicke Stoffen*.
© Royal Library of Belgium (Brussels). Rare Books Department, VH 28529 A.

7.1. Town Planning

A single glance at fig. 11, a plan for a town in *Materiae Politicae, Burgherlicke Stoffen*, reveals Stevin's highly organized way of thinking. His plan did not feature the picturesque but irregular medieval tangle of streets that was to be found in the town of his birth; it was strictly geometric. The severe, geometric character of the Renaissance found a devoted adherent in the Bruges scholar. Stevin opted for a rectangular structure, in which straight streets create identical rectangular sites on which houses could be built in blocks. He preferred to build blocks of houses, rather than individual dwellings, as this allowed for a satisfactory division of *hooven* and *lichtplaetsen* – gardens and courtyards. The plan also makes it clear that Stevin endeavoured to give all his public buildings a central location. Thus the *Vorstelick huijs* – the Princely House, or court, in other words – also contains the upper council chambers, the municipal school and the fencing and musketry academies, while under the roof of the *armhuys*, or poorhouse, almshouses for aged and infirm men and women, the hospital, specific quarantines for leprosy, plagues and poxes, accommodation for the blind, for poor orphans and for foundlings, a hospice and the lunatic asylum could also be found.

Clearly, in Stevin's day considerable provision was made for the welfare of those unable to support or care for themselves. Churches are located centrally (on the *grote marct*, or market place) and in every district. Specialized markets, such as the *beestemarct* (cattle market), *steenmarct* (stone market), *houtmarct* (wood market) and *coornmarct* (corn market), are also symmetrically located around the *grote marct* in the centre of the town. Stevin planned his streets to be 60ft wide with 10ft wide arcades on either side to protect pedestrians from inclement weather and the dangers of traffic. Beneath the street a sewerage system carried off the wastewater from the houses. The town was surrounded by stout bastioned walls, though expansion was foreseen at either end of its longitudinal axis.

7.2. House building

Van den Heuvel's study makes it clear that Hendrick was selective in the publication of his father's manuscripts. Indeed, he says as much in the introduction when he states that the work should be comprehensible to everyone, 'even those with no knowledge of or inclination for mathematics', thus anticipating a phenomenon not unfamiliar today: texts containing mathematical formulae produce a certain reticence, not to say aversion, in some readers. Hendrick was keen to make his father's work as accessible and familiar as possible to the general public.

In the art of house design, as in the science of warfare, Stevin clearly knew his classics. He begins his *huysoirdening* with a sketch of the layout of Greek and Roman houses, referring in particular to the Roman architect and engineer, Marcus Vitruvius Pollio, active at the time of Caesar and Augustus and author of *De Architectura*, which was regarded as the primary source of information on classical architecture. Stevin emphasizes that symmetry was the guiding principle of the Greeks and Romans, which is to say that 'the right and left parts of a body correspond in form, size and build'.

7.2.1. Symmetrically arranged cabinets, pantries and sculleries
When it comes to examining contemporary dwellings, Stevin attaches much importance to the places where light enters the building, which must be so arranged that every room receives adequate natural light. He starts from a ground plan for a model house (see fig. 12). We see a *Voorsael*, or hall, around which are arranged in perfect symmetry the *Eetcamer* (dining room), *Slaepcamer* (bedroom), *Vertreccamer* (living room) and *Keucken* (kitchen). Two staircases are indicated by the letter B. The four small rooms indicated by the letter C can

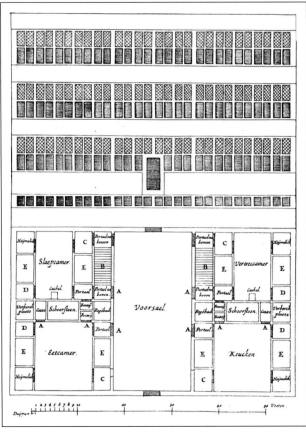

Figure 12: Stevin found symmetry of plan important in house building too. He was concerned with every part of a house, from attic to cellar. Nor did he forget the roof and facade. From: *Van de Huysoirdening* in *Burgherlicke Stoffen*.
© Royal Library of Belgium (Brussels). Rare Books Department, VH 28529 A.

be used as *schrijfpercken* (offices or studies), *cabinetkens* (small cabinets), *botelrie* (pantries), *morshoeck* (sculleries, where activities involving dirt can be carried out) or suchlike. The letter D indicates four storage spaces for wood or peat; E denotes four *cassen* or *betsteen* (cupboards). The *voorsael* is 34 × 17 ft and is intended as a 'common place for the entire household' as well as for the reception of strangers.

The upper floor and cellar are similarly arranged. Like a professional architect, Stevin devotes great attention to technical architectural details. He considers the direction in which doors should open (inwards or outwards) and lingers over locks and keys, stoves or fireplaces. He pays particular heed to water supply, lavatories and cellars, and argues that in addition to well water, rainwater should be used for such things as washing flax and cooking peas. He observes that in prolonged dry spells the water in the rain barrels becomes *wormich* and *stynckend* – wormy and stinking – as a result of all the dirt that washes out of the gutters along with the rainwater, and immediately explains how this can be remedied and the water purified by filtering. The *heymelicken*, or privies, are of great practical importance, and Stevin attaches one to each of the main rooms (alerting the reader to the possible smell). He also stresses the fact that a house is worthless without a good roof. With a view to actually collecting rainwater indoors (note the *regenbakken*, the cisterns provided), he favours a *trechterdak* – a funnel roof – covered with flat hard-baked tiles of good clay.

Having dealt with the technical, Stevin moves on to the aesthetic. In addition to the symmetry of a building's facade he also considers its decoration. Though classical architects frequently employed pillars, columns and pilasters, Stevin was not in favour of the non-structural use of such elements. 'Classical pillars that project from facades,' he said, 'such as those on Antwerp's town hall, or that are brought indoors (as a curtain between two bastions), obstruct the view. Sufficient decoration is achieved by the flat evenness of stones and the joints neatly wrought'.

Chapter Five

Economist *avant la lettre*

Simon Stevin lived at a time when increasing trade made financial management and accounting ever more important. In his professional capacity, first as a minor assistant in a merchant's office and later as personal advisor to Prince Maurice and 'Prince's Superintendent of Finance', he was well placed to acquire a thorough understanding of economics and subsequently to write on the subject. Stevin's work as an economist was significant for a number of reasons. He demystified the calculation of interest, making it easier for the ordinary man to understand, and he commended to the princely administration the system of double-entry bookkeeping.

1. Introduction

Stevin dealt with financial matters throughout his life. As a young man he worked as a cashier and bookkeeper in Antwerp, at that time northern Europe's greatest mercantile metropolis, into which merchants and merchandise poured from every corner of the globe. We do not know exactly what Stevin did in Antwerp, but we can assume that he worked in the office of a trading company and there became acquainted with mercantile practice. The commercial transactions of his stepfather Joost Sayon and his brothers Vincent and Jacob must also have made him familiar with the world of buying and selling. And he had known the mechanics of taxation since around 1577, when he was employed in the financial administration of the Brugse Vrije.

From the moment Stevin settled in the northern Netherlands, we can follow his activities in the economic field through his writings. R. Oomes (1996) suggests that the *Nieuwe Inventie van Rekeninghe van Compaignie* (*New Invention for Company Accounting*), published in 1581, is Stevin's earliest work (there is a copy in the Royal Library in The Hague). If so, this short volume on commercial practice would predate the *Tafelen van Interest* (1582), a work that challenged an ancient and established tradition. Many banking houses engaged in large-scale financial dealings of one kind or another. They were extremely powerful bodies, controlled by families like the Bardi and Medici of Florence, and the Welsers and Fuggers of Augsburg, who ruled financial empires and whose power was respected and feared by king, emperor and pope. At a time when the multiplication and division of integers was regarded as a complex mathematical operation, only experts were able to answer questions involving the calculation of interest. The larger well-established houses employed such experts to compute tables of interest, and the tables were jealously guarded as confidential information. This state of affairs continued unchanged into the sixteenth century, when mastery of arithmetic became (comparatively speaking) more widespread – an evolution that Stevin obviously sought to accelerate.

1.1. Stevin and the reorganization of the princely lands

Stevin's pupil and friend Prince Maurice was an able administrator of the family lands, as well as of the public monies he controlled as commander-in-chief of military operations. At some point – precisely when is uncertain but probably in the first years of the seventeenth century – Stevin was charged with the reorganization of the princely lands, so states his son Hendrick. On the title page of *Burgherlick Stoffen*, published in Leiden in 1649, Hendrick claims that his father had obtained the title of the 'Lord Prince's Superintendent of Finance'. With this appointment Stevin once again occupied a position of financial responsibility, though one that was far more important than his lowly jobs in Antwerp and Bruges. He seems to have held this post for the rest of his life.

As with most of the tasks he undertook for the prince, Stevin evidenced his activities by devoting a publication to the financial transactions involved in the management of lands and also of warfare, which he termed 'Extraordinary Financial Affairs'. His thoughts on the matter are contained in the *Wisconstige Gedachtenissen*; in part five, section two, we find *Van de Vorstelicke Bouckhouding in Domeine en Finance Extraordinaire* (*Princely Bookkeeping of Lands and Extraordinary Financial Affairs*), the lessons Stevin devised for the prince in this subject.

It is noteworthy that Stevin himself relates how he came to explain the principles of bookkeeping, which he does in the form of a lively dialogue, purportedly the faithful record of real 'conversations'. Maurice had complained that in some of the accounts his officials laid before him he found 'an obscure way of doing things'. This gave Stevin the opportunity to bring to the prince's attention the advantages of double-entry bookkeeping, a method of accounting first used in Italy. Stevin gives the first conversation the title *Oirspronck der beschrijving deser Vorstelicke Bouckhouding* (*Origin of the description of this Princely Bookkeeping*) and begins as follows:

> It so happened that one morning when I came to His Royal Grace, the first words he spoke to me were that that previous day he had been handed certain accounts that he found obscure, and in his opinion unnecessarily long.

The conclusion of the first conversation attests to the prince's thirst for knowledge, for he takes leave of his friend with the words, 'if the lessons in algebra are over we must make a start on bookkeeping; get some material ready for I am eager to begin.' It transpires that the prince had lessons from Stevin on this subject too and 'thoroughly understood the matter', at least the theory and practice of mercantile bookkeeping. From their second conversation we learn that the prince's interest was not so much in this kind of bookkeeping *per se*, but rather in its application to the management of lands and estates. In the second part of his tutorial, Stevin develops his principles of bookkeeping with regard to this kind of management and even the management of public finances, particularly the financing of military operations.

2. The Nieuwe Inventie

In 1996 a short article by R. Oomes appeared entitled 'An Unknown Work by Simon Stevin' (*'Een onbekend werk van Simon Stevin'*, in the newsletter of the Jan van Hout Society). The unknown work in question was a booklet printed in Delft in 1581, called a *Nieuwe Inventie van Rekeninghe van Compaignie* (*New Invention for Company Accounting*) and subtitled *inde welcke verclaert wordt een zekere corte ende generale reghele, om alle Rekenynghe van Compaignie ghewislic ende lichtelic te solveren, gheinventeert ende nu eerst int licht ghegheven door Simonem Stephanum* (*in which is explained a certain short and general rule for correctly and easily solving all business calculations, invented and now published for the first time by Simonem Stephanum*). This booklet of 24 pages, of which four are blank, has not hitherto been included in the lists of Stevin's works. One possible explanation for its omission may be found in the fact that only one copy (fig. 1) is known to exist (now in the National Library of the Netherlands in The Hague) and that the author does not refer to himself as being 'of Bruges', as Stevin did in practically all his later works. The booklet is described by W.P.C. Knuttel as pamphlet 581 in the catalogue of the pamphlet collection in the National Library (The Hague 1889–1920, reissue, Utrecht 1978).

2.1. Is *Nieuwe Inventie* Stevin's first publication?

The *Nieuwe Inventie* was dated by its author at Leiden in 1581, was printed in Delft, and deals with prevailing commercial practices in the first part of 1580 (before 15 July) in a further unspecified place. The work was dedicated to the burgomasters of Amsterdam, so perhaps that city may be regarded as the locus of events.

It is not possible to prove beyond doubt that this short work was Stevin's earliest publication. There are some facts, certainly, that give grounds for attributing it to the Bruges scholar, but other questions remain unanswered:

- In 1581, when the author dates his work, Stevin was in Leiden. He was entered in the municipal registers as Symon Stephani, the same name as that used by the pamphlet's author. However, there is nothing to indicate that in 1580 Stevin was working as a mercantile bookkeeper in Amsterdam. Then again, we have no idea what he was doing in the period between 1577 and 1581, so a stay in Amsterdam is not out of the question.

Figure 1: The *Nieuwe Inventie van Rekeninghe van Compaignie* is possibly the earliest work that can be attributed to Simon Stevin. It contains a description of commercial practices in use in the first part of the 1580s.
The Hague, National Library of the Netherlands, Pamphlet 581.

In the 1580s Stevin had contacts in Delft via the burgomaster Johan Cornets de Groot, which could explain the use of a printer in that city.

- The author draws up a balance sheet for a small company and determines the profit and loss of what would now be called the shareholders. Such an activity could, in principle, be in line with Stevin's bookkeeping work in Antwerp before 1577, though we do not know what his job actually entailed.
- That Stevin was very familiar with bookkeeping is clear from the lessons he devised for the prince and reproduced in the *Wisconstige Gedachtenissen*. Furthermore, the way in which the material is presented in both the pamphlet and the *Wisconstige Gedachtenissen* is very similar. In both works a practical example is taken as a starting point. The pamphlet's author begins with this example:

I assume that three persons, Heyndrick Pietersz., Franchois Jacobz. and Jan Michielsz. come to me on 15 July 1580 and say: The three of us have been in partnership for a certain time and each of us has bought, sold, received and paid out, and we have given money both to each other and to others.

Stevin's lessons for the prince are based on the detailed business accounts of one Dierick Roose for the year 1600. In a very short text entitled *Van Compaignieslot* (*On Closing Company Activities*), which was part of *Coopmans Bouckhouding op de Italianse Wijze* (*Mercantile Bookkeeping in the Italian Manner*), a section of the *Wisconstige Gedachtenissen*, Stevin also turns his attention to the problem of company finances. Like the author of the pamphlet he gives an example, which he introduces thus: 'Take a partnership with three partners, the first named A, the second B, the third C, each dealing, receiving, and giving money and goods, both to each other and to others.' The similarity in the way the problems to be solved are introduced is striking.

- The pamphlet's author tells us that he has invented what is written in its pages. He mentions no source, nor does he refer to other works. Contrary to his usual practice when writing on topics in his many other areas of expertise, Stevin likewise makes little reference to predecessors in his essay on bookkeeping in the *Wisconstige Gedachtenissen*, though predecessors there certainly were, both in the Netherlands and beyond. Perhaps it can be inferred from this that both Stevin and the pamphlet's author were self-taught in the practice of bookkeeping, that they had gained their knowledge from hands-on experience. Stevin even says that he employed the method developed to describe company finances in the Wisconstige Gedachtenissen many times thereafter.
- When dealing with income and expenditure, the pamphlet's author uses a form of double-entry bookkeeping (see box 1) by drawing up tables of credit and debit for each of the partners. From these he subsequently determines their items of profit and loss. Stevin also strongly recommends this method of accounting when he describes mercantile bookkeeping to the prince. He also makes use of it in his essay *Van Compaignieslot* in the *Wisconstige Gedachtenissen*.

Box 1: Double-entry or Italian bookkeeping

Double-entry bookkeeping is a collection of conventions for recording financial transactions. In essence, double-entry bookkeeping represents an equation: debits and credits must balance. The proprietor of the business is the financier of the business but one with very special rights. The business is regarded as a separate entity from the proprietor. Accordingly, the business owes the proprietor the funds he has invested in the business. The principles of double-entry bookkeeping can be illustrated by a typical balance sheet of a professional person.

Debits

    Non-current debits
    Capital of Charles Johnson, owner   300
    Long-term loan   250
    Current debits
    Creditors   125
    Tax liabilities   150
    Total debits   825 euros

Credits

    Non-current credits
    Furniture and material   25
    Car   250
    Current credits
    Debtors   100
    In the bank   150
    Total credits   825 euros

The debit and credit totals are the same. The balance balances. Maintaining this balance is the single objective of double-entry bookkeeping.

Obviously, if an item is added to the credit side, an item of the same value must either be added to the debit side or the same amount must be subtracted from a different item on the credit side. The same principle applies when an amount is subtracted from one or the other side. Here are some examples:

- Charles Johnson pays an outstanding bill of 60 euros. As a result, his credit (bank balance) decreases but his debit (creditors) also decreases by the same amount.
- Charles Johnson buys a desk for 40 euros. The credit (bank balance) decreases but another item of credit (furniture) increases by the same amount.
- A debtor pays his bill of 100 euros. Thus the credit (bank balance) increases but another item of credit (debtors) decreases by the same amount

One may wonder what happens to this principle of balance when an expense such as a rent of 100 euros is paid. Obviously this means the current credit in the bank decreases by 100 euros, but which item of debit or credit must compensate this? Mr Johnson might, for example, have the use of an office for a week in return for his 100 euros, in much the same way as by paying the distribution company one has electricity at one's command for a certain period. As regards this kind of short-term cost, which provides advantages for periods that end before the end of the

> financial year, it is agreed that they are regarded as costs. If they cover a period that ends after the end of the financial year, the part that relates to the current financial year is regarded as a cost and the remaining part as credit. At the end of a financial year, all the costs are combined. They are then seen as a sort of counterweight vis-à-vis the owner's capital; they represent a loss of capital. Thus in double-entry bookkeeping at the year-end there is a decrease of credit by this sum and a decrease of debit (owner's capital) by the same sum.

- Let us turn for a moment to the pamphlet's content. The three partners in the pamphlet ask the author to disentangle their business activities and thus determine who has made a profit and who a loss. To this end, the author poses a series of questions:

   Firstly I ask them if they all accept that the items that each says he has paid out and received for costs and other things have been done well and truly... Then I tell them that each must give me the items he himself received and paid out on behalf of the company: something that every merchant, even if he is inexperienced in bookkeeping, can do.

   Further, I ask them to give me the balance of the company, that is the account of the remaining debtors and goods and creditors: to wit, all the items of the debtors with the remaining goods under debit and all the creditors under credit such as they are on the present day.

   Finally I ask them to declare what part of the profit or loss each of them should have according to the agreement they made between them.

On the basis of the information that each of the partners provides, the pamphlet's author creates tables of credit and debit. He calculates the company's profit in fine detail in two different examples. In the first example, the profit is equally shared out among the three partners. In the second he takes into account the agreement made between the partners when they began their business, determining that 'the ratio of the share of profit or loss is such: Heyndrick Pietersz. receives half, Franchois Jacobz. a third, and Jan Michielsz. a fourth'. Note that a half, a third and a fourth add up to 13/12. When each partner wants to have his agreed share, not only the profit but also a part of the capital must be remitted. The pamphlet's author provides an elegant solution by giving each partner only 12/13ths of his agreed share of the profits. The shares are correctly calculated to the thirteenth of a penny. The author's grasp of mathematics and the elegance with which he solves the problem strongly suggest that he had a scholarly background. By this criterion, Stevin certainly qualifies as a candidate for the pamphlet's authorship.

- In *Van Compaignieslot* in the *Wisconstige Gedachtenissen*, Stevin begins with a very similar question:

I ask them urgently to do three things: firstly, each partner should explain how much more he has received than he has paid out, or how much more he has paid out than he has received. Secondly, they must give me the difference between the sum of the present creditors and the sum of the present debtors in ready money and goods present. Thirdly, what share of profit or loss each can claim according to their agreement, for then I will say to the last detail what is due to each, without troubling to couch the commerce of each in bookkeeper's style, as is sometimes done.

It is clear that both the pamphlet's author and Stevin start from the same principles. This could indicate that they are indeed one and the same individual.

- In *Van Compaignieslot* in the *Wisconstige Gedachtenissen*, Stevin explains why he has drawn up the method or rule for companies:

First I will say briefly why I have drawn up the foregoing rule. A few years ago there was a partnership that did a great deal of business and had depots and branches in Venice, Augsburg, Cologne, Antwerp and London. When drawing up the balance of the company's affairs it appeared that one of the partners had not kept his accounts in a proper bookkeeping fashion. This meant that he could bring in little from his side towards the complex accounts and had to accept whatever the other partners proposed.

Thus it appears that a few years before the publication of the *Wisconstige Gedachtenissen*, Stevin must have been confronted with this partnership problem. He mentions the trading centres in which the company had branches. It is striking that Antwerp is on the list, but not Amsterdam. Perhaps the problem set forth in the pamphlet and the example Stevin cites in *Van Compaignieslot* allude to the same methodological problem, to be situated in Stevin's Antwerp period, therefore before 1581.

We can conclude that there is a reasonable possibility that Stevin was indeed the author of the pamphlet, but it is not absolutely certain.

3. The *Tafelen van Interest*

The *Tafelen van Interest, Midtsgaders De Constructie der selver, ghecalculeert door Simon Stevin Brugghelinck (1582) door Christoffel Plantijn in den gulden Passer te Antwerpen* (Tables of Interest, Together with the construction of same, calculated by Simon Stevin of Bruges [printed] *by Christopher Plantin in the Golden Compasses at Antwerp*) appeared in 1582. The work was written in Dutch and was dedicated to the aldermen of Leiden, where Stevin lived. A French translation was included in *L'Arithmetique* (1585). The Dutch version was reissued in 1590 with corrections. The French translation appeared once again in Girard's 1625 edition of *L'Arithmetique* and in his *Les Œuvres Mathematiques* in 1634. However, there are several differences in content between the various versions. References to the French mathematician Jean Trenchant (see box 2), who Stevin acknowledged as one of his sources, appear only in the Dutch version. Some authors have hinted that Trenchant's fervent Catholicism made him unacceptable to the Huguenots, who deleted all mention of him from the French version of the book.

Box 2: *Le grand parti*

The medieval fairs were great trading events that were held at specific places two, three or four times per year, lasting up to several weeks and attracting foreign merchants bringing fine exotic products from distant lands. They frequently coincided with saints' festivals, from which they derived their names. At the fairs, merchants developed solutions to many of the problems of long-distance trade. The most pressing was that of payment. With variable exchange rates and the difficulty of moving large amounts of hard currency safely, a way of arranging payments without using cash was needed. The solution was the bill of exchange. Soon merchants began to trade in this paper, speculating on the issuer's ability to pay or changes in the exchange rate. By the fourteenth century, the fairs had effectively become money markets, where huge loans could be arranged with the most important bankers in Europe.

In 1555, at Lyons, King Henry II of France borrowed money from a number of bankers in order to finance his wars. He paid 4% per 'fair' on the loan. There were four fairs each year at Lyons, so this effectively meant 4% per quarter or 16% per year. In the same year, before the All Saints fair, he received from certain bankers the sum of 3,954,641 écus, a loan that was known as *le grand parti*. To repay this sum the king was asked to pay 5% per fair for 41 fairs, when he would have repaid both principal and interest. The question now was which of the two forms of repayment was the most advantageous to the lenders: payment of 4% interest per quarter and the return of the principal at the 41st payment, or payment of 5% interest per quarter with the principal included. Calculating the cost of the first option is simple; that of the second is not. Jean Trenchant endeavoured to explain this complex transaction in terms that the ordinary man could understand in his book *L'Arithmétique departie es trois livres*, published in 1558. He showed which option would be the most profitable to the bankers by means of two tables of interest, from which it could be seen that the first option was the most advantageous.

This was a problem with which Stevin was familiar. In the foreword to his *Tafelen van Interest* he refers to his predecessor, Trenchant:

> *These tables, together with their construction and use, I will explain in due order to the best of my ability in this treatise. Not that I publish them as my invention, but indeed as amplified by me; for before me Jan Trenchant has written about them in the 3rd book of his* Arithmetic, *in the 9th chapter, section 14, where this author made one of these tables of 41 terms at an interest of 4 percent for every term, every term being of three months. And although he has not made these tables for such general use as we present them here (for he had in view the most profitable of two conditions which the bankers offered to Henry, King of France, in the year 1555, concerning a principal of 3,954,641 gold crowns, which was called* le grand parti, *when they gave the king the choice to pay 4 per cent simple interest every quarter year or 5 per cent, during 41 terms or quarter years, so that he would have paid capital and interest at the same time), yet we say, to his good and everlasting memory, that he is rightly called the inventor of these tables.*

Here we see yet again how extensively read Stevin was, and how well he prepared his own publication.

3.1. Demystifying the calculation of interest

In Stevin's day the computing of interest was shrouded in secrecy. He was very well aware of this, and in his foreword to the tables he makes his opinion on the matter very clear:

> *I have also learned that here in Holland such tables are to be found in writing with some people, but that they remain hidden as great secrets with those who have got them, and that they cannot be obtained without great expense, and principally the composition, which is said to be shown to very few people. Indeed, it has to be confessed that the knowledge of these tables is a matter of great consequence to those who often need them, but to keep them a secret seems to argue in some sense a greater love of profit than of learning.*

Tables in manuscript form existed before Stevin's publication. One of these early tables, composed in 1340, has survived in a copy dated 1472. This manuscript was compiled for the powerful Florentine banking house of the Bardi family by their agent Francesco Balducci Pegolotti as part of his handbook for merchants, the *Pratica della Mercatura*. Pegolotti was in Antwerp in 1315. His tables record the increase of 100 liras at a compound rate of interest of 1%, 1½%, 2% and so on up to 8%. Each of the 15 tables runs over 20 terms. It would be another half a century before printed tables appeared. A few were contained in Jean Trenchant's Arithmétique, published in 1558 (see box 2).

Stevin's book, which came out in 1582, contained the first complete tables of interest ever published. At a price of two stuivers it was a bargain. In typical style, Stevin begins his book with a number of definitions, each followed by a detailed, elegantly formulated explanation. A few examples:

DEFINITION 1: Principal is the sum on which interest is charged.

EXPLANATION: For example, when a man gives 16 lb in order that he may receive for it 1 lb of interest a year, then the 16 lb is called Principal. Or when a man owes 20 lb, to be paid in a year, and he gives 19 lb present value, subtracting 1 lb for interest, then the 20 lb is called Principal.

DEFINITION 2: Interest is a sum that is charged on the outstanding part of the Principal over a certain time.

EXPLANATION : For example, when it is said: 12 per cent a year, that is as much as an interest of 12 on a Principal of 100 over a year, so that Principal, interest and time are three inseparable things, i.e. Principal does not exist unless in respect of a certain interest, and interest does not exist unless in respect of a certain Principal and time.

DEFINITION 4: Simple interest is such as is charged on the Principal alone.

EXPLANATION: For example, when 24 lb is charged for interest on 100 lb in 2 years at 12 per cent a year, the 24 lb is then called simple interest. Or when a man owes 100 lb, to be paid at the end of two years at 12 per cent a year, and he pays present value, subtracting for interest on the principal alone 21 3/7 lb, this 21 3/7 lb is then called simple interest, in contrast to compound interest, the definition of which is as follows:

DEFINITION 5: Compound interest is such as is charged on the Principal together with what is outstanding.

3.2. Examples to illustrate the theory of simple and compound interest

A series of similar definitions and explanations is followed by a series of examples to demonstrate the theory of simple and compound interest. Stevin introduces two types of interest, defined thus: 'profitable interest is such as is added to the Principal' and 'detrimental interest is such as is subtracted from the Principal'. It is clear that the first type is the interest received by investing a sum of money, whereas the second represents the payment of interest on a borrowed sum. All but one of the tables that give the book its title are tables of detrimental interest. They are based on sums borrowed for periods of 1 to 30 years at interest rates of 1–16 per cent and the so-called penny runs from 15–22. An interest rate of 'm penny' signifies that one penny interest is calculated on m penny capital; this is therefore equal to a percentage of $100/m$. But, Stevin points out, compound profitable interest cannot be worked out in quite the same way. Rather than adding a whole new series of tables, which might 'create confusion rather than clarity', he gives only one, that for the fifteenth penny, as an example. Stevin's reticence as regards profitable interest was perhaps not unconnected with the Church's dim view of compound interest. Some books on accounting refer to the immorality of this *coutume des juifs*. In that period, many moneychangers and moneylenders were Jews.

The definitions and explanations are followed by the tables themselves. In essence, Stevin solves the following problem:

Determine a constant value A_n that consists of $i\%$ compound interest and after n years has increased to a final value of 10,000,000; in other words, determine A_n in $A_n(1+i)^n = 10^7$.

The problem can also be put like this:

If one wishes to borrow 10,000,000 at $i\%$ compound interest over n years, what remains of this capital after payment of the interest?
Note that by starting from 'some large number', Stevin always works with whole numbers; his decimal ideas appeared only in *De Thiende*, published three years later.

The values of A_n are given in a first series of tables for values of i running from 1 to 16 and n varying from 1 to 30 (see fig. 2 for the interest table of 2%). In the third column of each table, values occurring in the second column up to the nth value are added together. Stevin explains in detail the way in which he has constructed his tables in Proposition 3, Construction of the Tables.

Stevin's initiative of constructing tables of interest encouraged several others, particularly in the northern Netherlands, to follow his example. Of course this was also partly the result of the north's rapid commercial expansion and the dominating influence of Calvinism, which took a rather more tolerant view of accruing interest on capital than Catholicism or Lutheranism did.

The first to emulate Stevin was the schoolteacher Marthen Wentzel Van Aken. He was invited to write a book by a Rotterdam merchant who found Stevin's explanation too difficult. Wentzel's tables were published in 1587 and reprinted in 1594. Two years later, Ludolph van Ceulen published his interest tables as part of his *Van den Cirkel*.

In this field too we can state that Stevin broke through barriers: with the publication of his *Tafelen van Interest* he made the whole mysterious business of money more transparent and accessible to the ordinary man.

4. Bookkeeping

4.1. Bookkeeping for merchants

Stevin's work on bookkeeping consists of two parts, mercantile bookkeeping and princely bookkeeping. In both cases he sought to convince the prince of the usefulness of the Italian double-entry method (see box 1), a type of accounting Dutch merchants had already been using for 60 years. The earliest printed account of mercantile bookkeeping was written by the Franciscan Luca Paciolo, who travelled throughout Italy as a teacher of mathematics. In his *Summa de Arithmetica, Geometria, Proportioni et Proportionalita*, published in 1494, he dealt with several arithmetical subjects, including the method of bookkeeping then employed by the great merchants of Venice. This know-how reached the

```
                    36
                         Tafel van Intereſt van
                              2. ten 100.

                    1.       9803922.       9803922½
                    2.       9611688.      19415610.
                    3.       9423244.      28838834.
                    4.       9238455.      38077289.
                    5.       9057309.      47134598.
                    6.       8879715.      56014313.
                    7.       8705603.      64719916.
                    8.       8534905.      73254821.
                    9.       8367554.      81622375.
                   10.       8203484.      89825859.
                   11.       8042631.      97868490.
                   12.       7884932.     105753422.
                   13.       7730325.     113483747.
                   14.       7578750.     121062497.
                   15.       7430147.     128492644.
                   16.       7284458.     135777102.
                   17.       7141625.     142918727.
                   18.       7001593.     149920320.
                   19.       6864307.     156784627.
                   20.       6729713.     163514340.
                   21.       6597758.     170112098.
                   22.       6468390.     176580488.
                   23.       6341559.     182922074.
                   24.       6217215.     189139262.
                   25.       6095309.     195234571.
                   26.       5975793.     201210364.
                   27.       5858621.     207068985.
                   28.       5743746.     212812731.
                   29.       5631124.     218443855.
                   30.       5520710.     223964565.
```

Figure 2: The *Tafelen van Interest* contains tables of compound 'detrimental' interest. Shown here is the table of the interest on 10,000,000 units at 2% over a period of 30 terms.
© *Royal Library of Belgium (Brussels). Rare Books Department, CL 2665 LP.*

Low Countries, brought thither with the foreign merchants (fig. 3) who made Antwerp and Amsterdam their base. It was first expounded in Dutch in 1543, by Jan Ympijn Christoffels. He was soon followed by several other authors, each of whom, to a greater or lesser degree, contributed to the development of this method of accounting. The most important of them, Claes Pietersz. (also called Nicolaus Petri), was personally known to Stevin.

Figure 3: Many painters depicted money dealing. *The Moneylender and his Wife* by Quentin Metsys (1514) is in the *Musée du Louvre* in Paris (Inv. 1444). © 1990. Photo SCALA, Florence.

In this treatise, unlike many of his other works, Stevin is very sparing in his references to sources and the literature he has read. In his foreword to *Vorstelicke Bouckhouding* he ascribes his familiarity with the material to experience: 'I became well-versed in mercantile bookkeeping and cashiering; and later on, in the matter of finance (the one in Antwerp, the other in Flanders in the region of the Brugse Vrije)'. He mentions only one of his predecessors in the field, Bartholomeus De Renterghem. Paciolo he ignores completely.

At the start of this part of *Vorstelicke Bouckhouding* Stevin explains why he uses non-Dutch words (in *Oirsaeck der onduytsche woorden die in desen handel ghebruyckt sullen worden* (*Cause of the non-Dutch words that will be used in this business*)), even though he had determined to give a specially adapted Dutch form of every technical word he introduced. He knew that the prince would eventually introduce this type of bookkeeping and would make use of 'most able bookkeepers'. They, however, were accustomed to non-Dutch words. The switch to Dutch words would have created difficulties at first, and so non-Dutch words such as *Debet, Credit, Debiteur, Crediteur, Balance, Iornael, Finance* and *Domeine* are employed.

Stevin's contribution to bookkeeping, as to mechanics, hydraulics, fortification and certain aspects of mathematics, was of huge importance. In his 1927 publication Pieter De Waal expressed his admiration for Stevin's work, asserting that: 'The art of bookkeeping would be based on Paciolo and Stevin until the end of the nineteenth century'. Indeed, Stevin is regularly mentioned in standard works on economic development, being cited, for instance, by R. Haulotte and E. Stevelinck (1994).

4.1.1. A complete picture

It is not the intention here to expound the science of bookkeeping according to Stevin at length, but there are certainly a number of interesting details that should be mentioned. Just what Stevin himself contributed to the practice and what he derived from other authors is not easy to determine. But the fact remains that he presented a complete picture of bookkeeping in his day in an instructive way.

Stevin called the most important ledgers in his system the *Journael* (journal or general ledger), the *Schultbouck* (debits) and the *Memoriael* (memorandum), adding the *Cassebouck* (cash book) and *Oncostbouck* (expense account) as auxiliary aids. In the *Journael* a merchant's actual transactions are recorded, while current debits are entered in the *Schultbouck*. The *Memoriael* is a memorandum of affairs that need to be recorded but do not belong in either *Journael* or *Schultbouck* – small short-term loans, particulars of pending transactions, notes by employees concerning the work they have performed and so on. A *cassebouck* is required only in large offices that employ a cashier, otherwise the items can be recorded in the cash account in the *Schultbouck*. An *oncostbouck* is always useful, as it saves having to enter every single transaction in the main ledgers. At the end of the month, the items in this book are balanced and 'journalized'. Here, Stevin does not produce a real theory of bookkeeping as we have come to expect from him as regards other subjects.

The lessons Stevin devised for the prince were based on the detailed *Journael* and *Schultbouck* containing the administration of the commerce of a certain Dierick Roose for the year 1600. The *Journael* contains 40 entries in 16 subsections: one for recording capital, one for cash, four for merchandise accounts, seven for personal accounts, one for company expenses, one for household (private) expenses, and finally one for calculating profit and loss. The *Schultbouck* is paginated; the debit side of the account is always set out on a page with an even number, and the credit side on an odd-numbered page. Roose begins his *Journael* on 0 January with two journal entries that represent his opening balance. Stevin explains that he starts the year on day 0 not from 'idle pedantry' but to distinguish the start of the calendar year from the first day of trading.

Stevin takes the mystery out of bookkeeping with special explanations and examples. But above all he postulates the general rule that 'in all items of commerce the master never puts himself in debit or credit again', making it clear, thus, that the balance must balance. Here are some examples to illustrate the principle:

- If a merchant buys a diamond as a piece of jewellery, or a picture as a household effect, he must debit the capital account (which is to say, himself) because his assets are reduced by this purchase.
- If he buys a commercial article, however, he must open a special merchandise account and debit it.
- If he gives a stuiver in alms, which also reduces his assets, it must be debited from the household expenses.

Stevin also builds into his bookkeeping system a form of balance that allows profit and loss to be determined at any given moment, not just when the books are closed.

He concludes his treatise on mercantile bookkeeping with sections on bookkeeping in a trading house's branch offices and on the settlement of a partnership's affairs. With regard to the booking of merchandise sent to branches, he introduces the principle that goods still unsold remain the property of the owner. The agent in place must start an individual book for goods received and send copies of all his ledger entries to the master. Only when the agent has sold the goods can the master credit them in his own accounts. As to trading partnerships, Stevin recommends that each partner should keep his books as if he were the agent and the company's bookkeeper the master. However, even if a partner fails to follow this principle he can still come to a closing of the company (see section 2, above).

4.2. Bookkeeping for a prince and his lands

Having dealt with mercantile bookkeeping, Stevin moves on to the real object of his lesson – the introduction of double-entry bookkeeping into the financial affairs of the prince. Though he had several predecessors in the development of mercantile bookkeeping, his contribution to the bookkeeping of the prince and his lands was, as far as we know, original.

Stevin begins with a resolute tackling of the management of the prince's estates. In the administrative system that prevailed at the time, the various stewards, who were responsible for particular categories of goods and services, and the *tresorier*, or chief steward, to whom they handed over the monies they received, were allowed – indeed, were sometimes even obliged by the system – to wait for long periods (it could be years) before closing their annual accounts and remitting the balances. This made it impossible to have any sort of reliable control over available funds. In particular, the *tresorier* could use the amounts he received from the stewards as his personal property, sometimes for years. It is clear that Stevin was highly critical of such maladministration and that the prince, who was known for his thrift and careful financial management, wholeheartedly agreed that a different way of doing things should be found. Thus it was decided that Stevin's reformed system would come into force on 0 January 1604.

4.2.1. A new system
Stevin's new system aimed at creating the minimum disturbance to the stewards' routine activities. They were however asked to submit their accounts on a monthly rather than an annual basis, and they had to state what they had actually received and paid out, rather than what sums should have been received and paid, as had previously been the case. Every month, based on the statements sent in, the monies were handed over to the *tresorier*. It was now possible to assess the state of affairs at any time and establish the amount of existing capital. The

closing of the books was delayed until the end of the year following the bookkeeping year in question. This meant that overdue payments could still be entered in the year to which they belonged. Many aspects of modern estate management can be recognized in this system.

As chief steward, the *tresorier* had charge of the revenue from lands and 'foreigners'. He was also responsible for the salaries, pensions and expenses of the princely household, which were entered each month under various headings of *dispense* (expense). Here too Stevin argues for the double-entry system. He gives a complete description of seven types of *dispense*: of the kitchen, the stables, hunting and falconry, the Prince's Chamber, the Chamber of Accounts, the building and the gardens. In each department a *dispensier* was responsible for buying and selling, and for storing and distributing goods and a *tresorier* for the paying out and receiving of cash money. A *bouckhouder* (bookkeeper) was responsible for the administration, while a *controlleur* kept an eye on the details and a superintendent oversaw everything – a task which, according to Hendrick Stevin, was carried out by his father, Simon.

4.2.2. Five sorts of *dispense*
Stevin defines a specific part of the princely bookkeeping – the financial administration of military operations – as 'extraordinary finance'. Here, he foresees five sorts of *dispense*, for *Fortificatie, der Amonitie van Oorloch, der Artillerye, der Vivres en der Zeesaken* (fortification, ammunition, artillery, victuals and naval affairs). He also argues for the urgent introduction of the Italian method of bookkeeping in the magazines, where the administration in a long drawn-out war tends to become totally confused.

It is difficult to determine whether the lessons on double-entry bookkeeping for the administration of the princely household and army were actually translated into practice. Presumably not. It is scarcely conceivable that the supervisor of the prince's pantries noted what was consumed to the last drop and crumb in terms of debit and credit in a series of ledgers, or that the prince's linens, his guns, horses, hounds, hawks and all his other possessions should have been subject to a form of double-entry bookkeeping. In his enthusiasm, Stevin evidently overlooked the fact that double-entry bookkeeping does not necessarily lend itself to every form of administration.

4.2.3. Stevin's income
The system of double-entry bookkeeping was introduced in the administration of the prince's lands, however. Recently, a journal and a ledger (for the year 1604) were found in the Nassau Domains Council Archives in the National Archive in The Hague (see Zandvliet, 2000) and in these Stevin's system is used and explained. Preceding the accounts is an introduction, if not written by Stevin himself, then certainly composed under his supervision. As these books were to be the first in a series, and to receive the prince's special attention, much care was devoted to their appearance. They are bound in parchment and embellished with the arms of the Nassau family (see fig. 4). Inside is a list of officers of the

Figure 4: The parchment-bound ledger of 1604, introduced and written under the supervision of Simon Stevin. On the cover are the arms of Maurice of Nassau, painted in watercolour and highlighted in gold. *The Hague, National Archive, Nassau Domains Council Archives, 1.08.11, inv. no. 1440 (ektachrome: JTD).*

prince's household and the salaries they received. Stevin's name is among them and we can see that he received 600 guilders annually, as indeed we know from the prince's correspondence with the States General (see Chapter Two, box 6). This was a high salary. Stevin is fourth on the list – an indication of the importance of his position at Prince Maurice's court (see fig. 5). The list also shows that Stevin received an extra 450 guilders.

Other sources make it clear that Stevin's system of administration of the princely lands was not destined to last long. Indeed, Stevin himself advocated its simplification. His proposals appear in *Verrechting van Domeine mette de Contrerolle en ander behouften van dien* (*Administration of Lands by means of the 'Contrerolle'*), in *Materiae Politicae*, published by Hendrick but 'written by Simon Stevin of Bruges, the said Lord Prince's Superintendent of Finance'. The *contrerolle* seems to have been a register in which was noted, at the start of the

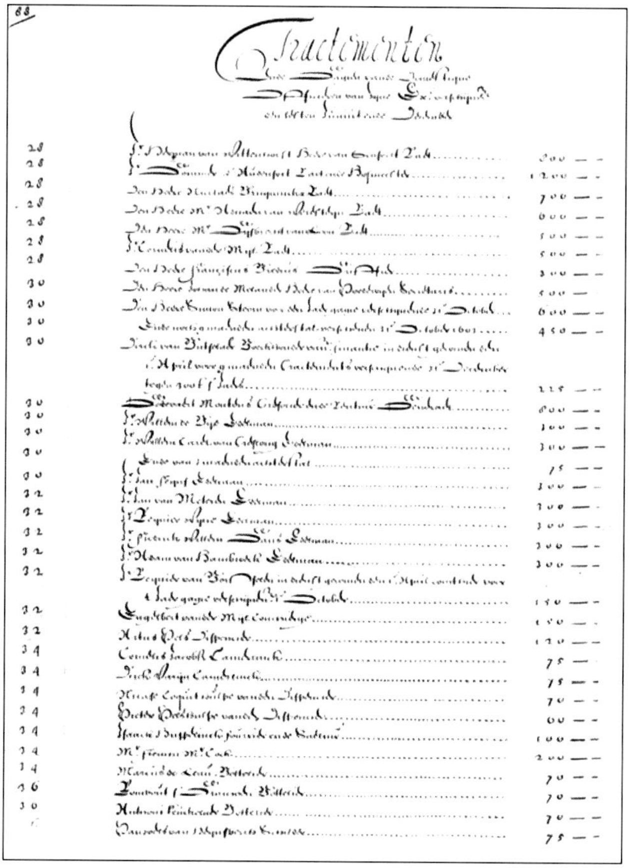

Figure 5: This journal of 1604 lists the salaries drawn by officials in Prince Maurice's service. It can be seen that Stevin received an annual salary of 600 guilders.
The Hague, National Archive, Nassau Domains Council Archives, 1.08.11, inv. no. 1439 (ektachrome: JTD).

financial year, every anticipated item of income and expenditure with the amount, where known. In each section, room was left to record the sums that were received and paid out in reality. The *Anhang van de Verrechting van Domeine* (*Appendix to the Administration of Land*) deals with things such as the financial administration of the prince's company of guards.

4.2.4. ...this sort of bookkeeping, or something very like it, was already in use in Rome in the days of Julius Caesar and long before...

Stevin concludes his essay on bookkeeping in the *Wisconstige Gedachtenissen* with the assertion that the Italian method of bookkeeping was already in practice among the 'Ancients'. He describes this in the 'presumption of the antiquity of bookkeeping' as follows:

> *When a good friend, who was versed in Ancient History, had seen the work on bookkeeping when it was first printed, he began to surmise that the practice began in Italy not some two hundred years ago as some people think, but that this sort of bookkeeping, or something very like it, was already in use in Rome in the days of Julius Caesar and long before, and that a few relics of those ancient times now lately came into the hands of those who reintroduced them. This I do not find unlikely, for it would be strange that such marvellous profound learning would be discovered as something new in the lay age (the age of the barbarians).*

What Stevin writes here is in line with the beliefs held by many of his contemporaries, that knowledge, gathered and rediscovered in the Renaissance, in fact originated with an earlier, forgotten civilization that flourished in what Stevin calls the *Wysentyt*, or 'age of the sages'. In contradistinction he terms the Middle Ages the *Lekentijd*, or 'lay age', noting in the margin that by this he means the *Barbari seculi*, or 'age of the barbarians'.

Chapter Six

Wonder en is gheen wonder

De Beghinselen der Weeghconst. De Beghinselen des Waterwichts: the most important works on statics and hydrostatics of the sixteenth century.

In 1586, just four years after he enrolled as a student at Leiden, Simon Stevin published De Beghinselen der Weeghconst *and* De Beghinselen des Waterwichts. *These were works of historic importance and their appearance made 1586 truly a year of wonders in the history of science. They were the first really significant innovative contributions to statics and hydrostatics since Greek antiquity, and they place their author in a line that runs from Archimedes – via Stevin, Galileo, Pascal and Huygens – to Newton.*

Even today it is (still) an intellectual delight to take up these works and follow Stevin's argument through their pages. Although the Bruges master set out to make his work as accessible as possible, this material (like the algebra that follows in Chapter Seven) requires a little additional concentration on the part of the reader. As Menaechmus reputedly pointed out to Alexander the Great when the ruler asked for a shortcut in geometry, there is no royal road – all must travel the same way.

1. Introduction

In 1586, at the Plantin printing house in Leiden then run by Frans van Ravelingen, Stevin published a work that would become a milestone in the history of science. It consisted of several volumes: *De Beghinselen der Weeghconst (The Elements of the Art of Weighing)* (see fig. 1), *De Weeghdaet (The Practice of Weighing)*, *De Beghinselen des Waterwichts (The Elements of Hydrostatics)* and the *Anvang der Waterwichtdaet (Preamble of the Practice of Hydrostatics)*. The first two volumes deal with what we now call statics, the last two with hydrostatics. In each case there is a theoretical part (*spiegheling inde beghinselen*) and a practical part (*daet*).

Stevin believed that theory and practice should go hand in hand. This was not always possible. In his engineering work he would frequently have to do without a strict theoretical foundation. We find the perfect combination of 'Stevin the engineer' and 'Stevin the mathematician and physicist' in his work *Van de Molens (On Mills)*, which is concerned with the technical construction of wind-powered water drainage mills (see Chapter Four, section 2). Stevin was no doubt motivated to design an improved water mill by the fact that he could acquire a

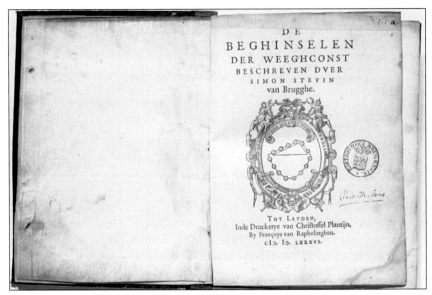

Figure 1: The title page of Stevin's historic masterpiece in which, after some 1,850 years, Archimedes's work in the field of statics was taken up and developed. Note the signature of the scholar Mulerius, to whom this copy of *De Beghinselen der Weeghconst* belonged, and who made several annotations.
Bruges, Public Library, B 265 (photo: Hugo Maertens, ektachrome: JTD).

profitable patent on it. Nonetheless, his was an indisputably innovatory approach, for he based the construction of a highly complex mechanism on a quantitative mathematical calculation. In order to make these calculations he had to take into account the laws of both statics (*De Beghinselen der Weeghconst*) and hydrostatics (*De Beghinselen des Waterwichts*).

Archimedes and Heron had both coupled applications to a theoretical understanding of statics and hydrostatics, more specifically in the operation of instruments and machines. Though Archimedes esteemed pure mathematical inductive thought far above its application, his name is, nonetheless, invariably associated with his law of the lever and the principle, named after him, which states that a body totally or partially immersed in a fluid is subject to an upward force equal in magnitude to the weight of the fluid it displaces – a discovery which, so legend has it, gave rise to the famous cry, 'Eureka!'

The significance of Stevin's work on statics lies in the fact that he not only studied and supplemented the work of Archimedes but also continued and developed it. Stevin distinguished himself from his predecessors, Jordanus Nemorarius and his school, Leonardo da Vinci, Commandino and others, by his systematic development of and innovative contribution to the theory and practice of weighing and the originality of his applications, such as the improvement to the water mill mentioned above. In this context he also made implicit and correct use of laws and concepts of physics – such as hydrostatic pressure and angular momentum – that were formulated only at a much later date.

1.1. The negation of perpetual motion applied in both statics and hydrostatics

It is important to place Stevin's work against the background of existing knowledge and the state of science and technology that prevailed in the sixteenth century. At that time nothing essential had been added to the work of Archimedes in the fields of mechanics and hydrostatics. Only in the school of Jordanus Nemorarius was the composition of forces studied on the basis of the Aristotelian theory of dynamics. Let us put Stevin in his chronological context: he died 22 years before Newton was born, he was 16 years older than Galileo and he lived decades before Christiaan Huygens. He was a contemporary of such creative spirits as Calvin, Palestrina and El Greco. It was only from 1550 or thereabouts that works by Apollonius, Archimedes, Pappus and others gradually began to become generally available in Latin translations.

Against this historical background, Stevin emerged with brilliant contributions to the fields of statics and hydrostatics. In statics he proved the law of the equilibrium of bodies on an inclined plane with his famous *clootcrans* or 'wreath of spheres'. In hydrostatics he discovered the fundamental and important hydrostatic paradox and also put forward a more general proof of Archimedes's principle than the great thinker of Syracuse himself had ever been able to do.

Stevin always introduced highly original 'thought experiments' (experiments performed in the mind). Notably (as the first ever to do so, and on intuitive grounds) he used the same principle – namely the impossibility of perpetual motion – both in the realm of statics with the *clootcrans* and in the derivation of Archimedes's principle in hydrostatics (see fig. 2). The negation of perpetual motion is an important natural law, which could justifiably be called 'Stevin's Law'.

What Stevin's argumentation in statics came down to, strictly speaking, was an application of the very important law of conservation of energy, which would only be formulated around 1840 – 260 years later, thus – by scholars such as Hermann von Helmholtz. As Dijksterhuis rightly remarks, Stevin ranks with Evangelista Torricelli and Christiaan Huygens who, after Stevin, would also intuitively make use of the principle long before it was explicitly formulated.

1.2. Practical problems provide the impetus for studies on weighing and hydrostatics

The problems that confronted Stevin in his development of statics and hydrostatics were directly linked to everyday practicalities. Problems connected

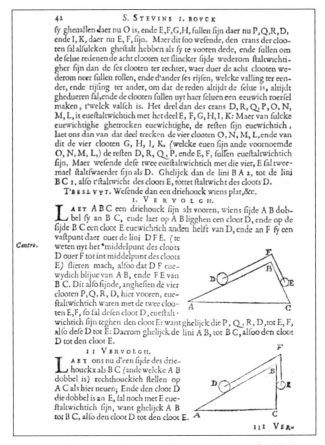

Figure 2: In *De Beghinselen der Weeghconst* Stevin refuted the existence of perpetual motion.
Bruges, Public Library, B 265 (ektachrome: JTD).

with the equilibrium of bodies and the combined action of forces in instruments and machines played a great part in many practical aspects of business and trade:

- In the case of the hand-held pair of scales, Stevin describes in *De Weeghdaet* 'a most perfect balance'. He establishes the weight of an object by creating equilibrium with a standard weight. Here, the lever with arms of unequal length automatically comes into play.
- In the case of diverse tools and instruments such as the pulley, windlass, screw and axle.
- In seafaring, where it is of crucial importance to understand and control the stability of ships and boats.
- In architecture, in the equilibrium of vaulting, for instance. Think of Beauvais Cathedral: construction began in 1225 and ended only three

centuries later. The cathedral had the highest vault in France. It collapsed. In the sixteenth century the crossing tower went the same way. This debacle was attributable to the Gothic masons' very imperfect understanding of statics.

1.3. Definitions, axioms and propositions: the axiomatic approach

It is remarkable how the axiomatic structure took shape in Greek geometry and how Archimedes applied it to statics and hydrostatics. The logical basis was coupled, certainly by Archimedes, with a deductive reasoning whose strictness would not be surpassed by later mathematicians such as Cauchy (1789–1857) or Gauss (1777–1855), known for their mathematical rigour.

It was clear, even to early mathematicians like Thales of Miletus in the sixth century BCE, that some basic statements and propositions could not be proved, but could nonetheless be considered as self-evidently true. These unprovable propositions are called axioms. The axiomatic method uses definitions in addition to axioms as starting points from which to derive theorems.

Stevin called axioms 'postulates. Since some matters of an elementary nature are common knowledge, and need not be proved...' (*Weeghconst*), noting that this was a method used by mathematicians: 'we shall, after the custom of mathematicians, before arriving at the propositions, postulate that the following things be granted'. Archimedes was the first in documented history to apply mathematics, and more specifically geometry, to physics, and Stevin refers to him in his address to the reader in *De Beghinselen des Waterwichts*: 'What was the cause that moved Archimedes to write that which he left to us in the *Book of the things supported in the water...*'

Stevin was thus familiar with Archimedes's work. On the one hand, he found it inspiring, yet on the other he sought to take up an independent position and follow new paths. He adopted his Greek predecessor's axiomatic structure, though he would not always equal his rigour. His distinction between axiom and definition was not always as sharp, for instance. That he inclined to a pragmatic approach was not strange of course.

In *De Beghinselen des Waterwichts* Stevin introduces the logical structure of his exposition thus: 'First the definitions', then 'The postulates' and 'Now the propositions'. Here and there in *De Beghinselen der Weeghconst* and *De Beghinselen des Waterwichts* passages can be picked out in which Stevin's argument is not entirely cogent by contemporary standards. We also encounter derivations in which the result (the proposition) is correct but the logical reasoning requires some refining.

These critical comments in no way detract from the historical significance of Stevin's contribution to physics. Concepts and understanding emerge with a certain recalcitrance in every branch of evolving science. Initial formulations often contain logical flaws. Even some of Galileo's arguments are not perfectly sound. And even in the case of Archimedes, cause for conceptual discussion can, in rare instances, be found.

2. The structure of *De Beghinselen der Weeghconst*

In the first book of *De Beghinselen der Weeghconst*, Stevin deals with two subjects. First he considers the lever (Propositions 1–18). Then, in the most pioneering section (Propositions 19–28), he analyses the inclined plane with the aid of the *clootcrans* or 'wreath of spheres' (see box 1), which enables him to propound his basic composition law for two forces. In Proposition 19, Corollary VI, this composition law is correctly formulated, even if the derivation is not entirely convincing. In Propositions 20–28 he illustrates and develops the knowledge gained in Proposition 19. In a later edition (in the *Byvough der Weeghconst* (*Supplement to the Art of Weighing*), a part of the *Wisconstige Gedachtenissen*, more specifically the *Eerste Deel des Byvoughs der Weeghconst, Van het Tauwicht* (*First Part of the Supplement to the Art of Weighing, Of the Cord Weight*)), Stevin further explains the composition law equivalent to the parallelogram of forces. The second book of *De Beghinselen der Weeghconst* is concerned with determining the centre of gravity (see fig. 3) in solid bodies. The application (*daet*) of the 'art of weighing' is dealt with in *De Weeghdaet* (*The Practice of Weighing*), in which Stevin considers instruments and machines.

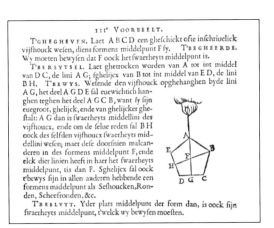

Figure 3: Stevin determined the centre of gravity in an empirical way. The method shown here is still followed in textbooks. From *De Beghinselen der Weeghconst*.
Bruges, Public Library, B 265 (ektachrome: JTD).

Box 1: The *clootcrans*

In Old Dutch a *cloot* is a ball or sphere; a *crans* is a garland or wreath. Hence *clootcrans* – 'wreath of spheres'. Stevin's proof using the *clootcrans* or 'wreath of spheres' in *De Beghinselen der Weeghconst* is an example of a derivation of a natural law that is intelligible to everyone:

> ...and the spheres will automatically perform a perpetual motion, which is absurd.

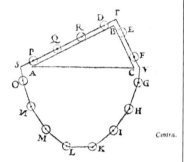

Figure 4: On this page Stevin presents his famous proof using the *clootcrans* or 'wreath of spheres'. With this experiment he was able to lay the basis for his derivation of the composition law of forces, virtually without mathematics. So pleased was he with this proof that he chose the 'wreath of spheres' as his 'trademark'. From *De Beghinselen der Weeghconst*.
Bruges, Public Library, B 265 (ektachrome: JTD).

Stevin considers a right-angled triangle ABC (see fig. 4), of which side AB is twice the length of side BC and base AC is 'parallel to the horizon'. He supposes that sphere D lies on side AB and sphere E, of equal weight and equal size, lies on side BC. He will prove: *Ghelijck dan de lini BA 2, tot de lini BC 1, also t'staltwicht des cloots E, tottet staltwicht des cloots D* – 'As therefore the length BA (2) is to BC (1), so the apparent weight of the sphere E is to the apparent weight of D'. Stevin uses the term *staltwicht*, a word not defined explicitly and for which there is no modern equivalent but which more or less means the component of the weight, which in the given situation is the only active one – the component along the inclined plane. It is translated here as 'apparent weight'. In his 'thought experiment', he now adds 12 identical spheres to spheres D and E to make up a wreath that can slide over three fixed points at S, T and V. *Laet ons maecken rondom den driehouck ABC eenen crans van veerthien clooten, evegroot, evewichtig, ende evewijt van malcanderen (...) al ghesnoert an een lini...* – 'Let us make about the triangle ABC a wreath of 14 spheres, of equal size and equal weight, and equidistant from one another (...) all of them strung on a line...' The number of spheres on AB is thus twice the number on BC. Stevin now seeks to prove that the *staltwicht*, or apparent weight, of spheres D, R, Q and P on AB is the same as that of spheres E and F on BC.

Let the first apparent weight (of spheres P, Q, R, D) be greater than the second (of spheres E and F). The four spheres O, N, M and L are in equilibrium with the four spheres G, H, I and K. Hence the group of eight spheres D, R, Q, P, S, O, N, M and L is heavier than the group of six spheres E, F, G, H, I and K, and, 'because that which is heavier always preponderates over that which is lighter, the eight spheres will roll downwards and the other six will rise.' That being so, each sphere takes the place of its predecessor, or as Stevin puts it, 'let D have fallen where O is now, then E, F, G, H will be where P, Q, R, D are now, and I, K where E, F are now'. Seen as a whole, however, the position of the wreath is unchanged by this substitution: 'the wreath of spheres will have the same appearance as before, and on this account the eight spheres on the left side will again have greater apparent weight than the six spheres on the right side, in consequence of which the eight spheres will again roll down and the other six will rise' – in other words, the spheres will perform a perpetual motion which, Stevin remarks, is absurd. Therefore, the wreath will remain still; the left and right groups of spheres 'balance' each other. If now the four spheres O, N, M and L are subtracted from the left, and from the right the four spheres G, H, I and K, which keep the other four in balance, the equilibrium will not be disturbed and the remaining spheres D, R, Q, P and E, F will be of equal apparent weight. But the two latter being of equal apparent weight to the four former, E will have twice the apparent weight of D, just as the length of AB is twice that of BC. The choice of the ratio of AB and BC is arbitrary, however, so that the proof is general.

Stevin now supposes (Corollary I to Proposition 19) that the weight of sphere D is twice that of sphere E (see fig. 6). Using the proof of the *clootcrans* demonstrated above, he arrives at the law of the inclined plane: 'Therefore, as the line AB is to BC, so is the sphere D to the sphere E'.

> It should be observed that, strictly speaking, in the absence of influences such as friction, air resistance and so on, perpetual motion is in fact possible and that classical mechanics is formulated specifically under these ideal conditions. However, it is perfectly possible to adapt Stevin's reasoning slightly and to show, in the context of classical mechanics (but then on the basis of the law of the conservation of energy), that Stevin's wreath of spheres will not begin to move if no initial impetus is given.

2.1. The centre of gravity of a solid body

The centre of gravity of a solid body plays a special role in the study of the equilibrium of complex bodies such as machines, buildings and boats, because a uniform gravity field effectively acts on an object's centre of gravity. Determining the centre of gravity is important in solving problems relating to balance. The complex shape of a construction or machine is then immaterial.

The Greeks were well aware of the concept of the centre of gravity. Archimedes even refers to still earlier writers, though their works have been lost. It is striking that Archimedes did not incorporate a consideration of the centre of gravity into the axiomatic structure of his theory of statics – even Stevin, the pragmatist, follows him in this. He regarded the theory of the centre of gravity as something separate and as something known.

2.2. The lever

It is not surprising that the lever had been studied at a very early date in history. Scales – a pair of scales is nothing other than a lever – were already in use in the distant past. The condition determining the equilibrium of a lever was first defined in the school of Aristotle, and it was determined very intuitively. Archimedes also developed a firmly based axiomatic study of the lever. Given that the solution was already known, his study was important primarily from a methodological point of view.

Archimedes's approach was completely different to that of the school of Aristotle. The latter associated the equilibrium of forces with the moving of their points of application, which occurs as soon as the lever is set in motion. Archimedes started from simple postulates such as 'a symmetrically loaded lever is in equilibrium'. He used the concept of 'the centre of gravity of a body'. In the course of history, these two complementary approaches eventually proved equally fruitful.

Stevin's study of the lever takes the Archimedean approach. Nonetheless, from the start Stevin regularly diverges from his Greek example, demonstrating his independent attitude in the construction of his argument. Stevin deals with the equilibrium of a body that is subject to parallel (vertical) forces. He shows many examples and applications and concludes this part of the first book of

De Beghinselen der Weeghconst thus: 'Up to this point the properties of vertical weights have been explained...'

Figure 5 illustrates a typical problem dealt with in the study of the lever in *De Beghinselen der Weeghconst*, that of *rechtwichten* – vertical weights. *Recht* means vertical, *scheef* means inclined. *Hefwicht* refers to an upward force; *daelwicht* to a downward force. Stevin states: 'A lowering weight and a lifting weight equal to it, acting at equal angles on equal arms, exert equal forces.' The point is that vertical upward forces can be replaced by vertical downward forces with an equivalent action. As Stevin puts it: 'Therefore E and D exert equal forces on equal arms AB, AC'. The gist of Stevin's observation (for the data given above) is that the vertical upward force that acts on C as a result of the addition of weight E is the same as the vertical upward force that acts on C as a result of vertical downward force D. Stevin's diagram, shown in fig. 5, is strictly limited to the essential elements needed for his proof. The link with the experimenter's hand clearly defines the point of suspension, given that A is also a fixed point.

2.3. From the inclined plane to the parallelogram of forces: the *clootcrans*

2.3.1. The inclined plane

In practical work situations, on building sites, in shipyards, in loading and unloading in the harbours, not only upward and downward forces but also pushing and pulling forces in oblique directions are involved. From Proposition 19 onwards, Stevin turns his attention to forces that act in what he calls *scheve*, or oblique, directions: 'in the following pages the properties of oblique weights will

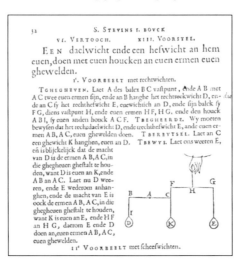

Figure 5: *Een daelwicht ende een hefwicht an hem even, doen met even houcken an even ermen even ghewelden.* – 'A lowering weight and a lifting weight equal to it acting at equal angles on equal arms, exert equal forces'. From *De Beghinselen der Weeghconst*.
Bruges, Public Library, B 265 (ektachrome: JTD).

be described'. His starting point is the analysis of the equilibrium of bodies on an inclined plane. For millennia the inclined plane has been effectively used – possibly even in the construction of the pyramids – as a 'machine' for hauling heavy loads upwards. This 'problem of the inclined plane' had already been solved in the school named after Jordanus Nemorarius in the fourteenth century using the Aristotelian approach. The solution was essentially correct, albeit rather laborious. Leonardo da Vinci also had an understanding of a body on an inclined plane. However there were also incorrect solutions in circulation. The Greeks failed to solve the problem. Archimedes did not address it at all, as far as we know. And the only known attempt in antiquity to calculate the force required to shift a weight upwards on an inclined plane (in a work by Pappus) failed. Stevin and Galileo solved the problem of the inclined plane at almost the same moment, undoubtedly independently of one another, and using different methods.

2.3.2. The inclined plane as the basis for the parallelogram of forces
Stevin aimed to determine the effective weight of a body on an inclined plane. He sought to define what we would call the 'component' of the weight of a sphere, in the direction determined by the slope of the inclined plane. In his derivation, Stevin introduced the notion of 'apparent weight' (*staltwicht*) for such a component of a force (see box 1). However, he did not always give the term exactly the same meaning. It is remarkable that Stevin was also able to use his results for the inclined plane as a basis for the parallelogram of forces. It is generally acknowledged that he introduced the parallelogram of forces.

2.3.3. The *clootcrans*: Stevin's visionary thought experiment
Stevin based the solution of the problem of the inclined plane on his famous proof of the *clootcrans* or 'wreath of spheres' (Proposition 19). Details of his brilliant experiment can be found in box 1.

In Proposition 19, Corollary I (see fig. 6), we learn that as side AB of the vertical triangle ABC is twice as long as side BC, so weight D must be twice as

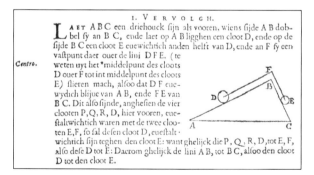

Figure 6: Corollary I of the *clootcrans* proof. If AB is twice as long as BC, sphere D will be in equilibrium with sphere E if the weight of D is twice the weight of E. From *De Beghinselen der Weeghconst*.
Bruges, Public Library, B265 (ektachrome: JTD).

great as weight E, for weights D and E to balance each other. Spheres D and E are then of 'equal apparent weight' (*evenstaltwichtigh*). Then follows the generalization: 'Therefore, as the line AB is to BC, so is the sphere D to the sphere E', which is applicable irrespective of the angle of inclination. This is the law of the inclined plane.

2.3.4. The parallelogram of forces

Building on the law of the inclined plane, Stevin arrived at the basic composition law of two forces. First he gave a derivation for the case of two forces working at right angles, followed by the general case. He illustrated the composition law with many clear examples. We encounter some of these in the first part of the *Eerste Deel des Byvoughs der Weeghconst, Van het Tauwicht* (*Supplement to the Art of Weighing, Of the Cord Weight*). There we find, on page 181, the diagram reproduced here as fig. 7. Stevin is dealing with the equilibrium of an object when three different forces act together on it, making use of the knowledge he has gained with his *clootcrans*. He calculates the forces that act along CH and CK, and the resultant force along CI keeps the weight G in equilibrium. This corresponds to the operation of the parallelogram of forces.

In fig. 8 we see another example from *Van het Tauwicht* clearly and graphically presented in which Stevin illustrates the composition law of two forces with arbitrarily chosen directions. The forces that act in the directions CE and CD (towards E and D, respectively) counteract the weight of the object AB. Stevin determines the magnitude of these forces by construction of a line segment HI parallel to CE.

In fig. 8, as in fig. 7, we see that Stevin uses line segments, along given directions, to represent forces. Thus he was in fact working with our modern

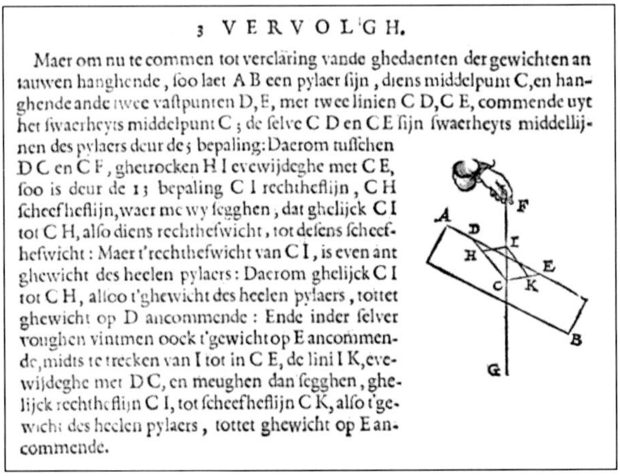
3 VERVOLGH.

Maer om nu te commen tot verclaring vande ghedaenten der gewichten an tauwen hanghende, soo laet A B een pylaer sijn, diens middelpunt C, en hanghende ande twee vastpunten D, E, met twee linien C D, C E, commende uyt het swaerheyts middelpunt C; de selve C D en C E sijn swaerheyts middellijnen des pylaers deur de 5 bepaling: Daerom tusschen D C en C F, ghetrocken H I evewijdeghe met C E, soo is deur de 13 bepaling C I rechtheflijn, C H scheefheflijn, waer me wy segghen, dat ghelijck C I tot C H, also diens rechthefwicht, tot desens scheefhefwicht: Maer t'rechthefwicht van C I, is even ant ghewicht des heelen pylaers: Daerom ghelijck C I tot C H, alsoo t'ghewicht des heelen pylaers, tottet ghewicht op D ancommende: Ende inder selver voughen vintmen oock t'gewicht op E ancommende, midts te trecken van I tot in C E, de lini I K, evewijdeghe met D C, en meughen dan segghen, ghelijck rechtheflijn C I, tot scheefheflijn C K, also t'gewicht des heelen pylaers, tottet ghewicht op E ancommende.

Figure 7: Equilibrium when three different forces are acting. From *Byvough der Weeghconst, Van het Tauwicht.*
© Royal Library of Belgium (Brussels). Rare Books Department, VH 8038 C.

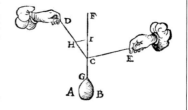

Figure 8: The parallelogram of forces. From: *Byvough der Weeghconst, Van het Tauwicht*.
© Royal Library of Belgium (Brussels). Rare Books Department, VH 8038 C.

concept of force vector; only the sense in which a force works is not graphically represented.

Stevin reminds us that cords are employed in all kinds of activities: '…amongst other things, cords are mostly used in practice'. The reader also learns that Stevin regards Proposition 27 in the first book of *De Beghinselen der Weeghconst*, which is equivalent to the parallelogram of forces, as his basic proposition for the composition of forces: 'The gist of the contents is as follows: in Proposition 27 of the first book of the *Weeghconst* it has been proved that if a prism is hanging in equality of apparent weight with two oblique lifting weights, as the oblique lifting line then is to the vertical lifting line, so is the oblique lifting weight to its vertical lifting weight.'

Lagrange, a pioneer of theoretical mechanics who spoke in the highest terms of the ingenuity of the argumentation in *De Beghinselen der Weeghconst*, thought that Stevin had derived the parallelogram of forces only for forces working at right angles; but in this he was incorrect. For instance, in the illustration to Proposition 27 in the first book of *De Beghinselen der Weeghconst*, shown here in fig. 9, Stevin deals with the equilibrium of a prism subject to arbitrary forces.

2.3.5. Bodies with a fixed point

Stevin also turned his attention to the analysis of the equilibrium of a body or machine with a fixed point (Propositions 20, 21, 23 and 24 of the first book of *De Beghinselen der Weeghconst*). A variation on this theme is his later study of bodies with a fixed axis, such as a windlass, or the axis around which a windmill's sails revolve.

Here again, Stevin was the first to address these important problems in a scientific way.

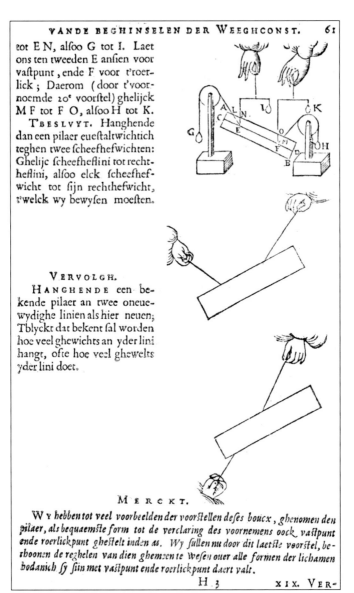

Figure 9: Illustration of the combination of two oblique forces. From *De Beghinselen der Weeghconst*.
Bruges, Public Library, B 265 (ektachrome: JTD).

2.3.6. Founder of the composition law of forces

Stevin had a thorough understanding of the composition law of forces and its applications. The literature describes him correctly as having introduced the parallelogram of forces. It is indeed striking that he was able to solve the problem of the inclined plane and the parallelogram of forces without any mathematical equipment to speak of, apart from extremely elementary geometry.

3. *De Weeghdaet*

In *De Weeghdaet* Stevin considers practical applications of the theories of statics he has expounded in *De Beghinselen der Weeghconst*. As he points out in the dedication of the work (to the burgomasters of Nuremburg), 'theory is lost labour when the end does not tend to practice'.

The applications Stevin sets out include the empirical derivation of the centre of gravity of a body, the analysis and sometimes also the making of a number of weighing instruments (balance, oblique balance and steelyard) and the operation of various kinds of lever. Of course, as Stevin points out, the lever has long been in use, and men have 'accomplished the construction of many common tools, to their great service and benefit. But because this was only done by experience, and not by thorough knowledge of the proportionality existing therein, many great new constructions often did not turn out well...'

Stevin then broadens his applications to take in the inclined plane and windlass. One of the applications he describes in *De Weeghdaet* is concerned with 'how much greater weight a horse harnessed to a wagon draws when ascending a height than on level land'. He determines quantitatively what force a horse must exert to keep a cart in equilibrium on an inclined plane (see fig. 10) and the triangle of forces KHI.

Below is a selection of some of Stevin's applications that are particularly interesting for their originality or their instructive character.

3.1. The *snaphaen*: a didactic gem

Stevin illustrates the lever with arms of unequal length in a most attractive pedagogic way (see fig. 11). He sketches 'a man [A] having on his shoulder B a lance CD...' The point of contact where shoulder and lance meet is the fixed point and there are two levers, one on the left and one on the right of this contact point. A cock hangs from the right lever so that a weight operates at K. Stevin asks what force the man's hand exerts on the lance. This is easy to calculate, given the knowledge gained in *De Beghinselen der Weeghconst*. Suppose, Stevin suggests, that the figure in the illustration were a *snaphaen* – a 'poaching soldier, with a poached cock I hanging at K...' A *snaphaen* is, among other things, a freebooter or highwayman. Could Stevin be making a wry allusion to the hated Spanish soldiery?

Some of the diagrams in Stevin's work are rather simple and sometimes they seem rather laboured, yet other figures, like the rascally *snaphaen*, are little graphic masterpieces. One wonders who drew them. Learning about the lever is much more attractive with the aid of the poaching soldier carrying off his stolen fowl on the end of his lance than with the neutral drawings usually found in physics textbooks.

WEEGHDAET. 31

t'fwaerheyts middelpunt der fchuyt, ende rechthouckich op t'plat P N: Daerom hoe dat G H ende P N de euewydicheyt naerder fijn, hoe lichter werck, ende hoe verder, hoe fwaerder.

IIIIe VOORBEELT.

UYT het voorgaende is oock kennelick, hoe veel ghewichts een peert in een waghen ghefpannen, meer treckt een hoochde op ftyghende, dan opt plat landt. Laet by voorbeelt A B t'plat fijn eens berghs, ende C D een waghen, weghende met datter op is al tfamen 2000 ℔, ende E F (inde plaets der ftrijnghen) fy de coorde, ende G fy t'peert, evestaltwichtich teghen den waghen. Laet oock ghetrocken fijn de * hanghende H I, ende I K, rechthouckich op t'plat A B, ende laet I H viervoudich fijn tot H K, ende is kennelick duer het 20 voorftel des 1^{en} boucx der beghinfelen, dat ghelijck K H tot H I, alfoo t'ghewicht der fcheefwaegh fooder een waer (in diens plaets nu t'peert is) tottet ghewicht des waghens, maer K H is t'vierendeel van H I duer t'gheftelde; des fcheefwaegs wicht dan foude van 500 ℔ fijn, te weten t'vierendeel van t'ghewicht des waghens; Daerom t'gareel oft riem oft fulcx alft waer, druckt t'peert foo ftijf voor den borft L, als een pack van 500 ℔ op fijn rugghe duwen foude, ende dat (wel verftaende alft voortgaet) bouen het duytfel dattet lijdt op t'plat landt treckende.

Perpendicularis.

T I S oock openbaer duer het 24e voorftel des 1e boucx, ende daer t'ghene wy hier vooren vande fchuyt ghefeyt hebben, dat als de ftrijngen euewydich fijn vande wech daer de waghen ouer vaert, dat de peerden dan de meeften ghewelt ande waghé doen, wel verftaende op eenen harden gantfch effenen wech, maer op eenen oneffenen hobbelighen eñ fandighen, fo voorderet de ftrijnghen achter wat leegher te doen dan vooren. T'welck den Hollantfchen voerlien duer deruating niet onbekent en is, diens waghens daer naer ghemackt fijnde, doen de ftrijnghen, langs t'zeeftrant varende, ende in derghelijcke euen harde weghen, achter hoegher

Figure 10: *Uyt het voorgaende is oock kennelick, hoe veel ghewichts een peert in een waghen ghespannen, meer treckt een hoochde op styghende, dan opt plat landt.* – [From the foregoing it is also evident how much greater weight a horse harnessed to a wagon draws when ascending a height than on level land.] The illustration is a visually convincing application of the composition law of forces; see the triangle KHI. From *De Weeghdaet*.
Bruges, Public Library, B265 (ektachrome: JTD).

Figure 11: The *snaphaen* illustrates the lever with arms of unequal length in a most attractive didactic way. From *De Weeghdaet*.
Bruges, Public Library, B 265 (ektachrome: JTD).

It should be observed in connection with fig. 11 that it is a woodcut, printed in reverse from a carved woodblock. The soldier wears his sword on his right rather than his left. The letters F (between the lance and the back of the hat but difficult to make out) and E have also been transposed, as the editors of *The Principal Works* note.

3.2. Cranes and windlasses according to Memling and Stevin

Using the understanding gained from *De Beghinselen der Weeghconst*, Stevin was the first to analyse yet another important instrument, the windlass. This

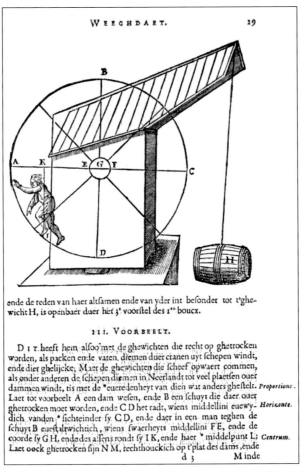

Figure 12: 'Given a windlass on whose axle hangs a weight...' The windlass is an integral part of the crane used on the quayside. From *De Weeghdaet*. Bruges, Public Library, B 265 (ektachrome: JTD).

machine, which by Stevin's day had already been in use for centuries, consists of a wheel that turns on a horizontal axis around which runs a rope. With it, heavy loads are raised or moved. Figure 12 shows a windlass as Simon Stevin analysed it in *De Weeghdaet*. It was used on quaysides as a crane operated by a treadmill. Figure 13.1, a detail from Hans Memling's *St John Altarpiece*, shows such a crane being used to unload barrels of wine in Bruges. The modern motor-driven crane works on the same principles.

Stevin contends that 'The drawing weight and the drawn weight of the windlass are proportional to the semi-diameter of the axle and the semi-diameter of the wheel...' The windlass can be compared to a lever with a short arm (the radius of the axle) and a long arm (the radius of the wheel). Like the lever, the windlass

Figure 13.1: A wooden crane powered by a treadmill on the Kraanplein in Bruges – a detail from Hans Memling's *St John Altarpiece*. Stevin studied this kind of machine in *De Weeghdaet*.
Bruges, Municipal Museums, Memling Museum,
© Reproductiefonds.

makes it possible to transmit a greater force to a body (here the wine barrels) than is exerted on the lever. The windlass used in Stevin's day is still encountered in the hoisting gear in a warehouse, for instance. Sometimes the windlass is installed on the topmost floor; sometimes it is attached to the hoisting beam outside (see fig. 13.2).

3.3. Stevin's 'Almighty'

One particularly remarkable proposition in *De Weeghdaet* is Proposition 10: '*To explain the construction and the properties of the Almighty*'. Stevin's

Figure 13.2: Tools such as the pulley, winch and lever, whose operation Stevin explained, were used for construction purposes in the Middle Ages.
© The British Library Board. All Rights Reserved (ADD 35313, f. 34).

Almachtich, or 'Almighty', is 'a device we so call on account of its exceptional power' and it was a machine of his own designing. In inventing it, he tells us, he was inspired by the *charistion*, ascribed to Archimedes. Legend has it that Hiero, king of Syracuse, had a ship built, exceptionally large and of ingenious form, as a gift for King Ptolemy of Egypt. However, when it was finished, all the citizens of Syracuse together could not get it into the sea because it was too heavy. It was then that Archimedes contrived his mechanical device, the *charistion*, which allowed King Hiero to move the ship to the sea single-handed. Proposition 10 commences with this legendary demonstration of the enormous practical use of the lever. Since a lever makes it possible to multiply forces, it must be possible to invent instruments that allow an ever-increasing multiplication of forces. 'Give me a place to stand,' Archimedes is said to have declared, 'and I will move the earth'. Stevin endeavoured to invent a machine that was even more powerful than the *charistion*. The Almighty is a sort of windlass driven by cogwheels. Dijksterhuis observes that it is something like the *baroulkos* described by Heron of Alexandria. Stevin makes quantitative analyses, for instance, to determine the ratio between the number of revolutions of the Almighty's axle and its crank.

At the end of *De Weeghdaet*, Stevin works out the distance over which the earth could be moved by turning the Almighty's crank 4,000 times a minute for 10 years (assuming that the requisite place to stand was available). Though such a calculation may seem fantastic, it illustrates anew that Stevin's mind was equal to this kind of hypothetical problem. Indeed, the mass of the earth he postulates is very realistic.

3.4. The *Byvough der Weeghconst*

In the *Byvough der Weeghconst* (*Supplement to the Art of Weighing*), which appeared 20 years after *De Beghinselen der Weeghconst*, Stevin set out to deal with six subjects:

- *Van het Tauwicht (Of the Cord Weight)*
- *Van het Catrolwicht (Of the Pulley Weight)*
- *Nande vlietende Topswaerheyt (Of the Floating Top-heaviness)*
- *Vande Toomprang (Of the Pressure of the Bridle)*
- *Vande Watertrecking (Of the Drawing of Water)*
- *Vant Lochtwicht (Of the Weight of the Air)*

In fact, the supplement contains only four of the six treatises that had been planned. The last two, the *Watertrecking* and the *Lochtwicht*, are missing. We may suppose that in the *Watertrecking* Stevin dealt with such things as water management and drainage, though that is not certain. It is clear from his works on harbour construction and water drainage mills that he frequently put the hydrostatic theories he expounded in *De Beghinselen des Waterwichts* to use in practice, and that he supplemented them with empirical knowledge as and when necessary. What he was lacking was a physically based theory of moving bodies, the so-called dynamics. This would later be formulated by Newton – the demonstrations carried out by Galileo and Torricelli were its precursors. It is a great pity that the *Lochtwicht* has not come down to us, always supposing it to have been written in the first place. The fifth chapter of the *Anhang van de Weeghconst* (*Appendix to the Art of Weighing*) shows that Stevin understood that a body in the air is subject to an upward force and that he also knew the magnitude of this force. He was certainly one of the first – maybe even the first – to arrive at this understanding. As it is, we can only speculate about the contents of the *Lochtwicht*, and hope that one day the treatise will turn up.

Of the four works that have survived, the first two – *Van het Tauwich* and *Van het Catrolwicht* – are closely related for they both refer to simple instruments that were, and still are, in everyday use. Stevin was clearly impelled by the desire to supplement and put into practice the theories he set out in *De Beghinselen der Weeghconst*. The texts on the 'cord weight' and 'pulley weight' can be classed as both theory and practice. In the Argument to *Van het Catrolwicht* Stevin writes: '... seeing that pulleys are frequently used in practice for pulling up large weights ... it may sometimes be useful to know beforehand what force is required to pull up a given weight'.

The third study in the supplement is *Vande vlietende Topswaerheyt*. This is an application of *De Beghinselen des Waterwichts*, and we shall return to it later when discussing Stevin's contribution in that field.

The treatise *Vande Toomprang* is both striking and original, and no doubt fascinating to the modern equestrian, as Hendrik Casimir says. Stevin studies the forces that operate when a horse wears a bit and bridle. One of his aims is to trace the connection between the shape of the bit and the severity of the bridle. In Stevin's analysis, the two cheeks of the bit are considered as levers. This allows him to use the insights gained from *De Beghinselen der Weeghconst*. The problem is not a simple one, however. For instance, in the case of the bit it is not evident what the fixed point is. We learn from the Argument to *Vande Toomprang* that Prince Maurice was keenly interested in this problem: 'His Princely Grace having from early childhood to this day continually practiced the Art of Riding ... nevertheless he never succeeded in gaining, either through words or writings, a thorough knowledge of the reason for the pressure of bridles...' Happily, Stevin had just the book to hand: '...this ... made him very anxious to understand the foregoing *Weeghconst*, hoping thereby to gain thorough knowledge of the matter. Which to his pleasure happened...'

3.5. Dynamics

Stevin's great contributions to physics were primarily in the fields of statics, hydrostatics and the equilibrium of forces. His contribution to dynamics was much smaller, although his composition law of forces also applies in dynamics. And some of his studies, on water mills for instance, also occasionally have applications to or points of contact with dynamics.

It was not until Newton that the laws of motion were fully articulated. Galileo laid the ground with his pioneering work on falling and rising bodies, *Discorsi e dimostrazioni matematiche intorno a due nove scienze* (c. 1638). De Soto, Charles V's confessor, was possibly the first to see, in 1545, that freely falling bodies have the same, constant acceleration. In 1618 Beeckman formulated the crucial problem of the quantitative connection between space and time for a falling body. Beeckman did not publish his work in book form, but wrote down his ideas in his now famous *Journal*, which came to light only in 1905. He initially set out his problem in a letter to Descartes, with whom he corresponded and worked for many years. It likewise appears from his *Journal* that in 1618, Beeckman had already formulated the principle of inertia that is so crucial to physics and which is often ascribed to Newton: whatever is moved once always moves in the same way until it is impeded by something extrinsic. In that same year, together with Descartes, Beeckman gave a first derivation of the law of falling bodies. They discovered that the distance a free-falling body travels in a void from a point of rest is proportional to the time of descent squared. Beeckman and Descartes arrived at the correct mathematical result for the law of falling bodies. However, this is not to say, as is sometimes assumed, that they had found a useful basis for the falling of bodies. The situation was very complex. Besides, Descartes and Beeckman's beliefs as to the motive force differed, and there was much disagreement about what it was that caused a body to fall. Whatever the case, we have to conclude that neither Beeckman nor Descartes managed to base their findings on the motion of falling bodies on a concept such as gravity.

Galileo came closer to comprehending gravity and the role it plays. His law of falling bodies dates from 1638. He had first derived the right result from the wrong premises. As early as 1604 he had made vigorous efforts to understand the motion of falling bodies. He came up against numerous difficulties and contradictions, to the extent, indeed, that he never published his planned work *De Motu*. Galileo kept systematic accounts of his many attempts to understand the motion of bodies. As a result, we can still see how he struggled with the new concepts that were evolving. He developed the notion of 'degree of velocity' and came very close to the concept of 'acceleration'. Yet it cannot be said that in his *Discorsi* Galileo succeeded in formulating perfectly consistent derivations for his contributions to mechanics – the law of free fall and the parabolic motion of a projectile.

Galileo explored, rather than exceeded, the limits of 'preclassical' understanding. This led him to new concepts that would enable his successors, and Newton in particular, to move beyond preclassical mechanics. It is also extremely difficult to determine what preclassical mechanics precisely included. How did ideas develop in the centuries between Archimedes and Aristotle, and Jean Buridan (c.1295– c.1358) and Nicholas Oresme (1323–1382) of the Paris school of William of Ockham, who also helped to develop the physical concepts and mathematical aids that Stevin, Galileo, Huygens *et al.* would use as starting points for their work on mechanics? Dijksterhuis points out that while philosophical-theological thought was experiencing a decline in the fourteenth century, in the Ockham school natural science was undergoing an evolution that would lead to the physics of the sixteenth and seventeenth centuries. Oresme's *De velocitate motuum*, on the 'cause' of the motion of falling bodies, shows how questions of mechanics were already being pondered in the fourteenth century.

3.5.1. Simon Stevin and the burgomaster of Delft drop two spheres from a height of 30 ft

Against this background of evolving dynamics, Stevin played a less conspicuous part in the fundamental and ultimately fruitful questioning which, via Kepler, Beeckman, Descartes, Galileo and others, would result in Newton's laws. Nonetheless, in the introduction to *De Weeghdaet* he does touch fleetingly on the motion of falling bodies, getting there from the somewhat separate approach of frictional forces.

At the start of his study of machines, Stevin observes that the *Weeghconst* introduces the forces that are required to maintain an equilibrium, but that it does not show which force is needed for a motion to actually begin. He is aware of the frictional forces that must be overcome in a real machine before motion can start. For instance, he postulates that the weight of a body, if it rests on an inclined plane, is not proportional to the frictional resistance the body experiences because of the inclined plane. This is an astute insight; Stevin was possibly the first to recognize frictional forces and attempt their description.

Stevin was probably motivated to carry out an experiment on falling bodies by his determination to refute Aristotle's assumption that the time a body takes to fall decreases as its weight increases. Assisted by his friend Johan Cornets de Groot, burgomaster of Delft, Stevin put it to the test. He dropped two lead spheres, one ten times heavier than the other, from a position of rest at a height of 30 ft and discovered that 'they fall together onto the board so simultaneously that their two sounds seem to be one and the same rap'. This experiment, which was probably carried out in Delft, took place no later than 1586.

It is often suggested that Galileo was the first to conduct this kind of experiment. He is said to have dropped cannonballs from Pisa's leaning tower between 1589 and 1592. But a study of Galileo's work and career reveals no indication that he ever carried out such an experiment. Furthermore, his experiments with falling bodies date from after 1586 – later than Stevin's. In fact, the question of which of the two was first is fairly pointless. Philoponus had already carried out this experiment in the sixth century. As had Giovanni Battista Benedetti in 1553 – Stevin was probably aware of this. The significance of these experiments was that they demonstrated that Aristotle, whose ideas were generally considered unassailable, had the wrong end of the stick, at least in this instance. Moreover, the experiments contributed to the development of the experimental method as a basis for science. In this area Galileo was a pioneer, among other things with his experiments with a ball rolling down an inclined plane, where he sought to establish the connection between the distance travelled and the corresponding time intervals.

4. *De Beghinselen des Waterwichts* and the *Anvang der Waterwichtdaet*

The danger posed by storm tides is all too well known in the Netherlands, and not surprisingly the character of Hans Brinker is very popular. Legend tells how the plucky young Haarlem lad kept a hole in the dyke plugged with his finger throughout the night and so saved his town from flooding. In Mary Dodge's *Hans Brinker or the Silver Skates, a story of life in Holland* (1865) it is Peter, son of the Haarlem lock keeper, who prevents the watery catastrophe. In the popular legend, however, it is not Peter but Hans who is 'the hero of Haarlem'.

There are even statues of Hans, on the dyke in Spaarndam near Haarlem and at Harlingen. That it is possible to hold back the sea with one finger, like Peter or Hans, follows from a natural law that Simon Stevin discovered.

There is no getting away from water in the Netherlands. Both its danger and its potential make study necessary. So it is not surprising that Stevin, the first physicist in the Low Countries, busied himself with practical studies of sluices, dykes and water mills. With a theory that he developed in *De Beghinselen des Waterwichts*, Stevin made his most important and original contribution to physics at the age of 38.

4.1. Two lines in the historical development of hydrostatics

- The Archimedean–Stevian theory, based on postulates specially developed for hydrostatics. Some of the axioms Stevin introduced were different to those of Archimedes, and they enabled him to push forward.
- The method of Galileo and his school. This was an attempt to describe liquids using concepts from the mechanics of rigid bodies

The first line, the approach of Archimedes and Stevin, gradually prevailed over the second. This is primarily due to the crucial role that the concept of 'pressure' plays in hydrostatics. Pressure was known to Archimedes only indirectly. Stevin was the first to calculate the force exerted by a liquid on the walls or bottom of the vessel containing it. Without actually using the term 'pressure' to describe it, that was, nonetheless, what he was calculating. The concept was elaborated later, first in the work of Pascal and Boyle, which built on Stevin's theories of hydrostatics, and later in Johannes and Daniel Bernoulli's works on hydrodynamics. The lack of the notion of 'pressure' in the Galileo-Torricelli tradition handicapped the evolution of hydrostatics in Italy. Nonetheless, despite this limitation, Torricelli was able to make progress in liquid mechanics.

Stevin makes a specific reference to Archimedes in the first sentence he addresses to the reader in *De Beghinselen des Waterwichts*: 'What was the cause that moved Archimedes to write that which he left to us in the *Book of the things supported in the water*, where he began to hit off nature wonderfully, I do not know…'

4.2. Stevin was the first to make any headway in hydrostatics since the time of Archimedes

Webster's Dictionary (1913) rightly states: 'The first discovery made in hydrostatics since the time of Archimedes is due to Stevinus.' In the field of hydrostatics he was the first to resume and continue the work of Archimedes, and indeed he was one of the first to read Archimedes's work in a Latin translation – something that would not in any case have been possible before Stevin's day, for it was not until around 1550 that Latin translations of Archimedes's Greek works were generally available (see Chapter One, box 1).

Stevin's contribution to hydrostatics opened up new horizons. The methods he used were innovatory, just as they had been in *De Beghinselen der Weeghconst*. A number of important results were altogether new, the most significant among them being what today we call the 'hydrostatic paradox'. Below, after a word about the structure of *De Beghinselen des Waterwichts*, we shall first have a reminder of Stevin's generalization of Archimedes's law, then a description of the thought experiment by which Stevin calculated the force that a liquid at rest exerts on the bottom and walls of the vessel that contains it.

4.3. The structure of *De Beghinselen des Waterwichts*

Following Stevin's usual pattern, *De Beghinselen des Waterwichts* consists of two parts: first the theory (*De Beghinselen des Waterwichts*), then the practice (*Anvang der Waterwichtdaet*). The theoretical section runs to 46 pages, the practical part to just nine. The *Anvang der Waterwichtdaet* contains three

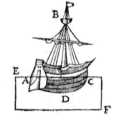

Figure 14: How deep does a ship lie in the water? From *De Beghinselen des Waterwichts*.
Bruges, Public Library, B 265 (ektachrome: JTD).

experimental demonstrations of the *Elements*. We may take it that Stevin uses 'practice' rather in the sense of a proof than of praxis. Both of these two main parts are prefaced by an address to the reader. In *De Beghinselen des Waterwichts* Stevin once again reminds us of the advantages of Dutch compared to Greek when engaged in scientific matters. To the reader of the short *Anvang der Waterwichtdaet* he explains that the three propositions he describes are hardly enough to merit the title of *Waterwichtdaet* (*Practice of Hydrostatics*), and the section is therefore called the *Anvang der Waterwichtdaet* (*Preamble of the Practice of Hydrostatics*). But, he assures us, we may 'expect the rest in due time'. As far as we know, however, Stevin never did expand the *Preamble* into a more complete *Practice*.

The first of the three propositions in the *Anvang der Waterwichtdaet* relates to the depth of immersion of a ship (see fig. 14), the second provides experimental illustrations of the hydrostatic paradox, while the third sets out 'To explain the cause why a man, swimming deep below the water, is not crushed to death by the great weight of water lying on him'. Stevin's experiment showing how upward pressure acts in a liquid merits special attention. Indeed, it is still demonstrated in many secondary schools today (see boxes 2 and 4).

Box 2: Stevin was there before Pascal

In the *Anvang der Waterwichtdaet*, Stevin proposes an experiment to demonstrate the upward pressure in a liquid. This experiment is still carried out in many secondary schools, in lessons about hydrostatics. Stevin describes his experimental set-up to prove that water exerts an upward force as follows:

> *In order to give a practical explanation about the examples in which the water exerts an upward thrust against the bottom, as in the 3rd corollary of the aforesaid 10th proposition, let ABCD be a water, and EF a closed tube, and G a disc of greater specific gravity than water, say of lead, as in the first figure (see fig. 15.1).*

Stevin now closes the bottom of the tube with a disc. Although this disc has a greater specific gravity than water it does not sink but clings to the tube because the water exerts on it an upward force (see fig. 15.2).

> *Let this disc G be laid against the hole F, in such a way that it fits closely thereto, and if then the tube together with the disc be put in the water ABCD – I take as far as H, as shown below – the disc G will not, in accordance with the common nature of lead, sink into the water, but cling to the tube and exert against it the same pressure as a weight of equal gravity to the water having the same volume as the prism whose base has the size of the hole F and whose height is HI, minus the difference between the weight of the disc G and the weight of the water having the same volume as the disc G.*

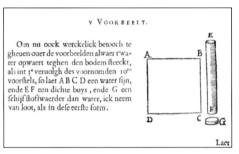

v Voorbeelt.

Om nu oock werckelick betooch te gheuen ouer de voorbeelden alwaer t'water opwaert teghen den bodem ſteeckt, als int 3e veruolgh des voornomden 10en voorſtels, ſo laet A B C D een water ſijn, ende E F een dichte buys, ende G een ſchijf ſtofſwaerder dan water, ick neem van loot, als in deſe eerſte form.

Laet

Figure: 15.1

DER WATERWICHTDAET. 61

Laet deſe ſchijf G gheleyt worden teghen t'gat F, alſoo datſe dicht daer op pas, ende de buys met de ſchijf dan alſoo t'ſamen in t'water A B C D ghеſteken, ick neem tot H toe, als hier onder, de ſchijf G en ſal naer den ghemeenen aert des loots, in t'water niet ſincken, maer ande buys blijuen hanghen, ende daer teghen ſoo ſtijf drucken, als een ghewicht eueſwaer an t'water dat euegroot is anden pilaer, diens grondt de groote des gats F, ende hoochde H I is, min t'verſchil des ghewichts der ſchijf G, tot t'ghewicht des waters an die ſchijf euegroot. Maer ſoo de ſchijf G niet dicht ghenouch teghen de buys en ſlote, ende datter eenich water indrong, ſoo ſal de ſchijf G daer ſoo langhe anhanghen, tot dat ſulck inghedronghen water t'voornomde ghewicht ouerwint.

Figure: 15.2

Maer want nu yemandt dencken mocht, het groot ſwaer water rondom de buys ſtaende, ſtijuer drucking der ſchijf teghen de buys te veroirſaken, dan een cleender water van de ſelfde hoochde, ſoo laet ons t'water neuen de buys rondom afcorten, dat is, dat de reſte water ſy in een vat van form als hier neuen, ende d'eruaring ſal bewyſen (verſouckende de macht des gheprangs in t'een en t'ander water, duer euewichten inde buys op G ruſtende) dit cleen water euen ſoo ſtijf te drucken als t'voornomde grooter, waer af de reden bouen grondelick beſchreuen is.

T'Beslvyt. Wy hebben dan duer daetlicke voorbeelden het 10e voorſtel der beghinſelen des Waterwichts verclaert, naer t'voornemen.

Merckt.

Wat het 11e voorſtel belangt, daer uyt is onder anderen kennelick, wat ghewicht waters datter druct, teghen elcke ſijde der duer van een ſluys, ende dierghelijcke: Oock dattet water ouer d'een ſijde alleenlick een ſtroobreet, daer teghen ſoo ſtijf prangt als t'water diens breede de Zee van Oceane ouer d'ander ſijde; welverſtaende als ſy euenhooghe ſijn. Van welcke dinghen wy om haer voornoemde claerheyt hier gheen beſonder voorſtellen en maken.

Hh 3 111 Voor-

Figure: 15.3

Figures 15.1–15.3: Stevin invented this simple experiment to illustrate his hydrostatic paradox. It is sill demonstrated in classrooms today. From *De Beghinselen des Waterwichts*. *Bruges, Public Library, B 265 (ektachrome: JTD).*

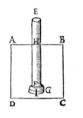

Stevin then links this proof to the hydrostatic paradox. The upward force the water exerts on the disc is the same whether the tube is immersed in a narrow or a wide vessel (compare fig. 15.2 with fig. 15.3).

> *But because someone might suppose that the large heavy column of water surrounding the tube would cause a greater pressure of the disc against the tube than a smaller column of water of the same height, let us take away the water all around the tube, i.e. so that the remainder of the water be in a vessel of a form shown opposite. Then experience will prove (testing the force of the pressure in either column of water by means of equal weights in the tube resting on G) that this small column of water exerts the same pressure as the aforesaid larger column of water, the cause of which has been thoroughly described above.*

That Stevin perfectly understood the hydrostatic paradox is demonstrated by his concluding observation:

> *NOTE: As regards the 11th proposition, from that it is evident, among other things, what is the weight of the water pressing against either side of the gate of a lock and the like. Also that the water on one side, even if it were only the width of a straw, exerts the same pressure against it as the water having the breadth of the Ocean on the other side, provided they are on the same level. Of these matters we do not draw up any special propositions, in view of the aforesaid clearness.*

CONCLUSION

> *We have thus explained, by means of practical examples, the 10th proposition of De Beghinselen des Waterwichts, as intended* (see section 4.5 for Proposition 10).

The experiments with the balance, which Stevin describes to illustrate his hydrostatic paradox and in which he also introduces the principle of the hydraulic press, are of exceptional significance (see figs. 16 and 17).

ABCD is a vessel full of water (see fig. 16) in whose bottom DC there is a round hole EF that is covered by a round wooden disc GH of a lesser specific gravity than water. IKL is another vessel full of water, of the same height as vessel ABCD but narrower, in whose bottom KL there is also a round hole MN, equal to the hole EF and covered by a disc OP of the same volume and weight as the disc GH. The experiment will now demonstrate that the disc GH will not rise, as wood normally does in water, but will exert on the hole EF the same pressure as a weight of equal gravity to the water which is equal in volume to the prism EFQR, minus the difference between the weight of the wooden disc GH and the weight of the water having the same volume as the disc. In order to prove this experimentally (*daet*), the disc GH can be attached to a balance whose weight S is of equal gravity to the aforesaid weight, 'and

DER WATERWICHTDAET. 59

dat water dan dat teghen dit: waer duer t'cranckfte voor t'fterckfte fal moeten wycken, dat is, t'water A B C D fal rijfen, ende van C D E F fal dalen; Maer dit fo wefende, haer oppervlacken en fullen niet euen hooch fijn t'welck opentlick teghen d'eruaring is. Daerom t'cleinfte water A B C D druckt euen foo ftijf teghen den bodem C D, als t'grootfte water C D E F.

III VOORBEELT.

Laet A B C D een vat vol waters fijn, in wiens bodem D C euewydich ligghende vanden fichteinder, een rondt gat E F is waer op ligt een ronde houten fchijf G H, ftoflichter dan water, ende dat gat E F bedeckende, en rondtom dicht fluytende teghen den bodem D C. Laet oock I K L een ander vat vol waters fijn, euenhooch mettet vat A B C D, maer cleinder, in wiens bodem K L oock een rondt gat M N fy, euen an t'gat E F, waer op light een fchijf O P, euegroot ende euefwaer ande fchijf G H; T'welck foo wefende, d'eruaring fal bethoonen dat de fchijf G H niet rijfen en fal, naer de ghemeenen aert des hauts in t'water, maer fal fo ftijf op t'gat E F drucken, als een ghewicht euefwaer an t'water dat euegroot is anden pilaer E F Q R, min t'verfchil des ghewichts der houte fchijf G H, tot het ghewicht des waters an die fchijf euegroot. Maer om fulcx duer de daet oock te fien, men mach ande fchijf G H een waegh voughen, diens ghewicht S euefwaer fy an dat voornomde ghewicht, ende de fchijf G H fal daer teghen euewichtich blijuen. Laet nu infghelijcx ande fchijf O P oock een waegh voughen, diens ghewicht T euefwaer fy an S, en de fchijf O P fal daer teghen oock euewichtich blijuen. Maer foomen S ende T yet fwaerder maeckt, fy fullen haer fchijuen doen rijfen, in der voughen dat de fchijuen G H, O P, duer fulcke euewichten beuonden worden eueftijf teghen haer bodems te drucken, waer uyt het voornemen blijckt, te weten het cleinder water I K L, euen fo ftijf teghen fijn grondt te drucken, als t'grooter A B C D.

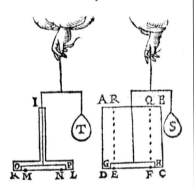

MERCKT.

Tis kennelick dat fo t'verfchil des ghewichts der fchijf als G H, tot het ghewicht des waters an haer euegroot, meerder waer dan t'ghewicht des
Hh 2   waters

Figure 16: This proof involving a balance also demonstrates the hydrostatic paradox. From *De Beghinselen des Waterwichts.*
Bruges, Public Library, B 265 (ektachrome: JTD).

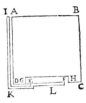

Figure 17: This example from the *Anvang der Waterwichtdaet* illustrates the principle of the hydraulic press.
Bruges, Public Library, B 265 (ektachrome: JTD).

the disc GH will be in equilibrium therewith'. Now likewise attach the disc OP to a balance whose weight T is of equal gravity to S, and the disc OP will also be in equilibrium. But if S and T are made a little heavier, 'they will cause their discs to rise', in such a way that discs GH and OP are found to exert the same pressure against their bottoms. Thus what was intended to be proved is evident, 'to wit that the smaller [column of] water KL exerts the same pressure against its bottom as the larger ABCD.'

Stevin had already established the principle of 'Pascal's scales'

Stevin's experiment just described, the third example in the *Anvang der Waterwichtdaet* of 1586, is in principle the same as Pascal's set-up with a pair of scales of 1660. This is still demonstrated in the classroom today to 15- and 16-year-olds. The priority given to 'Pascal's scales' therefore belongs to Stevin, for he had already established the principle in *Anvang der Waterwichtdaet*. It would thus be better to speak of 'Stevin's scales'.

Stevin's principle of the hydraulic siphon or hydraulic press is also often wrongly ascribed to Pascal. The principle is explained as example four of the *Anvang der Waterwichtdaet*. The original experiment is illustrated in fig. 17, so we can see how Stevin discovered the important principle of the hydraulic press. In example four we see a disc GH. When the thin tube IKL is filled with water to the same level as the vessel ABCD, the smaller volume of water in IKL will exert the same pressure on the base of the disc as the much greater volume of water in ABCD. The disc will then rise. In this way the 1 lb of water that fills the tube IKL is able to exert a greater pressure against the disc

GH than, for example, a column of water of 100,000 lb above it. It is sufficient that the water level in the thin tube IKL is made higher than in the vessel ABCD. Stevin is fully aware of the paradoxical nature of this law, 'which might be called one of the secrets of nature, if the cause were unknown'.

Throughout his works, and especially in the *Anvang der Waterwichtdaet*, Stevin describes various experiments – of which he was the inventor – in detail. Until Stevin's day it was completely unheard of to conceive of experiments as additional evidence for insights derived by means of 'logical deduction'. In this sense, the experiments Stevin devised were pioneering in methodological terms. *Daet*, by which Stevin generally meant praxis, is here demonstrably meant as 'experimental'.

Gradually, in the service of the prince and the Republic, Stevin became a true *homo universalis*. After 1586 he did little in the way of physics. When we look at the profound and pioneering character of *De Beghinselen der Weeghconst* and *De Beghinselen des Waterwichts* we can only wonder what else Stevin might have contributed to physics had he devoted all his considerable powers to the subject.

De Beghinselen des Waterwichts begins with a dedication: 'Simon Stevin wishes the states of the United Netherlands much happiness', thus underlining the importance of a knowledge of hydrostatics to a country like the Netherlands, where water management and shipping are of great concern: '...what continual practice these countries have with the water, more so than any others...'

Like *De Beghinselen der Weeghconst*, the work has an axiomatic structure. Eleven definitions are followed by seven postulates, succeeded by 21 propositions or theorems. Numerous examples support the argument and make the work accessible to the reader.

4.4. Generalized derivation of Archimedes's Principle

Although Stevin knew Archimedes's work and was inspired by it, he still developed an original approach to hydrostatics. In *De Beghinselen des Waterwichts* he employed two new basic principles. The first was his refutation of perpetual motion; the second and particularly original principle consisted of the imaginary 'stiffening' of a certain portion of the liquid mass without – he postulates – affecting the existing equilibrium in the liquid. The virtuoso thought experiments in which Stevin embeds his basic hydrostatic principles are themselves of great importance to the development of the methodology of physics.

Before dealing with the principle named after Archimedes, Stevin introduces a number of propositions relating to floating bodies so as to be able to apply them to a ship in the water. These preparatory suppositions lead to Proposition 5: '*A solid body of greater specific levity* [i.e. lesser specific gravity] *than the water in which it lies is of equal weight to the water having the same volume as its part in the water'*. Stevin demonstrates this with the aid of a geometric diagram (see figs. 18.1 and 18.2). This proposition is immediately applicable to a floating ship of whatever shape. In relation to this ship Stevin poses the question: given the volume beneath the surface of the water, what is the weight of the whole ship?

Again Stevin's argument is a pedagogic *tour de force*: it requires no previous knowledge on the part of the reader; the derivation is easy to follow. An extremely simple diagram consisting only of two squares is sufficient to describe a ship with an arbitrary complex shape and distribution of mass. Not only is the reference to a ship practical, the didactic design is also attractive and aesthetic. Here we see theoretical physics of great purity: the simplest possible modelling of a physical system, and the derivation from it of natural laws.

As one of the following applications of his new axioms Stevin gives an original re-derivation and generalization of Archimedes's Principle. He formulates it as follows (*De Beghinselen des Waterwichts, Proposition 8*):

> *The gravity of any solid body is as much lighter in water than in air as is the gravity of the water having the same volume.*

In a water-filled vessel a delineated quantity of water is considered, which is supposed to be in equilibrium with its surroundings (see fig. 18.3). Stevin calls the imaginary geometric surface of this volume of water *vlackvat* or 'surface vessel'. This is probably another of Stevin's neologisms, as it only appears in Dutch dictionaries in reference to him. Imagine the water removed from the surface vessel D; this surface vessel will then have a 'levity', or loss of weight equal to the weight of the water that was previously in the surface vessel. If the surface vessel is now filled with the solid A, the surface vessel with A in it will weigh as much as A minus the weight of the water that was removed from the surface vessel. Consequently, when A is placed in the water it suffers a loss of weight equal to the weight of the water it displaces.

4.5. Stevin's hydrostatic paradox

From Proposition 10 onwards Stevin introduces new discoveries in hydrostatics, discoveries that are of historical significance in the development of physics. Indeed, this is the most important creative contribution to the science in Stevin's entire oeuvre. Proposition 10 itself deals with Stevin's 'hydrostatic paradox'. It was only later that this surprising phenomenon acquired its name, through the agency of Robert Boyle (1627–1691), but it was certainly a discovery made by Simon Stevin.

In textbooks and on the Internet, Stevin's paradox is generally introduced with an arrangement of water vessels similar to that shown in fig. 19. It is stressed that as long as the principles of hydrostatics were not thoroughly understood, the behaviour of liquids was perceived as very mysterious. The

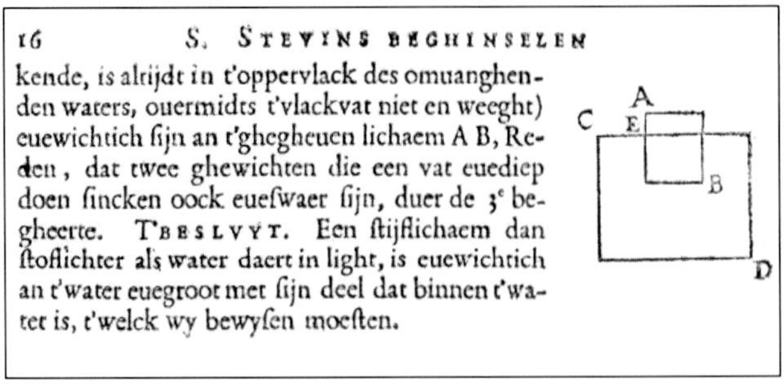

16 S. STEVINS BEGHINSELEN

kende, is altijdt in t'oppervlack des omuanghenden waters, ouermidts t'vlackvat niet en weeght) euewichtich ſijn an t'ghegheuen lichaem A B, Reden, dat twee ghewichten die een vat euediep doen ſincken oock eueſwaer ſijn, duer de 3e begheerte. T'BESLVYT. Een ſtijflichaem dan ſtoflichter als water daert in light, is euewichtich an t'water euegroot met ſijn deel dat binnen t'water is, t'welck wy bewyſen moeſten.

Figure: 18.1

DES WATERWICHTS. 17

VI VERTOOCH. VII VOORSTEL.

WESENDE twee oneueſtofſware wateren, ende een ſtijflichaem ſtoflichter dan eenich van dien: Ghelijck de ſtofſwaerheyt des ſwaerſten waters, tot de ſtofſwaerheyt der lichtſten, alſo de grootheyt diens ſtijflichaems binnen t'water in t'lichtſte water gheleyt, tot ſijn grootheyt binnen t'water in t'ſwaerſte gheleyt.

T'GHEGHEVEN. Laet A B een water ſijn, ſtofſwaerder dan t'water C D, ende E F ſy een ſtijflichaem ſtoflichter dan eenich dier twee wateren, t'welck eerſt gheleyt in t'water A B, ſoo daelter onder t'water het deel G F, maer t'ſelue lichaem E F gheleyt in t'water C D, t'welck daer ſy H I, ſoo ſinckter onder t'water het deel K I. T'BEGHEERDE. Wy moeten bewyſen dat ghelijck de ſtofſwaerheyt des waters A B, tot de ſtofſwaerheyt des waters C D, alſoo de grootheyt K I, tot G F.

T'BEWYS. T'water des waters A B euegroot an G F, is eueſwaer mettet lichaem E F, ende t'water des waters C D euegroot an K I, is eueſwaer mettet lichaem H I duer het 5e voorſtel, maer t'lichaem E F ofte H I is al een ſelfde lichaem duer t'ghegheuen, daerom t'water des waters A B euegroot met G F, is eueſwaer an t'water des waters C D euegroot met K I; Maer weſende twee eueſware wateré, ghelijc haer grootheyt tot grootheyt, alſo ouerandert haer ſtofſwaerheyt tot ſtofſwaerheyt, als nootſakelick volght uyt de toeghelaté 5e bepaling, daerom ghelijck de ſtofſwaerheyt des waters A B, tot de ſtofſwaerheyt des waters C D, alſoo de grootheyt K I, tot de grootheyt G F.

T'BESLVYT. Weſende dan twee oneueſtofſware wateren ende een ſtijflichaem, &c.

Figure: 18.2

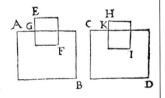

vessel shown in fig. 19 consists of two connected chambers that are open at the top and have apertures of equal diameter at their bases. If water is poured into one of the chambers it also flows into the other chamber until it reaches the same

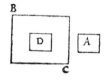

```
18      S. STEVINS BEGHINSELEN

    T'BEGHEERDE. Wy moeten bewyfen dat A in t'water B C ghe-
leyt, aldaer foo veel lichter fal fijn dan inde locht, als de fwaerheyt des
waters met hem euegroot.  T'BEREYTSEL. Laet D een vlackvat
vol waters fijn, euen ende ghelijck an A.  T'BEWYS. T'vlackvat D
vol waters, en is in t'water B C licht noch fwaer, want het daer in alle
gheftalt houdt diemen hem gheeft, duer het i' voorftel, daerom t'water D
uytghegoten, t'vlackvat fal t'ghewicht des waters lichter fijn dant in fijn
eerfte ghedaente was, dat is, van foo veel volcommentlick licht: Laet ons
nu daer in legghen t'lichaem A, t'felue
fal daerin eften paffen, om dat fy euen
ende ghelijck fijn duer t'gheftelde. Ende
t'vlackvat metter lichaem A alfoo daer
in, fal weghen t'ghewicht van A met
fijn voornoemde lichticheyt, dat is t'ghe-
wicht van A min t'ghewicht des waters
datter eerft uytghegoten was, maer dat
water is euegroot an A. Daerom A in t'water B C gheleyt, is daer in foo
veel lichter dan inde locht, als de fwaerheyt des waters met hê euegroot.
    T'BESLVYT. Yder ftyflichaems fwaerheyt dan, is foo veel lichter
in t'water dan inde locht, als de fwaerheyt des waters met hem euegroot,
t'welck wy bewyfen moeften.
```

Figure: 18.3

Figures 18.1–18.3: Stevin uses simple sketches to introduce propositions relating to floating bodies. From *De Beghinselen des Waterwichts*. Bruges, Public Library, B 265 (ektachrome: JTD).

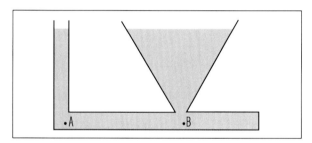

Figure 19: Textbook illustration of Stevin's hydrostatic paradox. The greater volume of water in the chamber above B is in equilibrium with the smaller volume in the tube above A, at least when the level of water in both chambers is the same.

level in both chambers. This phenomenon seems paradoxical. One would intuitively expect that the greater volume of water in the cone-shaped chamber above B would cause a greater force per surface unit to be exerted on base B than the water in the tube above base A, and that the level of the water in tube A would therefore be higher than in cone B. That this is not the case is an example of Stevin's hydrostatic paradox.

4.5.1. The Archimedes–Stevin tradition

It is striking that in *De Beghinselen des Waterwichts*, having laid the foundations and tried them out on Archimedes's principle, Stevin immediately gets down to business and determines and calculates the force exerted by water on horizontal or inclined submerged plane surfaces (which amounts to pressure). In this way he also establishes the basis for what is sometimes called the 'Archimedes–Stevin tradition'. This tradition is at the root of modern hydrostatic and hydrodynamic theory.

Stevin's derivation of the fundamental principles of hydrostatics is so important and his explanation so fine that we shall follow him through it. In Proposition 10 Stevin proves that:

> *On any bottom of the water being parallel to the horizon there rests a weight equal to the gravity of the water, the volume of which is equal to that of the prism whose base is that bottom and whose height is the vertical from the plane through the water's upper surface to the base.*

Stevin first applies this proposition to a rectangular parallelepiped (see fig. 20.1). It has been rightly observed that strictly speaking this should rather be termed an axiom than a proposition to be proved. Nevertheless, Stevin's attempt to deliver the proof for the rectangular parallelepiped is greatly inspired. In Corollary I of Proposition 10, he demonstrates that a solid body of greater 'levity' than water (that is, floating on water) neither lightens nor weights the bottom – provided the water remains at the same level before and after the introduction of the floating body. This is illustrated by the top diagram in fig. 20.2. Stevin calls a body with greater 'levity' a body that has a weight smaller than that of the water in which it is immersed.

In the reasoning Stevin develops in Corollaries II–IV, he introduces his highly original concept of the imaginary 'stiffening' of certain portions of the water filling a vessel with vertical walls, illustrating his argument by means of the diagrams shown centre and bottom in fig. 20.2. His starting point is that a body, of whatever shape and with a mass equal to the water in which it is immersed, will be in equilibrium in this liquid, no matter its position. If the water level is always maintained, with the imaginary stiffening of a part of the liquid, the force exerted on the bottom EF will remain unchanged.

As Stevin puts it in Corollary II:

> *Let there again be put in the water ABCD a solid body, or several solid bodies of equal specific gravity to the water. I take this to be done in such a way that the only water left is that enclosed by IKFELM. This being so, these bodies do not weight or lighten the base EF any more than the water first did. Therefore we still say, according to the proposition, that against the bottom EF there rests a weight equal to the gravity of the water having the same volume as the prism whose base is EF and whose height is the vertical GE, from the plane AB through the water's upper surface MI to the base EF.*

Note that in fig. 20.2 the diagram on the centre right and the one below it contain a solid body, part of which is lower than the base EF.

In Corollary III, Stevin shows that the water also exerts an upward force on the base EF.

> *Let again ABCD be a water, and EF a bottom therein, parallel to the horizon. This being so, the water below the bottom EF exerts an upward thrust against it as great as the downward thrust which the water above the bottom EF exerts against it. For if this were not so, the weakest would give way to the strongest, which does not happen, for each keeps its appointed place, by the first proposition. Now let a number of solid bodies of equal specific gravity to the water be laid therein in such a way that the water IKEFLM thrusts against EF from below, as in the figure opposite. This being so, the water below the bottom EF exerts the same thrust against EF, i.e. against the solid body, as it did before against the water. But it exerted against the latter the same thrust as the upper part against EF, as has been said above, and the upper part exerted a thrust*

Figure: 20.1

DES WATERWICHTS. 21

een ghewicht ruſt, euen ande ſwaerheyt waters die euegroot is met den pilaer, diens grondt E F is, ende hoochde de hanghende lini G E, van * t'plat A B, duer t'waters oppervlack A N totten grondt E F: Waer uyt blijckt dat eenighe drijuende ſtof in t'water gheleyt, ſy en verſwaert noch en verlicht (welverſtaende als t'water inde ſelfde hoochde blijft) den grondt niet.

Plane.

II. VERVOLGH.

Laet andermael int water A B C D, legghen een ſtyflichaem, ofte verſcheyden ſtyflichamen eueſtofſwaer mettet water, ick neem alſoo, datter maer water en blijft als t'begrepen binnen I K F E L M; T'welck ſo ſijnde, deſe lichamen en beſwaren noch en verlichten den grondt E F niet meer dan t'water eerſt en dede: Daerom ſegghen wy noch naer luyt des voorſtels, dat teghẽ den bodẽ E F, een ghewicht ruſt euen ande ſwaerheydt waters die euegroot is metten pilaer, wiẽs grondt E F is, eñ hoochde de hanghende lini G E, van t'plat A B duer t'waters oppervlack M I, totten grondt E F.

III VERVOLGH.

Laet wederom A B C D t'eenemael water ſijn, ende E F een bodem daer in, euewydich vanden ſichteinder. T'welck ſoo weſende, t'water onder den bodem E F, ſtoot euen ſoo ſtijf daer teghen opwaert, als t'water bouen den bodem E F, daer teghen neerwaert ſtoot: Want by aldient ſoo niet en waer, t'cranckſte ſoude voor t'ſterckſte wycken, t'welck niet en ghebuert, want yder houdt ſijn ghegheuen plaets duer het 1e voorſtel. Laet nu eenighe ſtyflichamen eueſtofſwaer mettet water, alſoo gheleyt worden, dattet water I K E F L M, van onder anſtoot teghen E F, als hier neuens. Dit ſoo ſijnde, t'water onder den bodem E F, ſtoot nu ſo ſtijf teghen E F, dat is teghen t'ſtijflichaem,

Cc 3 alſt te

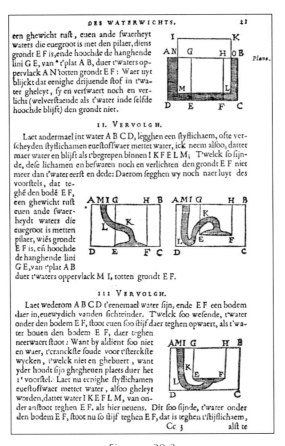

Figure: 20.2

Figures 20.1–20.2: In this masterly section from *De Beghinselen des Waterwichts*, pages 20 and 21 are concerned with the force that water exerts on the bottom. Stevin represents very clearly the ingenious thought experiment that led him to the hydrostatic paradox. *Bruges, Public Library, B 265 (ektachrome: JTD).*

> *against EF according to the present proposition. Therefore the lower part also exerts a thrust against EF according to the present proposition, that is, as we have said above, that against the bottom EF there still rests a weight equal to the gravity of the water having the same volume as the prism whose base is EF and whose height is the vertical GE, from the plane AB through the water's upper surface MI to the base EF.*

In Corollaries IV and V, Stevin develops his thought experiment. He keeps only that part of the water represented by the shaded parts of fig. 20.2. The force that is exerted on the base EF will not change because the rest of the water (outside the shaded areas) can be imagined to be "stiffened" without affecting the equilibrium. Proposition 10 is therefore valid for liquids of any shape, such as the shaded areas in fig. 20.2. Thus a force is exerted on the base EF that is equal to the weight of the water present in the prism whose base is EF and whose height is equal to the distance between the surface of the water and the base EF.

4.5.2. 'Stevin's Law'
In Proposition 10 and the following generalization, Stevin announced his discovery of a fundamental law of physics, and more specifically of hydrostatics. The realization that the force exerted by a liquid on a horizontal submerged surface (later made generally applicable to inclined surfaces) depends only on the area of this surface and the depth of its immersion, not on the volume of the liquid in the vessel (which later came to be known as the 'hydrostatic paradox') is one of Stevin's most masterly contributions. Not only should the great importance of this natural law, which could justifiably be termed 'Stevin's Law', be emphasized, but also the great originality of Stevin's demonstration.

In the following proposition Stevin builds on the law he has discovered in Proposition 10. He then extends his ideas to the study of the pressure exerted by a liquid on a wall of arbitrary shape and an arbitrary degree of submersion or incline.

Today we would formulate it thus: the pressure exerted on a flat surface immersed in water is equal to the weight of a prism whose base is equal to that flat surface and whose height is the vertical distance between the centre of gravity of the flat surface and the surface of the water.

It is remarkable that Stevin made the correct quantitative calculation of the pressure exerted by the water on an inclined as well as a vertical wall. The understanding of the forces that operate in a liquid is of capital importance to Stevin's applied work, such as the technology of water mills and sluices.

The significance of *De Beghinselen des Waterwichts* is further explained in boxes 2–4. Box 2 describes the experiment Stevin carried out to demonstrate the hydrostatics paradox, as well as the experiment, still used nowadays in many secondary schools, to demonstrate upward pressure in water. It also proves that Stevin invented the principle of the hydraulic press. Box 3 outlines the calculation from *De Beghinselen des Waterwichts* in which the pressure on both a vertical and an inclined wall is calculated. Together with 'Stevin's Law' (Proposition 10) and its derivation, this specific calculation is an early example of applied theoretical physics. Finally, Box 4 examines the stability of ships, dealt with in the *Byvough der Weeghconst* (1605) and illustrates that burgeoning Renaissance science had not yet managed to surpass the Greek examples in every respect.

Box 3: The pressure of water on vertical and inclined walls

The pressure exerted by water on both vertical and inclined walls of vessels is also determined in *De Beghinselen des Waterwichts*. Stevin was the first to correctly calculate the resulting force and its point of application.

Look at fig. 21 and imagine that the vertical wall ACDE of the vessel AB is moveable. Stevin asks himself which force must act on point K (to be found) in order to counter the force exerted by the water on the wall ACDE. He shows this force I as operating on K via a pulley L. He proves that this force is equal to the weight of the water in the prism ACHDE and that its point of application K (the centre of pressure or pressure point) is located at 1/3 of the height EA. As nearly always, Stevin gives the sense in which the force operates in words, and it is also apparent from the diagram.

Stevin then generalizes his discussion with the aid of successive examples, from vertical walls to inclined walls, from surfaces whose upper edge is level with the surface of the water to completely submerged surfaces of arbitrary shape, and so forth. In each example he defines a prism filled with water, so that its weight is equal to the force being sought. As Dijksterhuis observes, Stevin's results are equivalent to the statement that 'The force exerted on a plane surface of any form, location and orientation is equal to the weight of the column of liquid whose base has the same area as the surface and whose height is equal to the vertical distance between the centre of gravity of that surface and the free surface of the liquid.' First the system is described. AB is a vessel of water, a rectangular parallelepiped, with a 'bottom' or vertical wall ACDE. (In *De Beghinselen des Waterwichts* the word 'bottom' is used equally for horizontal and non-horizontal surfaces.) The water's surface is at ACGF. Stevin will prove that the force with magnitude I, equal to the weight of the prism ACHDE and operating in the direction KL, thus parallel to DH, is in equilibrium with the force exerted by the water in the vessel: 'the weight I is in equilibrium with the pressure of the water...' Let the wall ACDE be divided into equal rectangles (ACVR, RVXS, etc.). Stevin establishes that the magnitude of the force exerted by the water on such a rectangle always lies between two values, that of an inscribed and that of a circumscribed volume. Now he refines this division, creating eight equal rectangles instead of four, so that the difference between the forces exerted by the inscribed and circumscribed volumes decreases each time. In this way the force to be found always lies between upper and lower limits that are getting closer and closer together as the 'bottom' ACDE is divided into ever smaller rectangles: 'It is therefore manifest, through this infinite division of the bottom, that no weight so small can be given but it can be shown that the difference ... between the weight resting against the bottom ACDE and the weight of the half prism ACHDE is even less...' Then Stevin uses a syllogism to conclude that 'therefore the weight of the half prism ACHDE is equal to the weight resting against the bottom ACDE'.

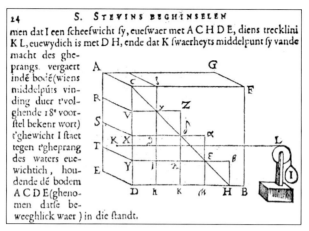

Figure 21: Construction used to determine the force water exerts on a vertical surface. From *De Beghinselen des Waterwichts*. Bruges, Public Library, B 265 (ektachrome: JTD).

Without going into detail it should be mentioned here that in determining the pressure exerted on inclined surfaces Stevin uses techniques and concepts that anticipate those of differential and integral calculus. We find a characteristic example of this (in fig. 21) in the division of the wall ACDE into a number of rectangles (ACVR, etc.). He will then limit the forces that act on each of the smaller surfaces – ACVR, RVXS and so on – by enclosing them between an upper and lower limit. This is made possible by looking at bodies of water such as RVZδXS and RVγυXS, which respectively exert a greater and lesser force on the wall of the body RVγδζXS. The refinement of this process by infinitely increasing the number of rectangles into which the vertical wall ACDE is divided, together with Stevin's more direct treatment of the concept of limit compared to that of Greek mathematics, is a step towards modern infinitesimal analysis.

In example four Stevin calculates the force I that operates on K (see fig. 21) by means of numbers, taking the foot as the unit. This is one of the first ever examples in which the methods of theoretical physics, in the spirit of modern analysis and with numbers, are seen in action.

Stevin performs an analogous masterpiece with his calculations for water mills. Among other characteristics he calculates the force that must act on a sail (per surface unit) to achieve equilibrium with the waterpower that is exerted on one of the paddles of the scoop wheel of a water mill. Here Stevin formulates a practically-orientated problem of physics in mathematical terms and in the process applies the principles expounded in both *De Beghinselen der Weeghconst* and *De Beghinselen des Waterwichts*. In other words, he makes a theoretical-physical model, which he also solves. At the same time he uses combinations of forces that today are called 'couples' or 'moments' and which would only be rediscovered much later.

Box 4: Why do ships float?

Understanding the stability of floating bodies is essential for shipping. A disaster such as the capsizing of the roll-on roll-off passenger ferry 'Herald of Free Enterprise' just outside the Belgian port of Zeebrugge in 1987 is a harrowing demonstration of what can happen when the laws of stability are ignored. Archimedes was the first to study the laws of hydrostatics. Even today the depth and genius of his study on floating bodies are impressive. Not only did he invent and lay the foundations for the science of hydrostatics in this work but he did so with considerable mathematical rigour and elegance. *On Floating Bodies* is the first work in which mathematics and physics go hand in hand, a Greek methodological innovation that still rules physics today. The second part of the work, in which Archimedes determines, among other things, the stable equilibrium positions of floating right paraboloids (see fig. 22.1), is a mathematical *tour de force* unmatched in antiquity and rarely equalled since.

Only one of Archimedes's simplest results will be mentioned here, Proposition 2 of Book II of *On Floating Bodies*:

> *A homogeneous body in the form of a right paraboloid floats in stable equilibrium in the water when its height is no greater than $3/2p$ (see fig. 22.1; with p as the parameter which describes the parabola).*

To arrive at his conclusion Archimedes examined two particular points: the centre of gravity of the submerged section of the right paraboloid segment (point B in fig. 22.2) and the centre of gravity of the section of the right paraboloid segment above the water (point Γ in fig. 22.2). If the combined action of the forces that operate in B and in Γ pushes back the paraboloid to a state of rest, then the paraboloid floats in a stable position.

Stevin deals with the stability of ships in the *Derde Deel des Byvoughs der Weeghconst, vande vlietende Topswaerheyt* (*Third Part of the Supplement to the Art of Weighing, Of the Floating Top-heaviness*). He describes this as 'of the top-heaviness of materials floating on water' (see fig. 23). Top-heavy means: 'disproportionately heavy at the top so as to be in danger of toppling' (New Oxford Dictionary of English, 1998). Stevin was motivated by a practical problem that was sometimes encountered during a siege:

> *It has sometimes happened that it was desired to make certain vessels, with ladders standing upright therein, about 20 feet high, for soldiers to ascend them. But since it was doubted whether this would not cause too great top-heaviness, so that the vessel might capsize and the soldiers fall into the water, a vessel was made, in order to be surer, with its ladder and accessories; thereafter it was tested in practice. This set me thinking whether it would not be possible to know this through static calculations of assumed forms and gravities...*

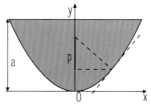

Figure: 22.1

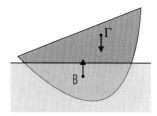

Figure: 22.2

Figures 22.1–22.2: Archimedes studied the equilibrium of floating bodies with the form of a right paraboloid segment in a way rarely equalled.

In his study, Stevin makes use of the concepts of centre of gravity of a floating body and the centre of gravity of the *waterhol*, by which he means the body of water the floating body displaces. He concludes that a ship floats in a condition of stable equilibrium if its centre of gravity is lower than the centre of gravity of the displaced fluid. In specific cases this is true, but it is not a generally valid argument.

Whereas in *De Beghinselen des Waterwichts* Stevin introduced new bases for hydrostatics, opened up new paths and discovered new laws, his treatment of top-heaviness does not attain the same level as Archimedes' analysis of floating bodies.

Christiaan Huygens

As regards the stability of floating bodies, it was Christiaan Huygens who took up Archimedes's torch. As a young man of 21, he wrote *De iis quae liquido supernatant Libri 3*. Although he never published his text, his biographer C.D. Andriesse (1994) calls it 'his master's thesis'.

When no more than 17, Huygens postulated a new physical principle for the 'minimum height of a centre of gravity '. In deriving it he was apparently inspired by the chain of spheres from Stevin's *clootcrans*. He discovered new criteria of stability, for forms other than those Archimedes had studied. Because Huygens never published *De iis...*, the results of his research had to be rediscovered later on, as indeed they were by Daniel Bernoulli (around 1738), Leonhard Euler and others.

The problem of stability of bodies floating in a liquid would only be solved in a more systematic way after Pierre Bouguer introduced the concept of metacentre in 1746. (The metacentre of a floating body such as a ship, for instance, is the point of intersection between a vertical line through the centre of buoyancy and a vertical line through the new centre of buoyancy when the body is tilted.) A floating body is 'in stable equilibrium' when its centre of gravity is below the metacentre.

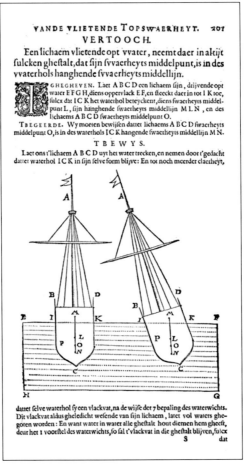

Figure 23: Stevin's diagram accompanying his analysis of the stability of ships. From *Wisconstige Gedachtenissen, Derde Deel des Byvough der Weeghconst, vande vlietende Topswaerheyt* (1605). Bruges, Public Library, B 1919 (ektachrome: JTD).

4.6. Situating Stevin's work on hydrostatics

It is important to emphasise that Stevin reformulated, rediscovered and further developed hydrostatics. He did not begin from the same axioms as Archimedes, but introduced two new axioms as starting points:

1. The impossibility of perpetual motion.
2. The new concept of imaginatively 'stiffening' a portion of the water volume. This 'stiffening' does not affect the equilibrium.

Neither principle is mentioned in the list of postulates in *De Beghinselen des Waterwichts* but they are, nonetheless, the basic axioms of Stevin's treatment of hydrostatics.

Ernst Mach (1838–1916) made systematic and explicit use of Stevin's 'stiffening' when dealing with hydrostatics in his renowned standard work on mechanics. He lauded Stevin's important methodological inventions: 'Look at Stevin's creations – the *clootcrans*, for instance. Here too a deep and broad insight is displayed'.

In the introduction to this chapter we referred to what was known about hydrostatics before Stevin began his studies. Dijksterhuis summarizes the situation that prevailed around 1580: 'hydrostatics ... not only had never risen above the level to which he [Archimedes] had raised it, but had even been unable to maintain its ancient degree of development'. The results reached by Archimedes, father of hydrostatics, had been forgotten. Basic hydrostatics before Stevin's day was limited to Archimedes's Principle. As Ernst Mach was aware, in *De Beghinselen des Waterwichts* Stevin gives a more satisfactory derivation of Archimedes's Principle than that of the great Syracusan himself.

Who, in that long period between Archimedes and Stevin, had explored the realm of hydrostatics?

4.6.1. The development of hydrostatics after Archimedes and before Stevin

Undoubtedly there existed a good deal of empirical knowledge about the properties of water, including flowing water, among the 'technicians' who throughout the centuries were occupied with building ships or canals. But a theoretical underpinning was apparently more difficult to achieve. Hydrostatics came along late in the development of Greek science, only to be lost again until the Renaissance. Leonardo da Vinci (1452–1519) wrote on the equilibrium of two liquids of different density in connected vessels. He was also active as a hydraulic engineer. His voluminous notes show that in this regard, as in many others, his considerations were mainly qualitative; he made no use of mathematical techniques to buttress his insights. As regards physics, he should be seen primarily as an inspirational transitional figure and as one of the first to attach importance to 'scientific observation', more so, even, than the Greeks.

The Venetian mathematician Giovanni Battista Benedetti (1530–1590) was a predecessor of Galileo. In hydrostatics he was one of the first Renaissance scientists to have a clear understanding of the upward force to which an object submerged in water is subject, as described by Archimedes. Benedetti took up a critical attitude to some of Aristotle's ideas, completely rejecting the principle of virtual velocity reaching back to Aristotle and his school, as used by Jordanus in the thirteenth century. Stevin also commented rather negatively on this principle, which did not prevent him from making use of it occasionally.

A virtual movement is an *imagined* movement (sufficiently small) of a physical system or part of one. The principle of virtual movement uses small imaginary movements, even though the system does not really move, even if it cannot move, to allow the application of a law – such as conservation of energy – and draw new conclusions from it. Historically, the concept of virtual velocity (or virtual movement per time unit) originated first.

The other tradition in hydrostatics, evolved by Galileo and Torricelli, began later, after Stevin's publication of *De Beghinselen des Waterwichts* (1586). The hydrostatics developed by Galileo (1564–1642) contained a rudimentary form of the principle of virtual velocity as a basic element. The Galileo–Torricelli tradition sought to borrow the principles of statics and apply them in hydrostatics. When Galileo studied dynamics and analysed water flowing down an incline, for example, he endeavoured to adopt the laws that operate when a ball rolls down an incline. At a certain point he perceived the importance of the fact that, whereas a lead ball may remain motionless on a horizontal surface, a liquid body cannot; it flows apart. This proved difficult to incorporate into his description. The main reason for the inadequacy of the Galileo–Torricelli tradition was its lack of the crucial notion of 'pressure'. The concept of 'pressure' in a liquid emerged quite naturally in the Archimedes–Stevin–Pascal tradition. The later works on fluid dynamics by Daniel Bernoulli and Leonard Euler, as well as all modern contributions to hydrostatics, are based on mathematical equations in which 'pressure' is a key concept. In this sense, they continue the Archimedes–Stevin tradition.

4.6.2. Stevin anticipates Pascal in some inventions

Various authors, from Ernst Mach to Dijksterhuis, have shown that Stevin's discoveries in hydrostatics have often and incorrectly been attributed to Pascal. Yet the situation is clear: not only the hydrostatic paradox and a number of consequences arising from it but also the principle of the hydraulic press were already correctly described in *De Beghinselen des Waterwichts*, which was published in 1586. Pascal was born only in 1623. Of course, this is not to detract in any way from the significance of Pascal's work. In the development of hydrostatics it was he who, after Archimedes and Stevin, took the next important step. He introduced the key concept of 'pressure at a point' in water, and recognized that pressure is not dependent on direction. He also formulated the law that bears his name, which states that the pressure of a fluid at rest is transmitted equally in all directions. The concept of pressure, the force that acts on a unit of surface of a wall, was nonetheless already implicit in Stevin's work. Stevin defined the force that acts on a surface, which is in effect 'pressure'. Still he had to define and calculate this force in every single case.

The knowledge Stevin expounded in *De Beghinselen der Weeghconst* and *De Beghinselen des Waterwichts* served him well in several of his later works. This is the case, for instance, in *Vande Spiegheling der Ebbenvloet* (*The Theory of Ebb and Flow*), in which he designed a kinematic theory of tides. Likewise when, as an engineer and inventor, he was occupied with water mills, sluices, hydraulic designs and architecture, this knowledge proved to be of great value.

5. Summary: the significance of Stevin's work in physics

Those who consider the discovering of a hitherto unknown *natural law* to be the pinnacle of a physicist's achievement will recognize in the hydrostatic paradox the most outstanding of Stevin's contributions to physics. For some 1,800 years – since Archimedes, in fact – no advance had been made in hydrostatics. Rather the reverse, for what Archimedes achieved in hydrostatics had largely been lost in the course of the centuries. That Simon Stevin did not introduce the core concept of hydrostatic pressure in a general sense in no way detracts from the unique character of his discovery of the hydrostatic paradox, an important natural law. Indeed, since Stevin calculated the magnitude of the force exerted by a liquid on an arbitrary surface, he was in fact already working with the quantity 'pressure'. Pascal's contribution consists of the fact that he postulated the general character of the concept of 'pressure' and its uniform transmission in a liquid.

It is surprising that the literature never refers to 'Stevin's hydrostatic paradox', but simply to the 'hydrostatic paradox'. Yet, when referring to discoveries of a similar significance – to fundamental natural laws – one speaks of Huygens' Principle (in optics) or Pascal's Law (in hydrostatics). Not only does the discoverer see his name associated with a law or formula, in these instances an appropriate accolade, but such a naming also gives the researcher a certain understanding of the historical evolution of his discipline – even if an attribution can sometimes be controversial.

5.1. Stevin was the first to use the negation of perpetual motion in deriving a natural law

In Stevin's proof with the *clootcrans* or 'wreath of spheres' in *De Beghinselen der Weeghconst* the original and fascinating method he used for his derivation is at least as important as the result he obtained, which in this case was already known via a different path. The thought experiment using the *clootcrans* is so fine and surprising that its aesthetic value and efficiency would be equalled only by Einstein's well-known thought experiments with moving trains, lifts and clocks, carried out in connection with his Special Theory of Relativity. Stevin has indeed taken a visionary step and achieved a masterly insight. In determining the equilibrium of the spheres he closed the wreath by suggesting a series of identical spheres strung on a line in the shape of a semicircle. Here he was intuitively falling back on a deep symmetry. He then combined these concepts by postulating the impossibility of perpetual motion, after which he effectively produced the composition law of forces.

Stevin was the first in the history of science to make use of the negation of perpetual motion in deriving a natural law. However, closer analysis shows that in the derivation with the wreath of spheres the principle of conservation of energy has to be invoked rather than the negation of perpetual motion. Dijksterhuis correctly observes that this in no way detracts from Stevin's fundamentally correct intuition.

In physics the methodological is enormously important, particularly when working at the profound level Stevin reaches here. With a strict formulation of the intuitive *clootcrans* proof the key ingredients appear to be:

1. The 'transposition' of the problem by the addition of imaginary spheres, which thus do not in reality play any part in the problem. In Stevin's thought experiment, this is a far from obvious transformation of the problem; it is somewhat deeper than the devising of an auxiliary construction in an exercise of plane geometry. Stevin's 'closing' of the wreath is sooner reminiscent of sophisticated methods used in modern theoretical elementary particle physics, where in certain fundamental problems fictitious or 'ghost' particles are added to the theory, in order to avoid certain unphysical approximations in the course of the calculation. At the end of the calculation the ghost particles disappear, just like the imaginary spheres of Stevin's wreath.
2. The use of a symmetry principle, in which the equilibrium of the four spheres O, N, M and L with the four spheres G, H, I and K is postulated. Strictly speaking, only the intuition that is at the basis of this symmetry consideration survives. In twentieth-century theoretical physics the great force of symmetry principles has been established, including the existence of some elementary particles, which were predicted with the help of symmetry principles and only later discovered experimentally. With his use of a symmetry argument – albeit an intuitive one – Stevin was again way ahead of his time.
3. In a rigorous formulation of the *clootcrans* proof it is the principle of conservation of energy that is involved, not, strictly speaking, the negation of perpetual motion. Stevin's considerations actually pertain to the limits of infinitesimal friction and he was one of the first – along with Huygens and Torricelli, and three centuries before the further study of these concepts – to intuit these conservation principles. Stevin's methods and thought experiments were not of the kind that could immediately be adopted by contemporaries, followers and succeeding generations (as was the case with the infinitesimal calculus developed by the Bernoullis and

Euler into a universal instrument). So profound was his intuition that it would be the nineteenth and even the twentieth century before some of his methods could be properly comprehended and developed.

Quite apart from the fact that no fewer than three original principles are combined here, the fact that the proof, which is the result of this combination, presupposes no previous knowledge on the part of the reader is also remarkable. This is unique in physics. The demonstration with the *clootcrans* is not only unsurpassed because of the parallelogram of forces; in the whole of physics it is a gem of aesthetics, clarity and cogency.

A leitmotif in Simon Stevin's oeuvre, and one that is at the very heart of the methodology of science, is his explicit coupling of theory and practice. He writes of *spiegheling* and *daet*. In most cases, *daet* meant 'practice'. But in *De Beghinselen des Waterwichts*, it rather means an empirical testing of theory. The ultimate test of the *spiegheling* is the *daet*. This decisive step in physics, which represented a break with the methodology of the Greeks, is usually attributed exclusively to Galileo Galilei. But Simon Stevin preceded the man from Pisa.

Chapter Seven

The link between Italian and French algebra

Simon Stevin's merits as a mathematician can best be assessed by examining his role in the history of algebra. He forms a link between the Italian algebraists – who at the start of the sixteenth century summarized the knowledge of the day, chiefly derived from Arab mathematicians – and the great French innovators of the second half of the sixteenth century and later. Repeatedly it appears that the latter must have known Stevin's work and his personal contributions to algebra.

1. Introduction

In assessing Stevin's contribution to the development of mathematics we can turn to his only work in Latin, *Problematum Geometricorum*, his only work in French, *L'Arithmetique* (in which are included translations of *Tafelen van Interest* and *De Thiende*), and to some parts of the *Wisconstige Gedachtenissen*. With the exception of *Tafelen van Interest* and *De Thiende*, which are discussed elsewhere in the present volume, the majority of Stevin's works have rather more the character of textbooks than of scientific publications. His work on geometry appears in *Problematum Geometricorum* and *Vande Meetdaet*, part of the *Wisconstige Gedachtenissen*, and it will not be specifically dealt with here.

Stevin's contributions to algebra were reworked in *Les Œuvres Mathematiques de Simon Stevin de Bruges*, published and provided with a commentary by Albert Girard Samielois in 1634 and printed by Bonaventure and Abraham Elzevier at Leiden. This book was instrumental in propagating Stevin's ideas in France.

The principal part of *L'Arithmetique* consists of the treatment of material that was already well known and often dealt with by Italian mathematicians such as Cardan and Tartaglia (see box 1). Nevertheless, Stevin can lay claim to originality in three respects:

- He arranges the material in a very personal way.
- His own findings are incorporated into the work; some of these will be examined below.
- He refutes certain deeply ingrained errors.

1.1. Stevin and the sixteenth-century notion of numbers

One of these deeply ingrained errors was the sixteenth-century notion of numbers. Still in thrall to the ideas of the Greeks, sixteenth-century mathematicians recognized only natural numbers larger than 1 as 'numbers'. They described every extension of the concept of number as absurd, irrational and inexplicable. Stevin challenged this view, arguing that:

- 1 is also number (*Que l'unité est nombre*);
- a number is not a discontinuous quantity (*Que nombre n'est poinct quantité discontinue*);
- there is nothing absurd or irrational about an expression such as $\sqrt{8}$; it is simply a number whose square is 8, just as 4 is a number whose square is 16. (*Que nombres quelconques peuvent estre mombres* (sic) *quarrez, cubiques, etc. Aussi que racine quelconque est nombre; Qu'il ny' a aucuns nombres absurdes, irrationels, irreguliers, inexplicables ou sourds.*)

All the French quotations of *L'Arithmetique* given above were of exceptional significance to Stevin; he had them printed in large capitals.

In the following section are a number of examples from *L'Arithmetique*. They were chosen because they are still very practical in modern mathematics. They illustrate Stevin's way of representing polynomials, of solving quadratic, cubic and quartic equations, and his ability to determine roots of equations of the form $f(x)50$ in a numerical way, a technique that is still used today in numerical algebra in the world of computers.

> ### Box 1: Quadratic, cubic and quartic equations
>
> It is often said that the Babylonians (c. 400 BCE) were the first to solve quadratic or second-degree equations. In reality, they developed an algorithmic approach or working method to solving problems that in modern terminology would give rise to a quadratic equation. Essentially their method consisted of completing the square (see box 2). Around 300 BCE, Euclid developed a geometrical approach to finding one of the roots of a quadratic equation. Euclid had no notion of 'equation', but he solved it in essence nonetheless. Hindu mathematicians took the Babylonian methods still further. Brahmagupta (598–665) produced an almost modern method that admits negative as well as positive solutions. He used abbreviations for the 'unknown', usually the first letter of a colour, and sometimes several different unknowns occurred in a single problem. The Arabs were unaware of the advances the Hindus had made, so they did not admit negative solutions and had no abbreviations or symbols for their unknowns. However, around 800 Al-Khwarizmi set out a classification of different types of quadratic equations.

He dealt with algebra in his treatise *Hisab al-jahr w'al-muqabala* – the word 'algebra' comes from *al-jahr* – though he gave only numerical examples of each type of quadratic equation.

Abraham bar Hiyya Ha-Nasi (Barcelona 1070 – Provence 1136) was a Hispano-Jewish mathematician and astronomer, also known by his Latin name, Savasorda. He published his *Hibbur ha-Meshihab ve-ha-Tishboret* ('*Treatise on Measurement and Calculation*'), which was translated into Latin in 1145 by Plato of Tivoli as *Liber embadorum*. This was the first book published in Europe to give the complete solution of the quadratic equation.

Around 1500 a new age in mathematics dawned in Italy

In Italy around 1500, mathematics entered a new age. The year 1494 saw the publication of the first edition of the *Summa de arithmetica, geometrica, proportioni et proportionalita*, now simply referred to as *Summa*, written by Luca Paciolo (Sansepolcro 1445–1517). Paciolo studied mathematics under several teachers in Venice, Rome and other places. From 1477 he travelled extensively and spent time at the universities of Perugia, Zadar, Naples and Rome, where he taught mathematics. In 1494 in Venice he published his *Summa*, a summary of mathematics known at that time, though it contains little in the way of original ideas. Paciolo did not deal with cubic equations but he began a discussion about quartic or fourth degree equations. He maintained that certain quartic equations could be solved by quadratic methods, while others, he believed, could not be solved with the mathematical knowledge available at that time. He was right.

Scipione dal Ferro (Bologna 1465–1526) held the Chair of Arithmetic and Geometry at Bologna University and he would certainly have encountered Paciolo, who taught at Bologna in 1501–1502. It is generally accepted that dal Ferro had found an algebraic solution for cubic equations. As far as we can judge now, though, he was only able to solve cubic equations of the form $x^3 + mx = n$. It is curious that dal Ferro must already have solved this equation in 1515 or thereabouts, but he kept his work a secret until just before his death in 1526, when he passed it on to his student Antonio Fior.

Fior was a mathematician of average attainments and he was certainly less adept than dal Ferro at keeping secrets, for it was soon rumoured in Bologna that the cubic equation could be solved. Prompted by these rumours, Niccolo van Brescia, better known as Tartaglia (see fig. 1), managed to solve equations of the form $x^3 + mx^2 = n$ and he made no secret of his success. On 17 January 1535, Fior challenged Tartaglia to a public contest in solving cubic equations. The following rules were agreed:

- each man gave the other 30 problems to solve;
- each had 40–50 days in which to do so;

NICOLAVS TARTAGLIA GEOMETRA
Diuitias patriæ cumulat Tartaglia linguæ,
Euclidem Etrusco dum docet ore loqui.
Hic certam tractare dedit tormenta per artem,
Et tonitru, & damnis æmula fulmineis. F.s

Figure 1: Niccolo Fontana Tartaglia was born in Brescia in 1499 and died in Venice in 1557. He studied cubic equations and discovered a solution for the roots of every known type. The portrait is by the sixteenth-century engraver Philip Galle, who compiled a collection of portraits of famous scientists entitled *Virorum doctorum* (Antwerp 1572).
© Royal Library of Belgium (Brussels). Rare Books Department, VH 9097 A.

- the winner would be the contestant who had correctly solved the most problems;
- there would be a modest prize for each problem solved.

By 13 February 1535, Tartaglia had found the general method for solving all types of cubic equations and with this he solved all the problems set by Fior in less than two hours. All the problems were of the form $x^3 + mx = n$, however. News of Tartaglia's victory reached Girolamo Cardano, or Cardan (Pavia 1501 – Rome 1576), professor of mathematics at Milan, Pavia and

Bologna, who at that moment was preparing his *Practica Arithmeticae* (1539) for publication. Cardan invited Tartaglia to Milan and prevailed upon him to reveal his solution. Tartaglia agreed to divulge his formula, provided Cardan swore an oath to keep it secret until after Tartaglia had published it himself. Cardan agreed, but did not keep his promise. In 1545, he published the method in *Ars Magna*, the first Latin work on algebra.

When he applied Tartaglia's formula to certain cubic equations, however, Cardan came up with some strange results. For example, when he considered $x^3 + 15x^2 + 4$ he obtained an expression in which $\sqrt{-121}$ appeared. Cardan knew that one could not take the square root of a negative number, but he also knew that $x = 4$ was the solution to his equation. On 4 August 1539 he wrote to Tartaglia asking him to clarify the matter, which Tartaglia was unable to do. In *Ars Magna*, Cardan ultimately produced a solution using 'complex numbers' to tackle a similar problem; apparently he did not entirely understand his own technique, calling it 'as subtle as it is useless'.

After Tartaglia had shown Cardan how to solve cubic equations, Cardan encouraged his own student Lodovico Ferrari, to examine quartic equations. Ferrari managed to solve these equations with what is probably the most elegant method ever proposed for this kind of equation. Cardan published all 20 cases of existing quartic equations in his *Ars Magna*.

Square roots of negative numbers also occur in Tartaglia's method for solving the cubic equation. This problem was studied by Rafael Bombelli (Bologna 1526 – Rome *c.* 1572). His *Algebra* (1572), which gives a detailed account of the algebra of his day, introduces the important concept of 'complex numbers'. He was the first to write down rules for the addition and subtraction of complex numbers. He then showed how correct, real solutions could be obtained from the Cardan–Tartaglia formula for solving a cubic equation, even when the square roots of negative numbers were involved.

Special forms of notation were a rarity among these Italian algebraists. Paciolo used a very limited mathematical notation; Cardan described everything in words and used no symbols at all.

The Frenchman Viète introduced the concept of algebraic variables

The Frenchman François Viète (Fontenay-le-Comte 1540 – Paris 1603) introduced the concept of algebraic variables in his book *In artem analyticam isagoge*. He wrote it down using letters – capitalized vowels (A, E, I, O, U) for the unknowns. He also invented the concept of the parameter (a known constant quantity) and represented this with capitalized consonants. The equation we now write as $5bx^2 - 2cx + x^3 = d$ would be expressed in Viète's notation as:

> *B5 in A quad - C plano 2 in A + A cub aequatur D solido.*

Here we can recognize *A quad* for x^2, *A* for x and *A cub* for x^3. We have to realize that Viète still solved his problem geometrically and regarded x^3 as a cube and x as a line segment. It made no sense to him to add a three-dimensional object to a one-dimensional object and so he made his equation dimensionally 'correct'. In scientific parlance, he based his thinking on the principle of homogeneity, so that every term shares a homogeneity of dimension:

- In the first term the unknown is of a quadratic form and B is linear, which together gives dimension 3.
- In the second term the unknown is linear and therefore the word *plano* (plane, which is quadratic) is added to C so as to obtain a dimension 3.
- The third term has the correct dimension.
- D is followed by the word *solido* (from solid, fixed, which indicates a cubic quantity), whereby this term too acquires dimension 3.

Viète's notation was thus still far removed from the modern algebraic notation of equations.

Rafael Bombelli was the first to change this. Here are a couple of examples of Bombelli's notation:

For $5x$ we find $\overset{1}{\smile}$ 5

For $5x^2$ we read $\overset{2}{\smile}$ 5

Here, for the first time, a notation for unknowns is used that is not based on geometrical interpretations. It is at this point in the chronology that Stevin's *L'Arithmetique* must be situated.

Box 2: Completing a square with an example given by Al-Khwarizmi (*c.* 800)

This example is meant for the mathematical clever clogs. We start from Al-Khwarizmi's equation $x^2 + 10x = 39$. He uses both purely algebraic and

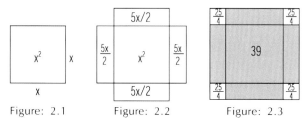

Figure: 2.1   Figure: 2.2   Figure: 2.3

Figure 2.1–2.3: The geometrical methods of finding roots for quadratic equations are based on completing a square. This method was first described by Al-Khwarizmi. Diagrams (1)–(3) show the three stages involved in working out the solution.

geometrical methods to find the solution. In algebraic terms he interprets the solution thus:

> To find a square and 10 roots that add up to 39. The question in this type of equation is formulated as follows: what is the square which, when combined with ten of its roots, will give a total of 39? The manner of solving this type of equation is as follows: take one half of the roots just mentioned. In the present problem there are 10 roots, and the half of this is 5, which multiplied by itself gives 25. This we add to 39, which gives 64. The square root of this last number is 8. From this we subtract half the roots, i.e. 5, which leaves 3. The number 3 therefore represents one root of this square, which itself is 9. Nine therefore gives the square.

At first sight this method of solution seems fairly abstract. The geometrical solution, which effectively completes a square, makes the algebraic solution much clearer. Historians of science tend to think that Al-Khwarizmi was familiar with Euclid's Elements.

Al-Khwarizmi starts by drawing a square with side x, which is in fact a geometrical representation of x^2 (see fig. 2.1). To the square we must add $10x$. This is done by adding to the original square four rectangles each with a width of $\frac{10}{4}$ and a length x to the square (see fig. 2.2). This last figure now has an area equal to $x^2 + 10x$, which is equal to 39. The square is now completed by adding four little squares, each with an area of $\frac{5}{2} \times \frac{5}{2} \times \frac{25}{4}$ (see fig. 2.3). The new square has an area of $4 \times \frac{25}{4} + 39 = 25 + 39 = 64$. The side of this large square is therefore 8. But the side has a length of $\frac{5}{2} + x + \frac{5}{2}$, so $x + 5 = 8$, or $x = 3$.

2. Geometrical numbers

Stevin used the term 'geometrical numbers' to indicate that a series of numbers consists of the successive powers and the successive roots of a certain base

188 'Magic is No Magic'. The Wonderful World of Simon Stevin

number. We can follow his line of reasoning in the construction of the geometrical numbers as given in the second part of the first book of *L'Arithmetique* under the title *des definitions des nombres geometriques*.

> Like the Ancients we can see that progressions of numbers such as 2, 4, 8, 16, 32, etc. or 3, 9, 27, 81, 243, etc. have remarkable properties. The first multiplied by itself gives the second, the second multiplied by the first gives the third, the third with the first the fourth, and so on. For 2 multiplied by itself gives 4, 4 by 2 gives 8, 8 by 2 gives 16, and so on. Likewise 3 multiplied by itself gives 9, 9 by 3 gives 27, 27 by 3 gives 81, and so on. The Ancients saw that it was necessary to give names to the numbers, by which they can be clearly distinguished; the first in the row is called Prime and denoted by ①, the second is called Second and is denoted by ②, and so on:
>
> ①2. ②4. ③8. ④16. ⑤32. ⑥64, &c.
>
> ①3. ②9. ③27. ④81. ⑤243. ⑥729, &c.
>
> In view of the fact that this first number is like the side of a square, and the second like its square, and the third like the cube of the first, etc, and that in this similitude of numbers and magnitudes many secrets of the numbers are revealed, the Ancients also gave these numbers the names of magnitudes, calling the first Side, the second Square, the third Cube, etc., and consequently all these numbers in general Geometrical Numbers (see figs. 3.1 and 3.2).

2.1. Geometrical numbers and geometrical figures

After the introduction, Stevin shows how well these numbers correspond to the size of certain geometrical figures. The original diagrams taken from *Le Premier Livre d'Arithmetique* illustrate his reasoning (see figs. 3.1 and 3.2). In the diagrams, the line segment A represents $a > 1$, Stevin's ①, which here is equal to 2; square B represents a^2; and cube C represents a^3. To represent $a^4 = a^3 a$, Stevin stacks cubes C a times on top of each other and gets the tower D, a rectangular block with a height that is twice the side of the base, in other words: D:C = B:A. To make $a^5 = a^3 a^2$ clear, he stacks up the cubes C a^2 times (note that here a^2 represents an abstract arithmetical number). The result is again a tower so that the ratios E:D = C:B are retained, and so on. In the second row the case $a = 1$ is rendered in the sequence N, O, P, Q, etc. and for $a < 1$ (in particular $a = ½$) we get the sequence Y, Z, AA, BB, etc. (note that in this case the volume of the three-dimensional figures becomes smaller).

In the same diagrams, Stevin also considers the fractional powers of the prime. He introduces the following notation:

$$W4 = \sqrt{\sqrt{4}} = \sqrt{2}$$

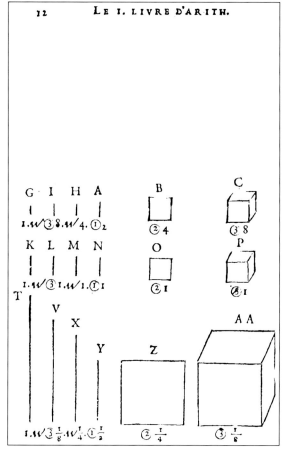

Figure: 3.1

and

$$W\;③8 = \sqrt[3]{\sqrt[3]{8}} = \sqrt[3]{2}$$

To represent $\sqrt{2}$, Stevin takes a line of unit length G and then constructs a line H as the mean proportional between G and A: G:H = H:A with G = 1 and A = 2 in the diagram. By analogous reasoning he gets G:I = I:J = J:A and I = $\sqrt[3]{2}$. Analogous structures occur when $a \leq 1$, where K and T are unity segments.

In reasoning thus, and representing powers and roots of a number a as geometrical diagrams, Stevin was taking the first step towards linking geometry and algebra. For finding roots he introduced a unity element (see line segments

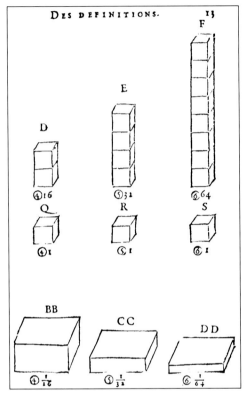

Figure: 3.2

Figures 3.1–3.2: The concept of 'geometrical numbers' played an important role in the development of analytical geometry. Simon Stevin introduced significant ways of representing them in *L'Arithmetique*.
Utrecht, University Library, Rariora duod. 17.

G, K and T in the diagram). However, if he had also used a unity element *e* for the whole powers he could have written the following ratios:

$$e : ① = ① : ②$$

$$① : ② = ② : ③$$

Had he done so, he could have introduced the correspondence between numbers and line segments that would have made analytical geometry possible. It would be a good half-century before Descartes made this discovery. In 1673 he published *La Géometrie* as an appendix to his famous *Discours de la Methode*. It is not too fanciful to suggest that Descartes knew Stevin's work through Girard's publication and also thanks to his contacts with Isaac Beeckman and Frans Van Schooten the Elder. It was, after all, through Beeckman that he was aware of Stevin's *Singconst* (see Chapter Eleven). Elaborating on his findings, Descartes would later devise what we now know as the Cartesian coordinate system.

3. Algebraic numbers and the representation of polynomials

In the second book of *L'Arithmetique*, entitled *L'Opération*, Stevin deals with the theory of algebraic numbers. The notation he introduces here represents an important phase in the development of algebraic symbolism.

Stevin defined an algebraic number as a polynomial in an indefinite quantity. He represents this, just as he does in *De Thiende*, by the symbol ① (though it should be noted that in *De Thiende* his symbols have a different meaning). He calls this indefinite quantity 'first quantity' (*prime quantité*). The higher powers of the indefinite quantity he calls 'second', 'third', etc. and represents them by ②, ③,... With the aid of these symbols, Stevin is able to express polynomials in a simple way. Our modern representation of the polynomial $3x^3 + 7x^2 + 2x + 3$ in Stevin's notation would be,

$$3③ + 7② + 2① + 3$$

Stevin's notation of polynomials was a vast improvement on the Cossic symbolism in general use in the sixteenth century and on the principle of homogeneity still used by many scientists (see box 1). In Cossic notation the *cosa, tanto* or *coss* (Stevin's 'first quantity'; our 'unknown' or 'variable' x) and all its powers were identified by their own symbols and called by their own names. Stevin's new exponential notation did away with the need to know all these names and symbols by heart.

3.1. Stevin's notation

Stevin was not the first to suggest such a simplification. The Italian mathematician Rafael Bombelli had preceded him in proposing an exponential notation, although in a slightly different form, in his famous work *L'Algebra*, published in Bologna in 1572 (see box 1), as indeed Stevin acknowledges in *L'Arithmetique*. Symbolic algebra was greatly improved by this innovation. By way of illustration, here is an example of the use of Cossic characters – which were usually added by hand and so differed slightly from author to author.

In 1565 Valentin Mennher set out a table showing the connection of an exponent value of x and the Cossic symbol:

0	1	2	3	4	5	6	7	8	9
N	φ	ξ	τ	ξξ	β	ξτ	bβ	ξξξ	ττ

As just mentioned, each of these symbols had a name. Starting with φ and reading from left to right they were *cosa, zensus, cubus, zensus de zensu, surdesolidus, zensicubu, bisurdesolidus*, and so on.

Take, for example, the product of two polynomials:

$$(3x-1)(5x+2) = 15x^2 + x - 2$$

In Cossic notation this becomes

$$(3\varphi - 1N)(5\varphi + 2N) = 15\xi + \varphi - 2N$$

and in Stevin's notation

$$(3①-1)(5①+2) = 15②+①-2$$

The problem with Cossic notation, quite apart from its unwieldiness, was that there was no clear connection between φφ and ξ. In Stevin's notation and in our modern terms, that connection is there:

$$① \; ① = ② \text{ or } x \, x = x^2,$$

which means nothing other than the rule that the product of powers with a common denominator is equal to the base to a power that consists of the sum of the exponents.

To complete his system of notation, Stevin introduced the symbol ⓪ for the number 1, calling it *Commencement de quantité*. He did not use it in writing a given polynomial, however. It occurs only when the encircled exponents are used to indicate the general form of a polynomial. For instance, ③②①⓪ means the same as $a_0 x^3 + a_1 x^2 + a_2 x + a_3$ in which the *a*s are indefinite coefficients.

Considered superficially, Stevin's notation can be overestimated. At first glance there seems to be little difference between ③ and x^3. How great and essential the difference really is can be seen when two more indefinite quantities occur in a problem – in other words, when a variable x is joined by a variable y. In such a case, Stevin uses the same symbols (① ② ③...) for what he called 'postposed quantities' (*postposees quantitez*) – the indefinite quantities occurring in addition to the original 'positive quantities' (*quantitez positives*) – but adds the Latin words *sec* (second), *ter* (third), etc. in front of them. Thus if 3② + 2① + 6 signifies $3x^2 + 2x + 6$, then $5y^2 + 3y + 7$ occurring in the same problem would read 5 *sec* ② + 3 *sec*① + 7.

Here too Stevin is a forerunner of the French algebraists. His representation of polynomials in different variables is far from elegant, however. François Viète was the first to use letters to represent both known and unknown quantities in his book *In artem analyticam isagoge* (see box 1). He used vowels for the unknowns and consonants for the knowns. The convention that uses letters taken from the beginning of the alphabet to represent known quantities, and letters from the end, such as x and y, for unknown quantities, was introduced by Descartes in *La Géometrie*. This convention is still in use today, though we take it so much for granted that we rarely give it a second thought. If asked for the solution to $ax = b$, no one responds by wondering which unknown must be determined.

4. Equations

4.1. Introduction

Whereas everyone can appreciate that Stevin's exponential notation for representing polynomials was an improvement on the unwieldy Cossic system, it is harder for the modern reader to see the solution of an equation as the determination of a fourth proportional, and thus as an application of the familiar rule of three. Stevin sought to solve an equation in this way because the rule of three was used to figure out the sort of problem that arose every day in the mercantile world with which, as we have seen elsewhere in this book, he was so familiar. Indeed, so important was the rule of three to the merchants and money-masters of the time that they called it 'the golden rule'. Practically every trading problem was formulated in this way:

'If 3 apples cost 60 francs, how much do 5 apples cost?'
What we are looking for is thus one number, out of four which are proportional two by two; for our example this means:

$$\frac{60}{3} = \frac{x}{5} \text{ or } \frac{3}{60} = \frac{5}{x}$$

For equations Stevin expressed the problem in the following form: 'If ② = 2① + 3, how much is 3① + 5?' He uses this approach when solving higher degree equations, but not when formulating them. The problem that we would nowadays put as 'find x so that $x^2 = bx + c$', is formulated by Stevin, in *L'Arithmetique, de l'operation, Probleme LXVIII*, as follows:

> *Estant donnez trois termes, desquels le premier ②, le second ① ⓪, le troisiesme nombre algebraique quelconque: Trouver leur quatriesme terme proportionel.*
>
> Given three terms, the first of which is ②, the second ① ⓪, and the third an arbitrary algebraic number: find their fourth proportional.

However, when solving a specific equation such as

$$② = 2① + 24,$$

the first three terms of the proportion prove to be

$$②, 2① + 24, ①.$$

Stevin always placed the highest occurring power of the unknown in the first member of the equation.

4.1.1. Stevin and coefficients

For Stevin, as for all his predecessors, coefficients were always known numbers; symbolism was reserved exclusively for the unknown and its powers. Viète and

Descartes expanded this idea (see box 1). Stevin was unable to find a general method for solving an equation of a given degree. He divided equations according to the highest occurring power of the unknown x and classified them according to their 'degree'. Before Stevin this classification was done in a different way, dependant on the number of terms in the equation. For instance, the relation $ax = x^2$ is a so-called simple equation, while $x^3 + bx^2 = cx$ is a complex equation. This classification began to be dropped from the late sixteenth and early seventeenth century: we find the first divisions according to degree in Stevin (1585), Viète (1590), Girard (1629) and Descartes (1637). Here too, Stevin had clearly influenced the French mathematicians.

In a quadratic equation, expressed in Stevin's notation as ② ① ⓪, three different cases had to be distinguished. There is the principal form

$$② = ① + ⓪;$$

the other two cases are created by replacing a plus sign with a minus:

$$② = - ① + ⓪, \quad ② = ① - ⓪.$$

The forms, constructed with the two elements ② ⓪ or ② ① are not considered separately. The first is the general type of equations with *quantitez derivatives des primitives*, which is to say that they are converted into a linear equation by a substitution $x^2 = y$ The second becomes linear if divided by ①; however, the fact that by doing so a root 0 is obscured goes unnoticed. In Stevin's day a 0 was rarely regarded as a root.

There are three principal forms of a cubic equation, but each must be subdivided according to the symbols of the terms in the second member, so that the total number of cases is 13. Obviously, in quartic equations the number of principal forms and subdivisions is still greater.

Below, Stevin's ideas on how to solve equations are explained and a typical example of a quadratic equation is worked out. Stevin always followed the same series of four steps to solve an equation, whatever the degree: *construction, démonstration arithmétique, démonstration géométrique* and *origine de la construction*:

a. Construction: carrying out the various operations for the calculation of the root.
b. Arithmetical demonstration: verification of the root by substitution.
c. Geometrical demonstration: geometrical proof that the root, considered as a line segment, satisfies the equation.
d. Origin of the construction: derivation of the method of calculation followed in (a).

In *L'Arithmetique* Stevin makes specific reference to his sources, as we expect from him by now. In doing so, however, he was rather the exception than the

The link between Italian and French algebra   195

rule. Copyright and royalties were as yet undreamed of, and knowledge of scientific literature was the prerogative of a privileged few. It is therefore instructive to see how Stevin rated his predecessors, whom he dubbed *des inventeurs de ces regles de trois des quantitez*.

Stevin attributes the first application of the rule of three in solving higher degree equations to a number of unknown authors and to 'Mahomet', son of Moses the Arab, who we know today as Al-Khwarizmi. For quartic equations he mentions Louys de Ferrare, more familiar to us now as Lodovico Ferrari of Bologna. He observes that Diophantos (*c*.250 CE) would have known the methods of Al-Khwarizmi (*c*.800 CE) – his chronology being a little shaky here. On the other hand, he is very well up on the Italian algebraists, naming Luca Paccioli (Luca Paciolo), Scipio Ferreus de Boloigne (Scipione dal Ferro), Nicolas Tartalia Bessian (Niccolo of Brescia, better known as Tartaglia), Antonio Maria Florido Venetien (Antonio Fior) and *disciple dudict* Scipio, Cardane (Cardan). He also refers to the book, published in Italian, of Raphael Bombelle (Rafael Bombelli). The correct enumeration and the references to the rivalry between the various mathematicians (see box 1) leads one to suspect that Stevin had been in personal contact with at least some of them. The fact that he mentions Bombelli's work in the Italian language and refers to Bombelli himself as *un grand arithmeticien de nostre temps* may indicate that Stevin spent time at a university in Italy during the 'missing years' between leaving Bruges and arriving in Leiden. Thus far, however, investigation in that direction has been without result.

4.2. The quadratic equation

For the quadratic equation

$$② = 2① + 24,$$

the four steps cited above are followed (Stevin deals with such an example in *L'Arithmetique, Livre II, Probl. LXVIII*, where his example is expressed as ② = 4① + 12; however, in order to reproduce a clearer diagram we, as Dijksterhuis (1943), have opted to consider ② = 2① + 24 (see fig. 4):

a. The half of 2 is 1; its square is 1; the given number ⓪, which is 24, which, when added to 1, makes the sum 25; its square root is 5; the first-mentioned 1, added to it, makes the sum 6. I say that 6 is the required fourth proportional.
b. Substituting 6 for ① in the proportion

$$\frac{②}{(2①+24)} = \frac{①}{①}, \text{gives } \frac{36}{36} = \frac{6}{6}.$$

c. In fig. 4, if AB = ①, then square ABCD = ②; if AE = 2, then AEFD = 2①, therefore EBFC = ② − 2① = 24.

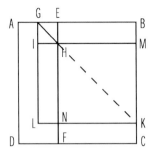

Figure 4: One of the most important objects of study in the algebra of Stevin's day was finding the roots of higher degree equations. They were determined using geometrical principles. This diagram illustrates the solution to the quadratic equation $x^2 - 2x - 24 = 0$.

Following the construction (a), we make AG = GE = 1 and complete the square GEHI = 1. GH is produced to meet BC in K; completing the square GBKL, we have EBCF + GEHI = GBKL therefore square GBKL = 24 + 1 = 25 and BG = 5. Apparently AB = AG + BG = 1 + 5 = 6.

In general, for the solution of the equation we have $x^2 = px + q$, AG = GE = $p/2$, AEFD = px, EBCF = q and therefore

$$BG = \sqrt{\left(\frac{p}{2}\right)^2 + q} \Rightarrow AB = \frac{p}{2} + \sqrt{\left(\frac{p}{2}\right)^2 + q},$$

which exactly describes one of the solutions now familiar to us. The second solution follows by putting a minus sign in front of the square root. Stevin's geometrical demonstration does not follow this solution.

d. In order to explain how the solution has been found (see also box 2), Stevin writes the equation in the form

$$x^2 - 2x = 24$$

and completes the member to a square of a linear function.
It follows that $(x-1)^2 = 25$, from which $x - 1 = 5$, hence $x = 6$.

4.2.1. Algebra and geometry

We have looked at the complete derivation in order to illustrate the line of reasoning followed by the mathematicians of the sixteenth century. One may well wonder why it should be necessary to carry out the steps in (a) or to make the geometrical demonstration in (c) when one knows the method described in (d). But we have to remember that in Stevin's day algebra had only just begun to struggle free from the close connections with geometry fostered by mathematicians since Euclid. As we have seen, to Stevin ① still represented a line segment; he could not accept the demonstration of the exactness of the solution until it had been verified geometrically that ① = 6 satisfies the geometrical relation ② = 2 ① + 24.

We have already noted that Stevin found only one of the roots. That did not hamper him, however. In his day, an equation was deemed to be solved as soon as a single root of it had been determined. That it might be satisfied by more roots than the one found was seen rather as a peculiarity than as an essential contribution to the solution. Stevin also makes no mention of the possible existence of negative roots, which of course may occur in the general solution. At that time, negative roots were not recognized as solutions, because they did not tally with the idea that geometrically a root represented a line segment.

In 1629, several years after his reissue of Stevin's work, Girard published *L'invention nouvelle en l'algèbre*, in which he articulated his thoughts on the fundamental theorem of algebra, namely that an equation of degree n has n roots (*Toutes les équations d'algèbre reçoivent autant de solutions, que la denomination de la plus haute quantité le demonstre*). Thus he paved the way for finding all the roots of nth degree equations, though he did not state that the solutions involved complex numbers. In 1673 Descartes expressed this problem as follows: 'for every equation of degree n one can imagine n roots, but these imagined roots do not correspond to any real quantity'. General rules for solution had to wait for Viète. He was the first to devise analytical rather than geometrical solutions.

4.3. Cubic equations

To solve this kind of equation Stevin followed the Tartaglia–Cardan theory. In his own terminology he sets out the problem for the solution of the basic equation $x^3 + mx = n$ as follows:

> *Estant donnez trois termes, desquels le premier ③, le second ① ⓪, le troisiesme nombre algebraique quelconque: Trouver leur quatriesme terme proportionel.*
>
> *Given three terms, the first of which is ③, the second ① ⓪, and the third an arbitrary algebraic number: find their fourth proportional.*

He also followed the description of Tartaglia's method in his four-step technique, which we demonstrated for the quadratic equation. He also noted in a separate section entitled *De L'imperfection qu'il y a en ceste premiere difference* (*L'Arithmetique, II*) that in the solving of cubic equations, roots of negative numbers must sometimes be extracted, mentioning that to this end he had studied other authors' methods of solution, among them those of Cardan and Bombelli. But he decided against including these in his work, feeling that they were not really leading anywhere. Stevin preferred a certain rule to the workings of chance, stoutly opining that there are 'enough legitimate matters to work on without any need to get busy and to lose time on uncertainties', though he allowed that 'those who like such examples can do with them what they please.' Stevin was following the dictates of his common sense, as he did in his numerous other activities, but it has to be said that in the long run the seeming irrationality of Bombelli's formula has proved of greater use to mathematics.

4.4. Quartic equations

It can be seen at once that Stevin's handling of the quartic or fourth degree equation is much less complicated than his treatment of quadratic and cubic equations. The reason is not hard to find: the quantity ④ could not be interpreted geometrically.

To some historians of science it is a matter for rejoicing that the Greeks did not create a four-dimensional geometry. Had they had a four-dimensional method of representing quantities, then in the case of quartic equations, too, the algebraic method would not have been the only one available. In the event, Stevin had no option but to use an algebraic procedure. It seems strange that having done so, he did not promptly jettison the centuries-old geometrical tradition for the second and third degrees. For the quartic equation he relied on the work of Ferrari as reproduced in the publications of Cardan.

4.5. A numerical method

Stevin did not invent the algebraic algorithms discussed so far, but he brought a certain clarity and order to them in his work. And he added some new discoveries relative to the solving of equations. One of these is dealt with in his *Appendice Algebraique*. Here Stevin describes a numerical method for obtaining an approximation of a root of an equation of any degree. In modern terminology we would say that the solution is found by means of the bisection method: if the numbers $f(a)$ and $f(b)$ have a different sign, the equation $f(x) = 0$ has at least one root between a and b. As Stevin did not reduce his equations to 0, his formulation is different. In the example

$$x^3 = 300x + 33915024,$$

with which he demonstrates his method, he substitutes for x the successive powers of 10 and finds that the first member takes for $x = 100$ a smaller value, and for $x = 1000$ a greater value than the second. Thus there is a root between 100 and 1000. Next he tries the series of values 100, 200, 300 and so on, and succeeds in enclosing the root between 300 and 400. Going on in this way he finds the value to be 324. Stevin then changes the example to

$$x^3 = 300x + 33900000$$

and finds that there is a root between 323 and 324. In the same way as before he continues with the series of numbers 3230/10, 3231/10, and so on. Here Stevin is making use of ideas that appear in *De Thiende*, albeit he is still working with fractions (a ratio of a numerator and a denominator) and not with decimal numbers to find the tenths. Then, still in the same way, he goes on to find the hundredths and is able to approximate every root as closely as desired. In modern terms one could say that his numerical solution in the limit converges to the exact solution.

4.5.1. Stevin and the concept of function

In the second example the equation has only one root. Stevin says nothing about how the method should be applied if there is more than one root. The insight that underlies this method of approximation is very important. The two members of the equation are no longer conceived as unknown constant numbers but as functions of a variable; the solution is seen as the answer to the question, for which value of the variable do the two functions in the two members of the equation assume the same value? In Cardan we find a similar reasoning. It should be noted that the word 'function' was not part of Stevin's vocabulary, though he certainly worked with the concept. In the same period, Galileo employed the idea of 'function' to describe complex observed curves that were connected with physical processes. Much later, in 1734, Leonhard Euler (1707–1783) was the first to describe a function as an expression formed by allowing any operation (such as addition, subtraction, multiplication and raising to a higher power) to operate on variable quantities and constants.

What exactly has Stevin contributed to mathematics? We have seen his choice of the decimal system, his introduction of calculating methods for decimal fractions or numbers and his proposals for applying the decimal system to measurements, money and so on. We have also pointed out his striking treatment of the 'limit' concept and the introduction of a numerical method for finding an approximation of a real root of a function.

5. The Rule of Algebra

The final chapter of book two of *L'Arithmetique* is devoted to *la reigle des faux des nombres algebraiques, dicte reigle de ALGEBRE*. This is Stevin's envoi to the description and solving of (polynomial) equations. He uses the word 'algebra' or rule in its narrow sense, as one of the rules by which arithmetic or mathematics is served. Today the term algebra describes a wide domain encompassed – together with analysis, geometry, group theory, logic and so on – by mathematical science.

5.1. ...the Inexhaustible fountain of infinite Arithmetical Theorems...

The rule of algebra brought out the poet in many a sixteenth-century mathematician. Stevin too could not resist giving lyrical expression to his deep admiration for it:

> *Now we have come to the last problem of this book, which is on the highly singular and admirable Rule of Algebra, the Inexhaustible fountain of infinite Arithmetical Theorems, Revealer of the mysteries hidden in numbers.*

The rest of the chapter consists of 27 problems, applications of the rule. In fact, these are nothing more than what we would now call paraphrased equations with one or two variables. The 'art of the rule' consists of formulating equations on the basis of the requirement expressed in words. Here, for example, is *Question II*:

> *Partons 5 en deux parties telles, que leur produict soit 6.*

This kind of problem can be solved by means of an adapted quadratic equation. Stevin being Stevin, he was not content to let the reader off with the 27 similar problems he illustrated in the chapter. He had one last gigantic application in reserve – his French translation of the first four books of the *Arithmetica* of Diophantos. It was not Stevin's aim to give an exact rendering of the original work; he looked on Diophantos's propositions as so many problems to which the rule of algebra could be applied. So this part of his work should not be regarded as a literal translation but rather as a paraphrase in the notation introduced in *L'Arithmetique*. It should be noted that Stevin did not work from Diophantos's original Greek text, but from a Latin translation by the Heidelberg professor of Greek, Wilhelm Holtzmann, known as Xylander, published in 1575. We now know that Xylander worked from a Greek copy that was riddled with mistakes.

Chapter Eight

Stevin's contribution to the Dutch language

Ghemerct de groote rijcheyt onses taels – *'Considering the immense richness of our language'. Stevin was a virtuoso architect of language. He coined new words and incorporated original translations of Latin words into Modern Dutch, and his influence on the language can still be felt today. He sought to make his work accessible to a wide readership, not only to those few with an adequate command of Latin. This concern for the language was in keeping with the increasing use of the vernacular throughout Europe during the Renaissance.*

1. Stevin and the building of the Dutch language

Simon Stevin had a considerable influence on the Dutch language and he is generally acknowledged as one of the founders of scientific and technical Dutch.

He did not, of course, start in a vacuum. Renaissance and humanist influences had inspired a great flowering of regional awareness in the sixteenth-century Low Countries, while Spanish oppression had awoken a collective consciousness. Increasing numbers of literary and scientific works were composed in the vernacular, and the need for a standardized written language became imperative – for up to that date everyone had written in his/her own dialect. The need for standardization was reinforced by the Reformation and by the desire to make the words of the Bible available not only to the prelate but to the ploughboy. In the sixteenth and seventeenth centuries many lexicons and primers appeared, describing spelling and vocabulary and setting out the rules of grammar. Among those who promoted and expanded the standardization of the Dutch language in the mid-sixteenth century were the printers Jan Gymnich of Antwerp and Joos Lambrechts of Ghent.

There was also a need for an expanded vocabulary. Now that Latin was being replaced by the vernacular in science, new words had to be found for all kinds of scientific concepts. Stevin played a significant part in this. He was one of the most important architects of Modern Dutch (see fig. 1) in the late sixteenth and early seventeenth centuries. In the sciences he was preceded by the botanist Rembert Dodoens, who published his famous herbal, the *Cruydeboeck*, in 1554.

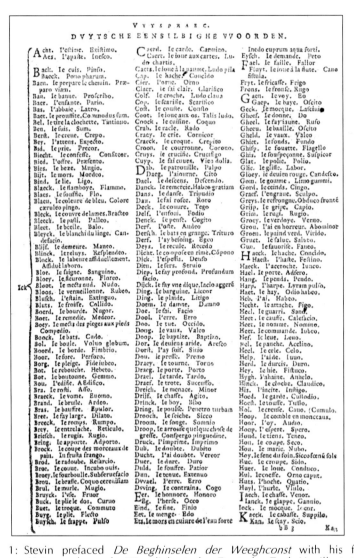

Figure 1: Stevin prefaced *De Beghinselen der Weeghconst* with his famous *Uytspraeck vande Weerdicheyt der Duytsche Tael*. He illustrates the conciseness of Dutch, which he regards as an intrinsic value, with a long list of monosyllables. From: *De Beghinselen der Weeghconst*. Bruges, Public Library, B 265 (ektachrome: JTD).

2. Early printed editions of works on medicine and the natural sciences

Middle Dutch and sixteenth-century scientific texts on the natural sciences mostly deal with natural sciences in the very broadest sense of the word. The

Middle Dutch *Der Naturen Bloeme*, an encyclopaedia of nature composed in rhyme by Jacob van Maerlant (1235–1291?), deals with the life and behaviour of men, quadrupeds, birds, 'wonders of the sea', fish, snakes, reptiles, insects, plants, herbs, water sources, minerals, precious stones and metals. *Der Naturen Bloeme* is based on the *Opus de Natura Rerum* by Thomas of Cantimpré (1201–1272), a Dominican active in Leuven. Van Maerlant's work contains no original science, but it is one of the earliest disquisitions on the natural sciences to be written in the vernacular.

Another important Middle Dutch manuscript that includes scientific and medical texts dates from the mid-fourteenth century. It was bought in London in 1818 by the Belgian politician and bibliophile Charles Van Hulthem (1764–1832), permanent secretary to the Royal Academy during the reign of William I, and it is now known as the Van Hulthem Manuscript. It contains a huge number of Middle-Dutch literary works by diverse authors. All kinds of subjects are dealt with, such as *Natuurkunde van het Geheelal*, a variety of medical topics, recipes, the medicine of Avicenna and Jan Yperman's *Boec der Medicina* and *Cyrurgie*.

Folk medicine, astronomy and arithmetic were the scientific subjects most commonly dealt with in Middle Dutch literature: folk medicine because no one escaped disease and death, arithmetic because it was essential in trade and commerce, and astronomy chiefly in the context of astrology and superstition.

Most of the works mentioned are compilations, often containing translations of classical texts or of French and Italian works. These played a role not only in the emergence of scientific research in the Netherlands, but also in the development of a uniform language. They are thus indispensable sources for the study of the history of the Dutch language.

3. Stevin's presence in the *Woordenboek der Nederlandse Taal*

The *Woordenboek der Nederlandse Taal*, or *WNT*, is the authoritative Dutch dictionary. Clicking on 'Stevin' in the CD-ROM version results in no fewer than 488 references. Some of these are attributed to Hendrick Stevin, but the very great majority refer to Simon Stevin. This is not to say that in every instance Stevin is accredited with being the earliest author in whose works these words can be found, though this is undoubtedly true in a number of cases. Some are almost certainly neologisms, words coined and introduced by the Bruges scholar himself. To define a word as a neologism is a delicate matter, however. There is always the chance that someone will discover a still earlier source in which the word appears. To find out whether a word we encounter in Stevin's work is really a neologism freshly minted by him, or a semantic neologism, or a word that was already in use in restricted scholarly circles and whose use he sought to increase, we have to compare Stevin's books with earlier books and manuscripts.

The *WNT* regards Stevin's works as an authoritative source. Stevin's prominent presence in the dictionary is attributable to a number of causes:

1. In a number of subjects Stevin is one of the earliest, if not the earliest, authors in the Dutch language. That is certainly true of *De Beghinselen der Weeghconst* and *De Beghinselen des Waterwichts*. Although, as regards arithmetic, for instance, simple textbooks already existed before Stevin, this was not the case for statics and hydrostatics.
2. In the fields of navigation, astronomy, bookkeeping, mathematics (arithmetic apart), technology, architecture and the arts of war, on all of which there were certainly publications in Dutch at the end of the sixteenth century, Stevin's works were also influential. His great prestige ensured that his linguistic contributions to these works also received much attention. Eminent men of letters such as Constantijn Huygens even referred to Stevin in their poems, which goes to show that in literary circles too he was recognized as an authority in several disciplines.
3. Stevin devoted much care and attention to the choice of the appropriate word for every concept. In some cases he would use existing words, giving them an expanded or more precise meaning; in others he deliberately coined new words, sometimes by translating a Latin or Greek concept and sometimes by forming a new compound word or derivation.

4. Neologisms and semantic neologisms

4.1. A comparison of Stevin's work with the dictionaries compiled by Plantin and Kiliaan

To date, Stevin's biographers have made no definitive study of the exact extent of his contribution to the Dutch language (other than in the field of arithmetic). Where he introduced neologisms and where he did not has, for instance, not as yet been determined. Dijksterhuis (1943) divided Stevin's words into several categories. He took the appearance or absence of a word in the famous dictionaries compiled and published by Christopher Plantin and Cornelis Kiel (known as Kiliaan) as one of the criteria for neologisms. These lexicons also provided a guide to the meaning of a word. Plantin's dictionary, the *Thesaurus Theutonicae Linguae* (in Dutch the *Schat der Neder-duytscher spraken*), dates from 1573. In 1574 Cornelis Kiel, one of Plantin's corps of scholarly correctors, brought out the *Dictionarium Teutonico-Latinum*, with a second edition in 1588. Dijksterhuis used the third, elaborated edition of 1599, which was entitled the *Etymologicum Teutonicae Linguae, sive Dictionarium Teutonico-Latinum*. He notes that Kiliaan did not use Stevin as a source.

These are the categories Dijksterhuis employed:

I. Words where Stevin made an annotation in the margin and which are mentioned by Plantin and/or Kiliaan:
   a. in the same sense Stevin used them
   b. in a sense different to Stevin's
II. Words where Stevin made an annotation in the margin and which do not appear in either Plantin's or Kiliaan's dictionary.
III. Words where Stevin made no annotation in the margin but which were nevertheless probably coined or introduced by him. This category was subdivided into:
   c. words that are found in Plantin and/or Kiliaan's dictionary (in senses other than Stevin's)
   d. words that are not found in either of these dictionaries

These criteria are not completely foolproof. The case of arithmetic, on which we do have indepth studies, makes this clear. Even so, the criteria provide a sound orientation as regards Stevin's role in the introduction of particular words. There are at least 250 words in categories I(b), II and III.

4.2. Arithmetic books in Dutch from 1445

The conclusions concerning Stevin's linguistic contributions drawn by Dijksterhuis in 1943 on the basis of a comparative study using the lexicons produced by Plantin and Kiliaan are due for some refinement. This has been shown recently by Marjolein Kool (1992) in the case of arithmetic. Arithmetic was indispensable in the daily practice of trade and commerce. Merchants in trading centres like Antwerp wanted their children to be able to calculate. It is therefore not surprising to find arithmetic already being taught in schools in the fifteenth century. The arithmetic that was gradually gaining ground was new and experimental, for it increasingly made use of Arabic numerals, which were much handier to calculate with than the cumbersome Roman numerals. Teachers of arithmetic had to invent Dutch arithmetical terms. The earliest Dutch textbook to describe the use of Arabic numerals dates from 1445. This manuscript, the *Allgorithmus is een aert in den welken siin ghevisiteert IX figuren...*, is preserved in the University Library of Basle.

4.2.1. Marjolein Kool's study
Marjolein Kool has made an indepth study of the arithmetical terms in Stevin's works and of whether or not they can be found in earlier sources. She starts from 22 'arithmetical sources' or textbooks that had appeared before 1585. In addition to the just-mentioned manuscript *Allgorithmus*, these include *Die maniere om te leeren cyffren na die rechte consten Algorismi. Int gheheele ende int ghebroken,*

printed in 1508 and now in the Royal Library in Brussels. Among the teachers of arithmetic she cites are Christianus Van Varenbraken, whose arithmetic book appeared in 1532; Gielis van den Hoecke, who produced several books including *Een sonderlinghe boeck in dye edel conste Arithmetica, met veel schoone perfecte regule* in 1537; and Adriaen van der Gucht, who published his *Cijferbouc* in Bruges in 1569.

Kool concludes that of the 61 arithmetical terms Stevin uses, 31 had already been used in the same sense in one or another of these 22 textbooks. One such term is *noemer*. According to Dijksterhuis this word belongs to category II. However, *noemer* had already occurred in *Maniere* in 1508. Kool cites examples of other already-existing words (words that appeared in the arithmetic books referred to, at least) that Stevin drew on, including *rekenen* (1510), *somme* (1463), *telder* (1580), *vergaderen* (1463), *viercant* (1537), *wortel* (1532), *worteltrecken* (1537), *mael* (1558), *aftrecken* (1445), *cijffer* (1510), *deelder* (1537), *deelen* (1445), *deelinghe* (1569), *ghebroken* (1508), *ghetal* (1445) and *helft* (1445).

Neologisms that Stevin very probably introduced in arithmetic and which Kool does not find in earlier works are *uytbreng* (the product of a multiplication), *teldaet* (practical arithmetic), *parich* (even), *onparich* (uneven), *soomenichmael* (quotient), *teerlincxwortel* (cubic root), *thiendetal* (a number expressed according to the decimal system), *ghebreeckende* (zero) and *viercanten wortel* (square root).

Semantic neologisms introduced by Stevin in arithmetic include *afcomst* (arithmetical calculation), *begheerde* (wanted), *beghin* (number 'left of the decimal point'), *beghinsel* (1. arithmetical calculation; 2. number 'left of the decimal point'), *mael* (quotient), *sijde* (root), *teerlinck* (cube), *telconst* (theory of arithmetic), *thiende* (the decimal place value system), *werckstuk* (problem) and *werf* (quotient).

In arithmetic the number of new words Stevin created (both neologisms and semantic neologisms) is clearly smaller than his pre-1992 biographers had indicated. It should also be mentioned that some of 'his' words are no longer used, and that a number of his neologisms in arithmetic are very close to words that already existed around 1580. For instance, *viercanten wortel* is a Stevin neologism, but the term *viercantighe wortel* was already in use in 1580. Similarly *parich*, which is a neologism, though *paer* already occurs in Kiliaan (1574).

Stevin was the first to formulate a number of concepts in arithmetic, which led to semantic neologisms such as *thiende*, *beghin* and *beghinsel*. These words appeared in some earlier arithmetic books, albeit with other meanings. It has rightly been pointed out that Stevin built up a very well-thought-out arithmetical vocabulary, in which there was virtually no ambiguity, and which contained almost no synonyms. With the publication of *De Thiende*, his work on the decimal system that was of immense historical importance, the influence of the vocabulary he employed was great. Not only did the new words and meanings he introduced become common currency, but existing arithmetical terms restricted to the specialized professional circle of arithmetic teachers also came into more general use.

4.3. What can be learned from the *Woordenboek der Nederlandse Taal*?

The *WNT* cites Stevin as the earliest source of a number of words. In addition to Plantin and Kiliaan, the *WNT* uses a broad base of sixteenth-century texts as sources. If the *WNT* cites an author as the earliest source, then this is certainly significant.

Arithmetic apart, in several of the areas in which Stevin was active, there were few, if any, texts in Dutch. This is particularly the case for statics and hydrostatics, disciplines in which nothing essential had been contributed or published in the Low Countries prior to Stevin. The Dutch terms he introduced here (with the Latin term in the margin) must be considered as neologisms or at least semantic neologisms, having been 'filtered' in Plantin and Kiliaan. Stevin undoubtedly chose words from the professional world, to which he assigned new meanings.

In physics and astronomy there was no Dutch equivalent of the arithmetic textbooks. Words relating to navigation, technology, the military sciences, bookkeeping, music, law, building and so on would as a rule be included in dictionaries such as Plantin's and Kiliaan's, because they were more accessible to the general public than the more abstract concepts of physics. The 'filtering' using the Plantin and Kiliaan lexicons is thus a reasonable method for all the disciplines in which Stevin worked, except arithmetic.

The sources used by the *WNT* go back to 1500 or earlier. The use of the Plantin and Kiliaan lexicons, supplemented by the *WNT*, is useful to characterize the status of a word used by Stevin as a neologism or semantic neologism and, in the case of words that are not derived from highly specialized fields, may serve as a guideline. As the case – albeit exceptional – of arithmetic illustrates, however, this strategy is not entirely decisive.

4.4. Stevin's ideas about Dutch

The motivations for Stevin's strong feelings about Dutch must obviously be judged in their historical context. But quite apart from that, Stevin ascribed a number of qualities to the Dutch language that, he felt, raised it above all other languages.

Firstly, he emphasized the conciseness of Dutch. He illustrated this with hundreds of examples of monosyllabic words, first-person present tenses of verbs, and so on. Stevin contended that there were far more monosyllabic words in Dutch than in Latin, French or Greek, for example. He regarded the ability to express a thought or concept in a word of one syllable as an intrinsic excellence in a language.

A second quality that Stevin esteemed very highly was the facility with which compound words can be formed in Dutch. 'The object of languages is, among other things, to expound the tenor of our thought (...) also in such a way that in every respect they properly admit of composition'. With great inventiveness, he illustrated how excellently words can be combined in Dutch compared to Greek or Latin, giving several examples, such as *water* (water) and *put* (well). These two words (see fig. 2) can be combined to give either *waterput* (water-well) or *putwater* (well-water, as distinct from rainwater or river water). In the same way,

S. STEVINS

daertoe gheuonden hebben, dat ick, met al de ghene die van d'oirſaeck niet meer en weten, ons alſvooren noch met recht mueghen verwonderen duer wat middelen dat mach gheſchiet ſijn. Merckt bouen al dit noch een beſonder, ende weerdighe eyghenſchap, by hemlien conſtelick inde T'ſaemvoughing veroirdent, ia ſulcke, dat gheen Griecx, noch Latijns * Letteraer, ſoodanighe uyt die talen perſſen en ſal, al wrong hy tot ſweetens toe: Te weten dattet laetſte der ghecoppelde altijdt * Grondt is, ende t'voorgaende * Ancleuing; Als wanneermen ſeght, ***Putwater***, ſo is water grondt, ende put ancleuing, want onſen ſin dan voornamelick tot water ſtreckt, om t'welck t'onderſcheyden van ſtroomwater, reghewater, &c. men vougter Put voor: Maer als wy dit verkeeren, ſegghende ***Waterput***, dan is den ſin (hoewel het de ſelue woorden ſijn) al een ander, want Put is dan grondt, ende Water ancleuing, ouermidts de voornamelicke meyning dan is van een Put, om welcke t'onderſcheyden van een Meſput, Calckput, &c. men ſtelter water voor. Alſoo oock is ***Glasveinſter***, een veinſter van glas, maer ***Veinſterglas***, is glas niet daermen uyt drinckt, maer plat daermen veinſters af maeckt. Wederom ***Olinuet***, is een nuet des gheſlachts daermen olie uyt perſt, maer ***Nuetolie***, is olie van nueten. Sghelijcx ***Iachthondt***, is een hondt daermen mede iaecht, maer ***Hondiacht***, een iacht niet met voghelen, dan met honden, &c. Wat den ghenen belangt die noch van meyning mueghen ſijn, het Griecx in deſen gheualle voor het Duytſch te gaen, wy achten dat ſulcx gheſchiet duer dat hem het Duytſch onbekent is, of datter verſtandt der oirdeling ghebreeckt, of dat hy hartneckich ſy, of eenich der ghelijcke beletſel heb, niet weerdich een woort daer af meer te roeren.

Grammaticus. *Subiectum.* *Adiunctum.*

TEN derden moeten wy ſegghen vande bequaemheyt deſes taels tot de leering der Conſten, waer af wy (bouen dien ſulcx nootſaecklick volght uyt het voorgaende toeghelaten) de volghende Weeghconſt, ſulck ſy is, tot voorbeelt ſtellen; welcke ghy, ghemerckt de groote rijcheyt onſes taels, uyt welcke alles veel beter behoort ghedaen te ſijne, daertoe miſſchien niet weerdich en ſult achten, te meer datter duyſenden by ons ſijn, dieſe veel beter, ende met beuallicker woorden beſchrijuen ſouden: Maer niet teghenſtaende al dit, ſo iſt doch ſoo ghedaen, dat gheen van al d'ander Gheſlachten der volcken wie hy ſy, t'ſelue, ſoo veel des ſpraecx grondelicke beteeckening, ende uytbeelding der Saeck belangt (ick en wil niet ſegghen, ſoude connen verbeteren, daer gheen vreeſe voor te hebben en is, want hemlien ghebreeckt Stof * welcke gheweert ſoo wort oock gheweert de daet) ſoude connen ſoo nauolghen. want waer wildy ſpraken halen daermen duer ſegghen ſal, Eueſtaltwichtich, Rechthefwicht, Scheefdaellini, en dierghelijcke daer de Weeghconſt vol af is? ſy en ſijnder niet, de Natuer heeft daer toe aldereyghentlicxt het Duytſch veroirdent.

Sublata materia tollitur effectus, *4. v 2. ſtrijtreden der Bewyſconſt.*

TEN

Figure 2: Stevin analyses the flexibility of *Neder-duytsch,* noting the facility with which compound words can be formed and giving examples such as *Waterput* and *Putwater, Veinsterglas* and *Glasveinster,* or *Iachthondt* and *Hondiacht.* From: *De Beghinselen der Weeghconst.* *Bruges, Public Library, B 265 (ektachrome: JTD).*

glas (glass) and *venster* (window) can make either *glasvenster* (a glass window) or *vensterglas* (glass for a window pane, rather than a glass to drink from). Likewise *jacht* (hunt) and *hond* (hound) can be combined to giver either *jachthond* (hunting hound) or *hondjacht* (a hunt with hounds, as opposed to falcons, for instance).

Thirdly, Dutch is a pre-eminently suitable language for the teaching of the arts, according to Stevin, 'for,' he argued, 'where would you find any languages in which one can say *Evestaltwichtich, Rechthefwicht, Scheefdaellini* and the like, in which the *Weeghconst* abounds? They do not exist'. (Those concepts can be circumscribed as: 'Evestaltwichtich' – 'of equal apparent weight', 'Rechthefwicht' – 'vertical lifting weight', 'Scheefdaellini' – 'oblique lowering line'.)

Fourthly, Stevin praised enthusiastically and at considerable length the power of the Dutch language to move and sway emotions, citing the persuasive oratory of preachers in the 'Dutch countries'. Should such a preacher 'get it into his head that a broom was to be the bride', he remarked, 'he will induce the congregation to come to the wedding'.

It is hard to recognize the usually so sober engineer Simon Stevin in these lyrical passages, which under the title *Uytspraeck vande Weerdicheyt der Duytsche Tael* (*Discourse on the Worth of the Dutch Language*) preface *De Beghinselen der Weeghconst*. However, it should be remembered that most of the works by Stevin and his contemporaries were introduced by a high-flown salutation or poetic exordium.

Of course, Stevin's rhetorical defence of Dutch must be set against the historical background. The Renaissance scholars of the north, who followed their Italian precursors in discovering and building on the intellectual achievements of the ancient Greeks, felt the need to distinguish themselves and call upon a long tradition, as the Italians – with their Greek and Latin roots – were obviously able to do.

The use and development of their own language also gave the Dutch their own identity. They were fond of referring to the Age of the Sages, a remote past in which the sciences related to mathematics had flourished. This was an age that considerably predated classical antiquity, an age of perfect knowledge and understanding, which had subsequently been lost. All that the ancient Greeks and the Renaissance scholars north or south of the Alps, were doing, in fact, was reconstructing that lost past. Dijksterhuis stresses that the notion of the Age of the Sages is irrational and fantastic, but it suited the ideas of the time. Thus Stevin added to the then popular explanations of the Antwerp scholar Joannes Goropius Becanus, who maintained among other things that Dutch was the oldest language in the world, that in the Age of the Sages there would have existed a perfect knowledge of mathematics and the natural sciences, and that the lost knowledge of that ideal golden age could soon be regained by introducing Dutch as the universal working language of science.

Stevin's extremely valuable contribution to the Dutch language is in no wise diminished by his motivation, which though rather bizarre to us would have seemed perfectly rational to his contemporaries. He also argued that science should be studied and practiced in Dutch, for otherwise many people who were gifted but

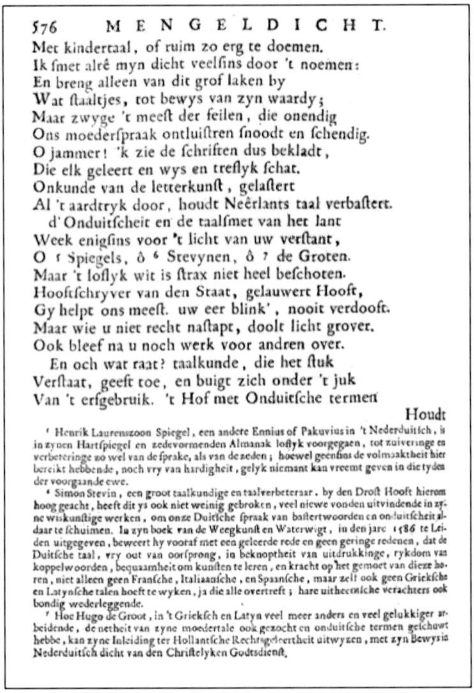
576 MENGELDICHT.

Met kindertaal, of ruim zo erg te doemen.
Ik ſmet alrê myn dicht veelſins door 't noemen:
En breng alleen van dit grof laken by
Wat ſtaaltjes, tot bewys van zyn waardy;
Maar zwyge 't meeſt der feilen, die onendig
Ons moederſpraak ontluiſtren ſnoodt en ſchendig.
O jammer! 'k zie de ſchriften dus bekladt,
Die elk geleert en wys en treflyk ſchat.
Onkunde van de letterkunſt, gelaſtert
Al 't aardtryk door, houdt Neêrlants taal verbaſtert.
d'Onduitſcheit en de taalſmet van het lant
Week enigſins voor 't licht van uw verſtant,
O [5] Spiegels, ô [6] Stevynen, ô [7] de Groten.
Maar 't loflyk wit is ſtrax niet heel beſchoten.
Hooftſchryver van den Staat, gelauwert Hooft,
Gy helpt ons meeſt. uw eer blink', nooit verdooft.
Maar wie u niet recht naſtapt, doolt licht grover.
Ook bleef na u noch werk voor andren over.
En och wat raat? taalkunde, die het ſtuk
Verſtaat, geeft toe, en buigt zich onder 't juk
Van 't erſgebruik. 't Hof met Onduitſche termen

Houdt

[5] Henrik Laurenszoon Spiegel, een andere Ennius of Pakuvius in 't Nederduitſch, is in zynen Hartſpiegel en zedevormenden Almanak loflyk voorgegaen, tot zuiveringe en verbeteringe zo wel van de ſprake, als van de zeden; hoewel geenſins de volmaaktheit hier bereikt hebbende, noch vry van hardigheit, gelyk niemant kan vreemt geven in die tyden der voorgaande ewe.

[6] Simon Stevin, een groot taalkundige en taalverbeteraar, by den Droſt Hooft hierom hoog geacht, heeft dit ys ook niet weinig gebroken, veel niewe vonden uitvindende in zyne wiskunſtige werken, om onze Duitſche ſpraak van baſtertwoorden en onduitſcheit aldaar te ſchuimen. In zyn boek van de Weegkunſt en Waterwigt, in den jare 1586 te Leiden uitgegeven, beweert hy vooraf met een geleerde rede en geen geringe redenen, dat de Duitſche taal, vry out van oorſprong, in beknoptheit van uitdrukkinge, rykdom van koppelwoorden, bequaamheit om kunſten te leren, en kracht op het gemoet van dieze horen, niet alleen geen Franſche, Italiaanſche, en Spaanſche, maar zelf ook geen Griekſche en Latynſche talen hoeft te wyken, ja die alle overtreft; hare uitheemſche verachters ook bondig wederleggende.

[7] Hoe Hugo de Groot, in 't Griekſch en Latyn veel meer anders en veel gelukkiger arbeidende, de netheit van zyne moedertale ook gezocht en onduitſche termen geſchuwt hebbe, kan zyne Inleiding ter Hollantſche Rechtsgeleertheit uitwyzen, met zyn Bewys in Nederduitſch dicht van den Chriſtelyken Godtsdienſt.

Figure 3: The Dutch poet Joannes Vollenhove (1631–1708) argued for the careful use of language in his poem dedicated to the 'Nederduitsche' writers (*Poezy*, 564–577). In his *Mengeldicht*, Stevin ('Stevyn') is mentioned and lauded alongside Joost van den Vondel, Hugo de Groot and P. C. Hooft. *The Hague, National Library of the Netherlands, 760 D 26 (ektachrome: JTD).*

did not know Latin would be prevented from improving their knowledge. This 'social' argument was evidently a significant one in Stevin's day too.

Stevin's dedication to the use of Dutch developed only gradually, and apparently under the influence of his surroundings. In his earliest works (*Tafelen van Interest* (1582) and, if the attribution to Stevin is correct, the *Nieuwe Inventie van Rekeninghe van Compaignie* (1581)), he used not only Dutch but also Latin (*Problematum Geometricorum* (1583)) and French (*L'Arithmetique* (1585)). After the publication of *De Thiende* and *Dialectike*, however, he clearly developed an increasing interest in the vernacular, and eventually came to use nothing but Dutch. In the first book of *De Beghinselen der Weeghconst* (1586) he made his views on the matter clear, and he returned to it in various later works.

No wonder Stevin was so highly esteemed by contemporary Dutch men of letters and chambers of rhetoric such as Amsterdam's *In Liefde Bloeyende*. The dedication to him in the florilegium *Den Nederduytschen Helicon* (1610) is well known. Even Bredero (1585–1618), the major Dutch poet of his generation, lauded Stevin's contribution to the Dutch language. Stevin's later influence on the Dutch language is illustrated by Figures 3 and 4.

523781.

Nederlandtſche

Woorden-Schat,

Waar in meeſt alle de

Baſterdt-Woorden,

Uyt

P. C. Hóófdt, H. de Ghróót, C. Huyghens, I. v. Vondel, en andere voortreffelijke Taalkundighe:

En

Konſt-Woorden

Uyt

A. L. Kók, S. Stevin, de Kamer in Liefd' bloeiende, en andere Duitſche Wijſghieren,

Verghadert,

Naauwkeurighlijk en met Kraft vertaalt worden.

De tweeden Druk, verbetert en veel vermeerdert.

t'AMSTERDAM,

By THOMAS FONTEYN, by de Deventer Hout-markt, inde ghekroonde Drukkery, 1654.

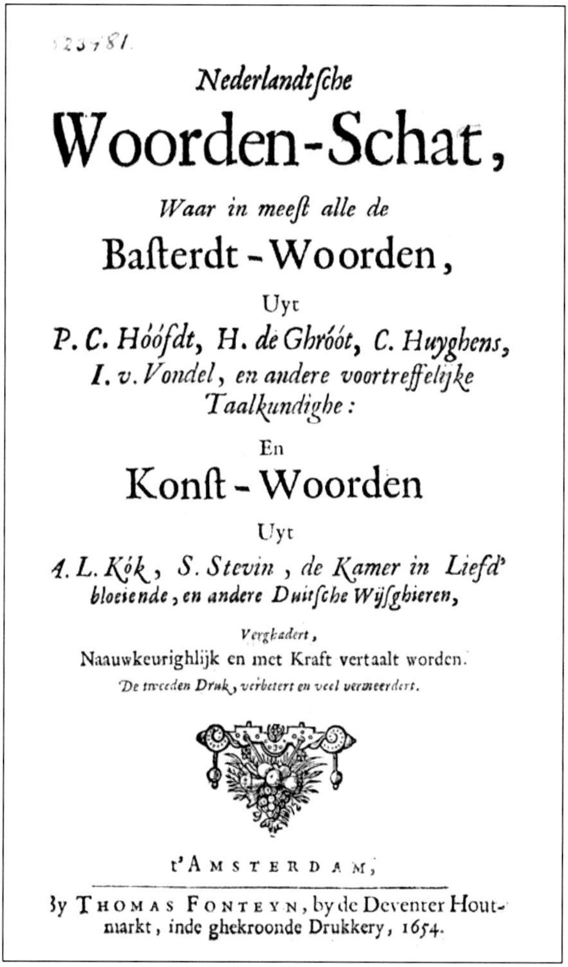

Figure 4: Stevin's influence on the Dutch language can be discerned in leading seventeenth-century linguistic works, as appears from the title page of the *Nederlandtsche Woorden-Schat,* published in 1654 by Thomas Fonteyn.
The Hague, National Library of the Netherlands, 3175 G 29 (ektachrome: JTD).

4.5. The interest in the vernacular as an international phenomenon

It was not just the scholars and men of letters of the Low Countries who sought to practice science and culture in their own language in the sixteenth century. Italy, France, Germany and England, for instance, were all moving towards a greater use of the vernacular.

One drawback to Stevin's decision to publish only in Dutch was that his ideas spread less rapidly in other countries than they might otherwise have done.

Several of his discoveries, most notably the hydrostatic paradox, were later repeated by other scholars. The translations of Stevin's books by Snellius, Hugo de Groot and others did not suffice to bring his work the international attention it deserved. (Mertens (1998) has produced a survey of the translation of Stevin's works into other languages.)

In this respect, the contrast with Christiaan Huygens, for example, is striking. Huygens lived for a long time in Paris, often published in French, and constantly corresponded and interacted with the leading scholars of his day. Huygens was indeed one of the most creative scholars in history, but Stevin's work would undoubtedly have had still greater resonance if he had written in Latin, for instance.

It should also be mentioned that Stevin was not a great writer of letters, even though they might have gained him greater international recognition. This was recently emphasized by Paul Bockstaele on the occasion of his discovery of one of the rare letters written in Stevin's own hand.

Chapter Nine

Perspective

Vande Deursichtighe

The painters of the Middle Ages gained their understanding of perspective empirically. Jan van Eyck is the pre-eminent example. Mathematically based perspective, ordered around a central vanishing point, was developed in Italy, and in 1600, Guidobaldo del Monte produced a pioneering work on the subject. In Vande Verschaeuwing (Perspective), *Simon Stevin made an original contribution to the study of perspective in the form of a didactic synthesis specially written for Prince Maurice, who had not been at all satisfied with the explanations he was given by painters.*

1. The historical context and pioneering aspect of Stevin's *Vande Verschaeuwing*

1.1. Early perspective

Perspective is not a straightforward matter. Nor has every culture felt the need to create the illusion of space and distance on a flat surface. The art of the ancient Egyptians, for instance, took no account of the effects of spatial recession. In Europe, perspective made its first appearance on Greek vases and in the frescos of Pompeii. But Byzantine art – religious and hieratic, and centred on the divine rather than the earthbound – returned to a convention of 'flat' images in which linear perspective played little or no part. Elements of perspective can be discerned in the paintings of Giotto (1266–1337) and his followers, but one feels no sense of inclusion in the depicted space.

In the first half of the fifteenth century, the principles of perspective were systematically and mathematically worked out, primarily in Italy. Architects and painters played a crucial part in this evolution. Filippo Brunelleschi (1377–1446) is often regarded as the 'inventor' of perspective. The architect Leon Battista Alberti (1404–1472) and the painter Piero della Francesca (1410/20–1492) described Brunelleschi's system in their works. The painters of the Renaissance endeavoured to follow their canons.

Uccello's (1397–1475) *Annunciation* (1425) is sometimes cited as the first painting in which perspective was used, an accolade that is also awarded to Masaccio's *Trinity* fresco (1425–28). Nonetheless, when it comes to actually admitting the beholder's eye into the pictorial space, neither painting was as advanced as the works of the Netherlandish painters of that period. Not even Piero della Francesca, whose work represented a high point of Italian Renaissance painting, achieved the level of the Low Countries school – certainly not when it came to the detailed representation of the effects of light. Flemish masters, particularly Jan van Eyck, were highly successful in creating the illusion of reality. Van Eyck worked in an essentially non-mathematical way. Today there is still debate about how he and his contemporaries dealt with perspective; witness the recent theories of David Hockney and Charles Falco, expounded by Rex Dalton, that lenses and other optical devices were being used in painters' studios as early as 1430.

The study of perspective, its historical development and its significance in the wider context of spatial concepts is a vast topic and well beyond the scope of the present volume. Our concern is with the evolution of the mathematical underpinning of perspective and how Simon Stevin contributed to that.

In 1558, Federigo Commandino was the first to give a geometric proof for the correctness of one of Alberti's perspective constructions. In 1585, Giovanni Benedetti produced a second. The more general mathematical formulation of the laws of perspective was developed by Guidobaldo del Monte in Italy and Simon Stevin in the Netherlands.

1.2. Stevin and Guidobaldo del Monte

In 1600 Guidobaldo del Monte published *Perspectivae libri sex*. This was the first work in which a mathematical – and in this case geometric – study of perspective was introduced. *Libri sex* was a pioneering work and represented a major step forward in understanding the geometry of perspective and in the development of projective geometry. It gave a method of making perspective images of objects with constructional elements in arbitrary directions, and demonstrated that in principle any problem of perspective could be solved by geometry. In 1605 Stevin published *Vande Verschaeuwing*. Though he makes no reference to Guidobaldo's work, he was probably directly or indirectly acquainted with it. His approach to perspective is purely geometric. He derived not only new theorems but also results that Guidobaldo had already obtained. For the already known propositions, Stevin generally devised more mathematically satisfactory proofs. His discourse is, as was his wont, pedagogically very attractive and pithy. Stevin's most innovatory contributions to mathematical perspective are to be found in Propositions 5 and 6 of *Vande Verschaeuwing*. They make it possible for the first time to use the 'glass', or picture plane, at an oblique angle to the floor. Here, in principle, Stevin has solved every problem of perspective. His demonstration of these propositions is remarkably elegant. The propositions continue to play a role in modern mathematics. Stevin's insights into the problem of inverse perspective were also penetrating and ahead of his time. In places the treatise is truly impressive.

2. A short description of *Vande Deursichtighe*

Stevin's pioneering contributions to the theory of perspective are to be found in part three of the *Wisconstige Gedachtenissen*, in a chapter entitled *Vande Deursichtighe. Inhoudende t'ghene daer hem in gheoeffent heeft den doorluchtichsten Hoochgheboren Vorst ende Heere Maurits Prince van Oraengien...* (*Of Optics. Containing that which was practiced by the excellent Highborn Sovereign and Lord, Maurice Prince of Orange...*)

Noted in the margin beside *deursichtighe* is *De perspectivis*. What did Stevin understand by these words? In the Summary he specifies:

> *Vande deursichtighe sullen drie boucken beschreven worden: T'eerste vande Verschaeuwing: Het tweede vande Beginselen der Spieghelschaeuwen: Het derde vande Wanschaeuvving.*
>
> [Of optics three books are to be described: the first of Perspective; the second of the Elements of Catoptrics; the third of Refraction]

To Stevin, therefore, *deursichtighe* was rather more in line with optics and *verschaeuwing* with the modern concept of perspective, which he defines as follows:

> *Perspective is the plane imitation of elevated objects, which gives the impression of elevation.*

In *Vande Deursichtighe*, as elsewhere, Stevin follows the axiomatic structure that starts from definitions and axioms and subsequently results in theorems. There are six of these.

Some of the definitions introduced follow from fig. 1:

- *We call the object that from which the perspective is made, and the perspective made, its image (4th definition).*
- *Floor* [the ground plane] *is the plane on which an object stands or lies (5th definition).*
- *Eye is a point which is assumed to perform the visual function of the eye (6th definition).*
- *Observer's line* [the prime vertical] *is a straight line from the eye to the floor, and its extremity in the floor is called foot* [centre of vision] *(7th definition).*
- *Observer's measure is a line equal to the observer's line (8th definition).*
- *Glass* [the picture plane] *is* an infinite plane between the eye and the object figure, in which this figure is assumed to show its image* [in the margin is noted * *Planum infinitum*] *(9th definition).*
- *Glass base* [the ground line] *is the intersection of the glass and the floor (10th definition).*

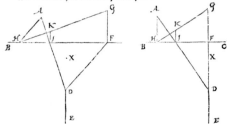

Figure 1: Basic construction determining the perspective image of a given point A. The six steps necessary to achieve this are described in the text. From: *Wisconstige Gedachtenissen, Vande Deursichtighe*. Bruges, Public Library, B 1919 (ektachrome: JTD).

- *Ray is the straight line that comes from the eye (11th definition).*
- *Meeting point* [the vanishing point] *is the point at which the produced images of different straight* parallel lines meet* [in the margin is noted ** Parallelis*] *(12th definition).*
- *And such lines thus meeting in the meeting point are called meeting lines (13th definition).*
- *A straight line drawn in the floor from the foot to the glass base is called floor line, and its point of intersection with the glass, floor-line glass intersection point (14th definition).*

When he comes to the postulates, Stevin states:

- That the natural point, its image in a natural plane glass and the natural eye are in a straight line.
- That a point, line or plane, given in the glass, also serve as their own images.

With fig. 1, which shows Stevin's diagram accompanying Proposition 5 (1st problem), we can illustrate some of the above definitions in more detail. Stevin's diagram lies in the 'floor' (which would nowadays be called the ground plane). At right angles to the floor we imagine the 'glass' (or picture plane), which cuts the floor along *BC*, the 'glass base' (or ground line). (Note that in the left half of fig. 1, *C*, which should be to the right of *F*, is missing.) *A* is the perspective point that here lies in the floor. The 'observer's measure' is the line *DE*. The 'foot' is point *D*. On the 'foot' an 'observer's line' is drawn that is equal to the 'observer's measure' *DE*. Stevin thus illustrates the 'observer's measure' (the distance from 'foot' to 'eye') with the help of the line *DE* on the plane of the paper (which corresponds to the floor). In imagination the 'observer's line', like the 'glass', is perpendicular to the floor, but is not illustrated in that way for to do so would already require a perspective drawing and that is precisely what is being sought here. The geometrical construction in six steps (*twerck* – the procedure) that Stevin now gives, in order to find the image (*schaeu*) of point *A* that is sought is direct and – especially because of its graphic nature – highly pedagogic.

It is a pleasure to carry out the six successive steps indicated by Stevin, which can be found above the diagram in fig.1, with a pencil and ruler. After the sixth operation, the point *K* is established and in the short proof is shown to be the image (*schaeu*) that was sought.

The general problem studied in *Vande Verschaeuwing* by means of these six operations is the determination of the perspective image in a given picture plane of a given object seen from an established point. Proposition 5 provides a basic result: it makes it possible to find the image of a given point that lies in the ground plane. For a point that lies above the ground plane, Proposition 6 gives the solution.

2.1. The six theorems

All the applications that Stevin worked out in the theory of perspective are based on six propositions or theorems.

- *Theorem 1. The straight line between two images of points is the image of the line that joins these two points.*

The remarkable thing is that Stevin saw the need to prove this proposition.

- *Theorem 2. If parallel lines are viewed through a glass parallel to the parallel lines, then their images are also parallel in the glass.*
- *Theorem 3. If parallel lines are viewed through a glass that is not parallel to the parallel lines, they meet in the same point of the ray that is parallel to the parallel lines. If the said lines are also parallel to the floor, their meeting point comes as high above the floor as the eye.*

This important proposition is made clear with the help of Stevin's diagram, shown in fig. 2. The glass is represented by ACK. The infinite line or carrier of line segment AB does not intersect the eye E and A lies in the glass. The image of AB is AM and it lies on the intersecting line of the glass and the plane determined by EAB. This intersecting line contains the point K in which the line EG (which is parallel to AB) intersects the glass. In other words the carrier of AM passes through K. The position of K depends only on the

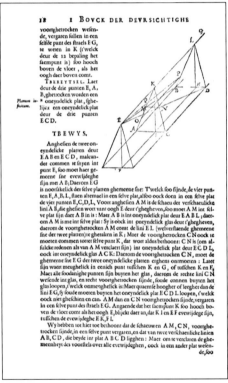

Figure 2: Analysis of the perspective image of parallel line segments. *E* is the eye. Stevin finds it obvious that *AK* is the image of the half-line that has *A* as its end point and that passes through *B*. From: *Wisconstige Gedachtenissen, Vande Deursichtighe*.
© Royal Library of Belgium (Brussels). Rare Books Department, VH 8038 C.

direction of AB, so that the image of whichever line parallel to AB will lie in a straight line that passes through K. K is thus a meeting point. If the parallel lines are horizontal, as is the case in the diagram in fig. 2, the meeting point of their images is at the same height as the eye.

It immediately follows from Theorem 3 that AK is the image of the half-line that has A as an end and that passes through B, a result that Stevin finds obvious and applies without making a separate proposition of it.

- *Theorem 4. If there are various groups of parallel lines, which are also parallel to the floor, but non-parallel to the glass, while one group of parallel lines is non-parallel to the other, their various meeting points are all equally high above the glass base.*

Implicitly, theorem 4 determines the horizon.

- *Theorem 5. If the glass revolves about the glass base as axis, and the observer's line about the foot, in such a way that it always remains parallel to a line which in the glass is at right angles to the glass base, the image of a point in the floor remains always in the same place in the glass.*

This proposition is Stevin's most important theoretical contribution to the theory of perspective. Its content is an interesting so-called invariant property. In some leading mathematical books on descriptive geometry, such as Gino Loria's 1907 work, Theorem 5 and its consequences are called 'Stevin's Proposition'. Loria writes that, with this Theorem 5, Stevin had discovered the fundamental theorem of central projection. Kirsti Andersen (1990) points out that the striking thing is not so much the demonstration but the fact that Stevin came up with the idea of formulating this theorem.

- *Theorem 6. This theorem generalizes Theorem 5 to the case that the point whose image is to be found lies not in but above the floor.*

Stevin was the first to satisfactorily solve the construction of the perspective image in an inclined picture plane.

With the help of the six theorems, which make it possible to determine the perspective image of an arbitrary point, every problem concerning the construction of a perspective image can, in principle, be solved. Stevin gives numerous examples in which he applies his method. He can reduce every construction to that of Problem 1, introduced above. To come round to perspective image *K* of point *A* via fig. 1, he successively constructs *DF*, *FG* (=*DE*), *AH* (//*DF*), *GH*, *AD* (→ intersection point I of *AD* and *GH*) and finally the image *K* (intersection of *HG* and the line perpendicular to *BC* in *I*). This basic construction is not always the simplest. Stevin gives still other examples, including the obvious cases of a square or cube (see figs. 3.1 and 3.2), for which he proposes speedier or more refined methods of construction.

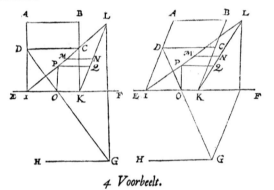

Figure: 3.1

Stevin's contribution to the perspective of curves is also important. He refers to the case that the perspective image of a circle is an ellipse. A circle in the picture plane can be the image of an ellipse in the floor. Sinisgalli calls this *teorema di Stevin: 'proiezione di una circonferenza in una ellisse'*. Sinisgalli (1978) has translated *Vande Deursichtighe* into Latin and Italian. In a searching study of the development of perspective theory from 1404 to 1605, he gives lengthy consideration to Stevin's work.

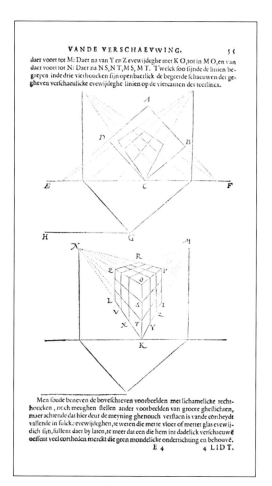

Figure: 3.2

Figures 3.1–3.2: In *Vande Deursichtighe* Stevin also suggests speedier perspective constructions; for example for a square (fig. 3.1) and the now classic perspective image of a cube (fig. 3.2). Figure 3.1 appeared in the scientific journal *Nature* (1998) in the analysis of Saenredam's 'geometric perception of space' in his paintings of church interiors. From: *Wisconstige Gedachtenissen, Vande Deursichtighe*.
Bruges, Public Library, B 1919 (ektachrome: JTD).
© Royal Library of Belgium (Brussels). Rare Books Department, VH 8038 C.

2.2. The inverse problem

Having described 'the method of perspective drawing', Stevin turns to the second subject in *Vande Verschaeuwing*, the 'finding of the eye'; in other words, to the problem of inverse perspective. Taking the example of a painting he poses this question: given a painting, where is the painter's eye? The idea is to find the optimum position, so that 'when the eye is placed there, the picture is seen in its perfection'.

It is well known in mathematics that there is no single solution to the general problem of inverse perspective, a fact of which Stevin is well aware. He knows that he must make additional suppositions about the objects depicted. For instance, on page 77 of *Vande Verschaeuwing*, he comments in a note that it might be wondered why, in 'finding the eye', he did not begin with examples in which the image is a point, line or triangle, and explains that the reason is 'that they do not admit of one certain solution, but of an infinite number of solutions'.

By gradually dealing with more general problems, Stevin would make a remarkable advance in the problem of inverse perspective. In its general applicability, it would have to await the mathematicians of the nineteenth century for a solution.

Following the explanation of how to 'find the eye' is a section entitled 'Detection of Errors', in which Stevin gives five useful rules for avoiding errors or detecting them in already existing perspective drawings.

3. Maurice, Stevin and Dürer

Finally, Stevin adds an appendix that deals with the perspective instrument (see fig. 4) he designed in consultation with Prince Maurice. He drew his inspiration from the apparatus designed by Dürer (see fig. 5). He describes the instrument thus:

> *I have read somewhere, and if I remember rightly, in Albert Durer [sic], how, wishing to explain what perspective actually is, he says that one ought to see the object through a pane of glass and imagine that what one sees in the glass is painted on it, for that is the true perfect image seen by the eye in that place.*

And further:

> *This description of the image (which induced us earlier to define a glass) appeared so suitable to his Princely Grace that he wanted not only to imagine such an image in a glass, but also to actually draw it on the glass, and to this end had a glass prepared, in the way shown in the accompanying figure, where A denotes the glass (...) pivoting about the hinge B, so that it can be set straight or inclined as desired, and fixed by means of the small screw C. The hole through which one may look is D, which can be pushed nearer to and further away from the glass and be fixed with the small screw E. The glass may also be set higher or lower and then be fixed with the small screw F.*

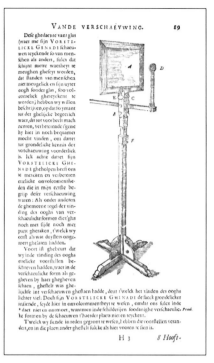

Figure 4: Inspired by Dürer's example, Stevin and Maurice designed this perspective instrument. The instrument consists of a pane of glass *A*, which can be adjusted to any angle. The artist looks through *D* at the object to be depicted, which is behind the glass. He then marks on the glass *A* the salient points of the object he sees through *D*. From: *Wisconstige Gedachtenissen, Vande Deursichtighe*.
© *Royal Library of Belgium (Brussels). Rare Books Department, VH 8038 C.*

Figure 5: Dürer – a master of experimental perspective theory – gives us a peep behind the scenes. He uses a window on which he has drawn a grid. He can then transfer the image he sees in the window to the paper, which also has a grid drawn on it.
© *Van Parys Media / AKG-images.*

It would be hard to equal the description of this perspective instrument and its operation. Yet again we see Stevin's extraordinarily lucid and concise line of reasoning. He is emphatic in his reference to the role of the prince, ascribing the idea of making the instrument to Maurice, who not only 'drew perspective images of men and other things' but who even, thanks to the instrument, was able to 'perceive and correct several imperfections' in Stevin's first version of *Vande Verschaeuwing*. Whether or not that was truly the case – a number of authors have speculated about Maurice's contribution in his pupil–teacher relationship with Stevin – the prince was certainly very keen to learn, particularly in the areas of mathematics and engineering. Stevin had already pointed out in his note to the reader at the start of *Vande Verschaeuwing* that this work had been undertaken at Maurice's request, for the prince had not been satisfied with the explanation of perspective that he had received from the painters he had questioned.

4. Stevin's followers: De Vries, 's Gravesande...

Although *Vande Verschaeuwing* can be called a masterpiece in the history of mathematical perspective theory, there are fewer references to it in the literature than might be expected. Kirsti Andersen has argued that *Vande Verschaeuwing* exercised an influence via 'followers' of Stevin, such as Frans Van Schooten the Younger (1615–1660) and Willem 's Gravesande (1668–1742). In turn, the latter inspired the mathematician Brook Taylor (1685–1731). It should not be forgotten that allusion to earlier works and authors was not yet common practice in Stevin's day. Other than in his books on algebra (see Chapter Seven), in which he did cite the work of his predecessors, Stevin does not even mention Guidobaldo del Monte, for instance, in his perspective theory.

The mathematicians and physicists of the Netherlands who followed Stevin were familiar with *Vande Verschaeuwing*. Isaac Beeckman cited it in his Journal, in July 1623, when discussing perspective in painting. The young Christiaan Huygens was tasked by his tutor in 1624 with studying perspective in the works of Hans Vredeman de Vries or Marolois or Stevin. He chose the last of these and found that after reading Stevin's treatise he had no problems with perspective theory. Frans Van Schooten the Younger, in his *Tractaet der perspective*, was influenced by both Guidobaldo del Monte and Stevin. Abraham de Graaf continued the Stevin tradition in *De geheele mathesis* (1676) and also built on Van Schooten's work. Willem 's Gravesande, an exceptional professor of physics at Leiden and one of the first to teach Newton's physics supported with proofs, also has a place in the Stevin tradition with his *Essai de perspective* (1711).

Beyond the Netherlands, reactions to Stevin's *Vande Verschaeuwing* can be found in the seventeenth century in general mathematical works such as those of Pierre Herigone (1580–1643) and Claude François Milliet Dechales (1621–1678). It is also interesting to note that Brook Taylor's *Linear Perspective*

(1715) and *New Principles on Linear Perspective* were clearly based on Willem's Gravesande and thus on Stevin. Taylor's work and consequently – indirectly – Stevin's *Vande Verschaeuwing* had a considerable influence on English perspective practice.

In all the abovementioned works the influence of Stevin's *Vande Verschaeuwing* can clearly be shown, even though the authors make no specific reference to it as their source. Andersen identifies the following chain of mathematicians who pushed forward the boundaries of mathematical perspective: Guidobaldo del Monte, Simon Stevin, *Willem 's Gravesande*, Brook Taylor and Johan Heinrich Lambert.

It is true that with its degree of difficulty and high mathematical content, Stevin's work was not really attractive to those who employed perspective, such as painters and architects. Works like those of Hans Vredeman de Vries, with their magnificent illustrations, or even Van Houten's *Verhandelingen van de grontregelen der doorzigtkunde*, were much more accessible. Naturally, these more practically orientated works had to comply with the laws of perspective as developed in the works of Guidobaldo del Monte and Simon Stevin.

5. Impact

Andersen points out that Stevin's approach to the concept of the perspective image and perspective theory was purely mathematical. To the Bruges scholar perspective was entirely a matter of geometry. All his definitions are essentially geometric and the core of perspective is, to Stevin, the image of a point. His predecessors had dealt with perspective mostly as a blend of geometry and ideas about seeing that were current at that time.

Going through *Vande Deursichtighe* carefully one is struck by its extremely lucid, structured and systematic construction. It is a masterly and most elegant pedagogic book. Stevin's understanding of the matter is profound and, in several places, he has fathomed subtle aspects of perspective theory. Here, as always, his argument is presented accessibly and clearly. Several authors have pointed to the very small number of pages (91) that sufficed for Stevin to work out the entire subject.

Stevin published *Vande Deursichtighe* as a chapter of his voluminous *Wisconstige Gedachtenissen*, not as a separate book devoted entirely to perspective. This probably explains why – despite the Latin and French translations that appeared as early as 1605 – its direct influence on architects, painters and others concerned with the theory of perspective was relatively limited.

Recently, however, there has been increasing interest in Stevin's work on perspective, even in *Nature*, one of the leading scientific journals. Authors such as Martin Kemp (1986), Alberto Pérez-Gómez and Louise Pelletier (1999), and Charles van den Heuvel (1994a,b, 2000) analyse perspective as a crucial element in art, architecture and science.

Figure 6: Saenredam's painting of the interior of St Bavo's Church in Haarlem (1648). 'The art of the geometric observer', as Martin Kemp called it, was based on Stevin's work on perspective.
Edinburgh, National Gallery of Scotland.

One of the painters who had a copy of *Vande Deursichtighe* was Pieter Saenredam (1597–1665), whose oeuvre consisted chiefly of church interiors (see fig. 6). Recently, Martin Kemp (1998) has shown how Saenredam used Stevin's work in his compositions. Even in his preparatory sketches he always drew the 'eye'.

Chapter Ten

An extraordinary didactic genius: Stevin's 'visible language'

Stevin's expositions are didactic masterpieces. Recently, academic research projects have been set up to take a closer look at the Bruges scholar's uncommonly effective pedagogic contrivances.

Stevin's 'visible language' and 'information graphics', which are characterized by wry allusions and lend themselves extremely well to computer simulations, are core concepts. They will be illustrated with examples, especially from Vanden Hemelloop (The Heavenly Motions), *the first 'textbook' to deal with both heliocentric and geocentric astronomy.*

1. Introduction

A considerable part of Stevin's oeuvre is essentially didactic. In the *Wisconstige Gedachtenissen*, for example, the majority of the lessons Stevin devised for Prince Maurice are brought together. However, it also contains original and innovatory treatises, such as the reissues of *De Beghinselen der Weeghconst* and *De Beghinselen des Waterwichts*. Even the chapters that consist mostly of a review and explanation of already existing knowledge, such as the treatise on perspective, *Vande Verschaeuwing*, contain original contributions to the subject. It is also noteworthy that Stevin's approach to existing knowledge is always a critical one, as in the case of the translation of Diophantos in *L'Arithmetique*.

Not every author has paid sufficient heed to the 'textbook' character of many of Stevin's writings. This has been encouraged by the mixture, sometimes diluted and sometimes concentrated, of original writings and familiar science in the *Wisconstige Gedachtenissen* and also in *L'Arithmetique*. Look, for instance, at *Vanden Hemelloop*, a fairly short but comprehensive text on astronomy. The French mathematician Dechales's negative criticism of this work was most unjust, for he judged it not as a textbook but in terms of a creative contribution to science. To Dijksterhuis, who cites the example of Dechales, belongs the merit of trying to distil Stevin's original contributions from his voluminous oeuvre.

It has sometimes been suggested that Stevin might have made still more original contributions to mathematics and physics had he not acted as Prince Maurice's tutor. But we should not lose sight of the fact that even in his early works, especially the mathematical ones, Stevin devoted many didactic pages and chapters to a critical assessment of the *status quaestionis*. And he presented even his most inventive discoveries within the framework of a more general volume such as *De Beghinselen der Weeghconst* and *De Beghinselen des Waterwichts*.

2. Stevin's information graphics

Stevin enhanced his exceptional pedagogic talent still further through his pioneering work in the area of graphic representations. A growing number of researchers are fascinated by Stevin's novel use of graphic images to help explain both physical phenomena and structures, and laws in general. Some have even set up projects based on Stevin as a pioneer of visual logic and syntax. In 1991 Krzysztof Lenk and Paul Kahn of Manchester University stated that in their quest for 'visible language' they returned repeatedly to the work of Simon Stevin and Johann Amos Comenius, 'two prominent European scholars and educators who were enormously effective in their use of visual material to complement their texts'. Lenk and Kahn's insights are well worth looking at.

Diagrams had already been widely used in the Middle Ages to give visual support to descriptions of physical and metaphysical aspects of the world, and by 1589 the language of diagrammatical notation was already well developed. Significant accomplishments in 'visual research' dating from the fifteenth and sixteenth centuries include:

- A more realistic representation of the visual order of the world as a result of the development of the theory of geometric perspective, to which Stevin contributed in his *Vande Verschaeuwing*.
- The creation of precise methods of visual recording, to which researchers from the Low Countries greatly contributed. Studies of human anatomy were superbly represented by Andreas Vesalius; in botany Dodoens, Clusius and Lobelius produced lifelike pictures of plants; and accurate maps were made by cartographers such as Gerardus Mercator.
- The insightful formulation of general rules of presentation at which Dürer arrived through his studies of the relation of proportions, shapes and particularities of nature. He compared the 'ideal' with what is actually found in nature. Illuminators and panel painters had already interpreted many of these insights, implicitly or explicitly, in a masterly way.

The visual pedagogy Stevin displayed in most of his works, such as *De Beghinselen der Weeghconst* and *De Beghinselen des Waterwichts* is indeed striking. Lenk and Kahn emphasize that new information can only be perceived and absorbed by the mind of the reader or student when it can be associated with some other information already existing in their experience. 'The structure or framework must be there before the new information can find its place.'

In *De Beghinselen der Weeghconst*, Stevin relies on the fact that the reader carries in his conscious mind the practical experience of weighing different objects. He also presupposes that the reader is familiar with the hand-held pair of scales, its arms, point of support and the small counterweights used in weighing. His aim is to explain the rules or laws by which this mechanism works. But he does not present his results in the form of a table of numbers, illustrated perhaps with purely geometric diagrams, which was the conventional method of that time. Stevin was aware that this kind of representation would likely be incomprehensible to the student who had not already grasped the principles being expounded.

Lenk and Kahn acknowledge that it was Stevin's unique contribution to place his visually abstract, geometric explanations in the context of objects that the reader would know from his everyday experience. The horse pulling a heavy cart up the hypotenuse of a triangle (Chapter Six, fig. 10) or the two men carrying a rectangle with a visible centre of gravity (see fig. 1) are fine examples.

Modern specialists and researchers working in the area of information graphics thus discover that Stevin draws on the student's ability to associate thoughts on several different levels at the same time. He sends information 'on several simultaneous levels', mixing realistic and abstract conventions in the process. He expects that the reader will connect the pieces of the puzzle in an appropriate way. Lenk and Kahn speak of a 'technique of overlapping conventions' producing 'information graphics' by means of which Stevin instructed the engineers and explorers of his time, and which still fascinate us today. Lenk and Kahn maintain that 'in this era of electronic media, when the manipulation of visual elements through technology has become facile, Stevin's method should be an inspiration. Many of his diagrams beg for animation and the kind of multi-window presentation made possible by today's hypertext and multimedia software applications.'

2.1. The *snaphaen* brings the concept of 'equal apparent weight' to cunning life

Another superb example of Stevin's visual language is the rascally *snaphaen* – the soldier carrying off his stolen fowl on the end of his lance – who features in *De Weeghdaet* (Chapter Six, fig. 11) and brings the concept of equal apparent weight to life. Examples from *De Beghinselen des Waterwichts* include the ship whose depth in the water Stevin seeks to discover (Chapter Six, fig. 14) and the diagrams and imaginary apparatus by which he visualizes the hydrostatic paradox (Chapter Six, figs. 15.1–15.3 and 20.1–20.2).

There may well be yet another aspect to the *snaphaen*. The visually familiar image of the Spanish soldier, used to illustrate the geometric elements of the lever, also undoubtedly embodies an ironic allusion to the loathed Spanish mercenary, here in the shape of a cock-poaching freebooter. This image is also delightful in graphic terms. Sadly, the identity of the woodcut's creator is unknown.

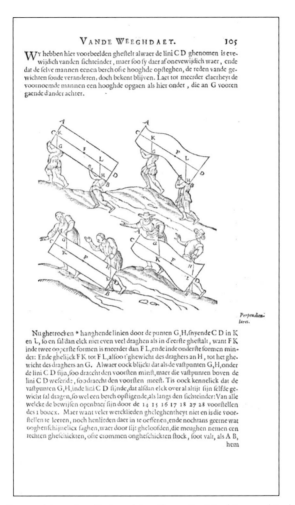

Figure 1: Stevin combines 'reality' and 'mathematics' in his graphic language. He gives abstract geometric explanations and relates them to objects and situations familiar to the reader from everyday life. From *De Weeghdaet*. Bruges, Public Library B 265 (ektachrome: JTD).

3. *Vanden Hemelloop*

Stevin's writings on astronomy are essentially didactic. They appeared under the title of *Vanden Hemelloop* (*The Heavenly Motions*) in 1608 as the third part of *Vant Weereltschrift* (*Cosmography*), which is itself the first of the five parts into which the *Wisconstige Gedachtenissen* (1605 and 1608) are divided. *Vanden Hemelloop* is a striking example of Stevin's extraordinary didactic mastery.

3.1. The exposition of Ptolemy and Copernicus

Vanden Hemelloop is a clearly structured and lucidly written text. The summary informs us that the book is divided into three main parts and of what follows. The approach is both remarkable and original. For a start it is striking that, as early as 1608, Stevin had realized that the heliocentric system as introduced by Copernicus in *De Revolutionibus Orbium Coelestium* (1543) was the 'correct' one. This is clear from his description of it, in which he explores very penetrating arguments in favour of the heliocentric theory. He announces Copernicus's assumption as the culminating third part of *Vanden Hemelloop* almost as a matter of course.

Even though Stevin's stance caused less controversy than Galileo's defence of heliocentrism would (from 1616 the Italian's support for this view of the universe led to his ill-fated encounters with the Inquisition), it was no less firmly grounded. There can be no doubt that Stevin arrived at his conviction of the correctness of the heliocentric theory quite independently of Galileo. In the Low Countries, Gemma Frisius (1508–1555), professor at Leuven and tutor of Mercator, had already been an advocate – albeit necessarily a clandestine one – of the heliocentric system, and it is very possible that his ideas and publications influenced Stevin. Note that Galileo first spoke in favour of the Copernican system only in 1611, three years after *Vanden Hemelloop* was published.

Stevin's championing of the heliocentric system may not have brought the wrath of the Inquisition down on his head, but it was nonetheless received with alarm by the religious authorities. This we know from a letter written in 1608 by Ubbo Emmius, first rector of Groningen University, in which – albeit without knowing much about the matter – he condemned Stevin's ideas and expressed great concern about his influence on Prince Maurice.

It has rightly been pointed out that Stevin was one of the first scientists to adopt and undergird the heliocentric system, and he can certainly also be regarded as the first to have written a textbook with a lucid explanation of it. From the moment it appeared, Copernicus's historic *De Revolutionibus* came under sharp theological attack, initially from the Protestant professors of Wittenberg. Nonetheless, *De Revolutionibus* was used and valued as a starting point in computations of the orbits of the planets. Now calculations were made 'as if' the sun was at the centre of the solar system. Stevin, along with others such as Benedetti (1585), Kepler (1596) and Gilbert (1600), was one of the few scholars who dared to interpret the heliocentric system literally.

The evolution of a heliocentric cosmology into the new paradigm could hardly be other than slow and problematic – witness Galileo's fate at the hands of the Inquisition. The standard works on astronomy, published around 1600, all took as their starting point the geocentric system as set down by Ptolemy. (A heliocentric system was postulated and defended by some among the Hellenistic Greeks, including Aristarchus of Samos, but in that period too the concept met with great opposition.)

Stevin not only accepted and produced a theoretical underpinning of the heliocentric system, which is remarkable enough in itself, but also explained the systems of both Ptolemy ('assuming a fixed Earth') and Copernicus ('assuming a moving Earth') and indicated the connections between them.

Have a look at *Vanden Hemelloop*. It says much for Stevin's profound didactic genius that as a non-professional astronomer he was able to compose a textbook that gives the reader a better introduction to the theories of Ptolemy and Copernicus than the original works themselves. Both Ptolemy's *Almagest* and Copernicus's *De Revolutionibus* are much less accessible texts. *Vanden Hemelloop*, on the other hand, is very systematic and lucidly structured, and is possibly still the best introduction to both the geocentric and the heliocentric theories (fig. 2).

Figure 2: The *Cortbegryp des Hemelloops*. Stevin's summary provides a clearly structured synthesis. He will, he tells us, explain everything as if the matter is completely unknown. From the *Wisconstige Gedachtenissen, vant Weereltschrift, vanden Hemelloop.*
Brussels, Royal Library of Belgium, Rare Books Department, VH 8038 C.

3.2. A didactic jewel

Stevin himself made no astronomical discoveries, so we shall limit ourselves here to a summary of *Vanden Hemelloop* – chiefly to emphasize its author's mastery of didactics. Even so, it will be seen that Stevin produced some original ideas in this work also.

The first book of *Vanden Hemelloop* deals with 'The Finding of the Planets' Motions and the Motions of the Fixed Stars by Means of Empirical Ephemerides, on the Assumption of a Fixed Earth'. Stevin uses the ephemerides computed by Stadius. Ideally, he would have started from real observations of the positions of the sun, moon and planets made over a sufficiently long period, but such observations were unavailable. He could, however, refer to 'ephemerides' (singular: ephemeris) or tables intended for astrological prognostics giving the calculated positions of celestial objects at regular intervals throughout a period, which were compiled by means of a number of known points that were interpolated. Stevin's skill lay in his ability to interpret these ephemerides as real tables or 'empirical ephemerides' in the lessons he devised for Prince Maurice.

The second book of *Vanden Hemelloop* is entitled 'On the Finding of the Motions of the Planets by Means of Mathematical Operations, Based on the Untrue Theory of a Fixed Earth'. It is clear from the title that Stevin did not accept the geocentric worldview as the correct one. In the book's Summary he reminds us that in the first book it had been perceived by experience how – seen from a fixed Earth – the planets move in eccentric circles and epicycles, of which he has noted the characteristics (apogees and so on). In the second book, 'containing the same mathematical operations', numerical values for the eccentricities and dimensions of the orbits are determined.

In the third book of *Vanden Hemelloop*, entitled 'On the Finding of the Motions of the Planets by Means of Mathematical Operations, Based on the True Theory of the Moving Earth', Stevin deals with the orbits of the planets according to the heliocentric system. His own opinion that the heliocentric image is the true one (*de wesentlicke stelling*) is repeatedly indicated, even on the title page (see fig. 3). Stevin accepted that the heliocentric system revealed the true structure of the universe principally on the grounds of its simplicity and 'naturalness'. For example, he found the fact that the velocity of a planet's revolution decreases the further it travels from the sun to be a good argument. He also cited characteristics of the orbits of the planets that are much less artificial in the Copernican heliocentric system than in Ptolemy's geocentric cosmos.

However, he also saw that for some calculations the geocentric system is simpler. It is obvious, for instance, that the orbit of the moon is simpler seen from the earth than from the sun.

Already in the first proposition of the third book, Stevin makes a significant improvement to the Copernican theory, as Anton Pannekoek (1961) points out (in *The Principal Works of Simon Stevin*, vol. III). Copernicus had had to assign three rotational motions to the earth: a diurnal rotation around its own axis; an annual orbital motion around the sun (in which he imagined the earth's own

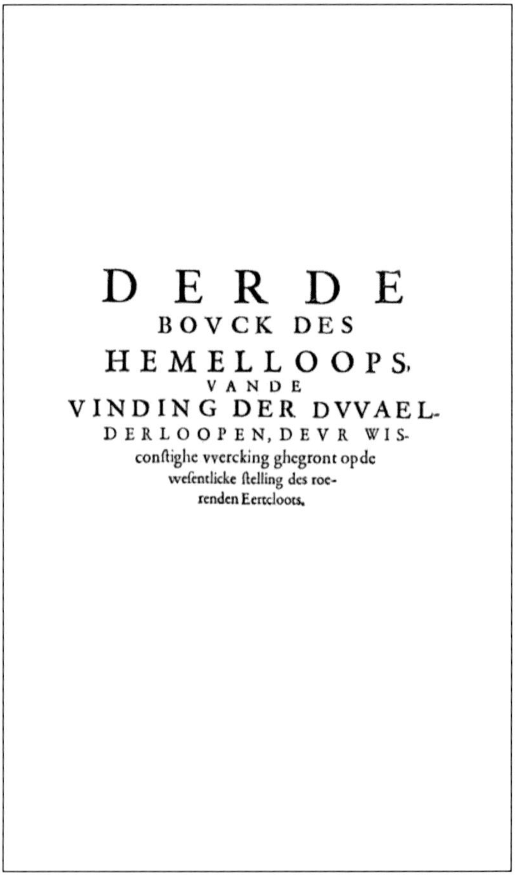

DERDE
BOVCK DES
HEMELLOOPS,
VANDE
VINDING DER DVVAEL-
DERLOOPEN, DEVR WIS-
conſtighe vvercking ghegront op de
weſentlicke ſtelling des roe-
renden Eertcloots,

Figure 3: Stevin endorses heliocentrism and undergirds it 'by Means of Mathematical Operations, Based on the True Theory of the Moving Earth'. From the *Wisconstige Gedachtenissen, vant Weereltschrift, vanden Hemelloop, vande Vinding der Dwaelderlopen...*
© Royal Library of Belgium (Brussels). Rare Books Department, VH 8038 C.

rotational axis connected with the earth–sun connection line); and a third, conical motion, also annual, that kept the axis of the earth pointing more or less in the same direction in space during its annual revolution.

Stevin managed to deal with the earth's own rotational axis as fixed. Even though he attributed its directional constancy to the fact that the earth itself should be regarded as a magnet, he correctly intuited it as a fundamental property. In this respect Stevin appears to have been inspired by William Gilbert's *De Magnete*, published in 1600. Copernicus was aware that the earth's rotational axis, with the exception of the precessional motions, maintained an almost constant direction, but he accounted for this only by the roundabout way of a third rotation. Stevin uses the term *seylsteenighe stilstandt* (literally the 'loadstony standstill') to denote the fixed direction of the earth's rotational axis. (*Seylsteen* is Old Dutch for compass. The *WNT*, the authoritative Dutch dictionary, cites Stevin as one of the sources for the word: see also Chapter Eight.) Just as a ship's compass points in a constant direction no matter how the ship's course changes, so the earth's rotational axis stays, in Stevin's image, in the same direction during the course of the earth around the sun.

The notion of planetary orbit as employed by Ptolemy and Copernicus and, subsequently, by Stevin does not correspond to the modern concept. The ancient astronomers thought that the planets were carried along on the surface of a turning sphere. Stevin specifically states that we should not think of the planets as moving 'like birds round a tower'. He was the very first to consider the connection between Ptolemy and Copernicus, and he did so in the most sophisticated way, with an insight that took in the finest nuances.

Vanden Hemelloop contains many didactic delights. With the aid of just two striking diagrams, Stevin introduces 24 definitions at the start of the book. Examples of new concepts are *natuerlicken dach* (natural day), *natuerlic jaer* (natural year), *Egips jaer* (Egyptian year), *Juliaensche jaren* (Julian years), *Dwaelders* (planets) and *Uytmiddelpuntichront* (eccentric circle). Stevin calls an eccentric circle a circle 'whose centre is outside the Earth'; a concentric circle or *middelpuntichront*, is a circle 'whose centre is also the centre of the Earth'. In figure 4 the line segment DE is called the *uytmiddelpunticheytlijn* or 'line of eccentricity'; an apogee he terms *uytmiddel-puntichronts verstepunt* or 'point farthest from the Earth' and a perigee is *naestepunt* or 'point nearest to the Earth'. FGH, which describes a planet's course, is an *inront*, or epicycle; ABC is the *inrontwech* or planet's orbit.

Figure 4 illustrates how points *ABC* (with centre point *D*) describe an eccentric circle when *E* is the earth. And then Stevin says, 'But if the point *D*

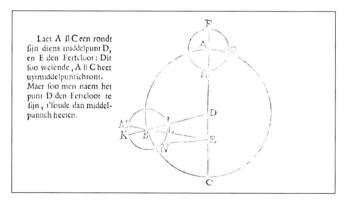

Figure 4: In this diagram, Stevin demonstrates an elegant mathematical transition between the heliocentric and geocentric models. From the *Wisconstige Gedachtenissen, vant Weereltschrift, vanden Hemelloop, vande Vinding der Dwaelderlopen*....

© *Royal Library of Belgium (Brussels). Rare Books Department, VH 8038 C.*

were taken to be the Earth, it would be called concentric'. Coupled with the diagram this sentence, at once concise and precise, makes the difference between *uytmiddelpuntichront* and *middelpuntichront* unambiguously clear. Stevin assumes that the pupil, in this instance Prince Maurice, needs no previous knowledge of the subject, as indeed he says in so many words. 'In the beginning of the description of the Heavenly Motions I will assume the matter to be altogether unknown' (see fig. 2). It shows how very concerned Stevin was with the exact definition of and approach to his pedagogic task. He continues, 'and then I will proceed with the discussion in the same order in which it seems to have evolved in actual fact'.

Figure 4 allows us to make the mental leap from a system with the earth E as the centre to a system with the sun as the centre. Here too Stevin combines two levels, the heliocentric and the geocentric, in a pedagogically effective way. Though this kind of image and association may now be generally familiar to the modern student, it was certainly not the case in Stevin's day.

Stevin's didactically refined expositions were coupled with a keen appreciation of nuances. In *Vanden Hemelloop* we encounter an example of this in the discussion of the transition from the geocentric to the heliocentric model, to which Stevin devoted much attention, being the first to do so. When the earth and the sun change places as central bodies in the model, difficulties arise. Copernicus seems to have ignored them, but they did not escape Stevin's attention. Indeed it is not enough, having swapped the sun for the earth, to say that the geocentric orbit has now become the heliocentric orbit. The geocentric orbit does not correspond to the orbit of the planets around the sun but to the orbit around the centre of the 'annual circle' of the earth. Stevin assumes that Copernicus perceived these problems but that he found their solution so obvious that he deemed a special proof unnecessary, 'apparently considering the matter so clear that it does not call for any proof'.

It is noteworthy that Stevin always starts from an independent, critical attitude, even where – as in *Vanden Hemelloop* – his objective is purely pedagogic. He thought deeply about the subject matter and grappled with subtle points, 'which for a more profound knowledge of the causes would, I thought, require greater certainty concerning certain things than thereby sprang to my mind'. In this way he also produced original contributions, as he did here, with the setting down of the geocentric–heliocentric transition, and with the writing of what had initially been conceived as a textbook.

3.3. Show and explain

In this chapter we have discussed and illustrated Stevin's pedagogic virtuosity with examples from *De Beghinselen der Weeghconst*, *De Beghinselen des Waterwichts* and *Vanden Hemelloop*. His innovatory combination of language (and mathematical symbols) and image (his visual logic and syntax) is very conspicuous. His apposite, lucid, restrained and virtuoso use of language is also striking. Stevin can be called a classic author of the Dutch language.

A selection has been made here. But examples of Stevin's 'visible language' abound throughout his work – from his patents (Chapter Four, fig. 1), to the pivoted sluice lock (Chapter Four, fig. 4), the analysis of the diverse types of harbours (Chapter Four, fig. 6), the ground plans in *De Sterctenbouwing*, among them the development of the bastion (Chapter Four, fig. 10), and the striking images in *De Beghinselen der Weeghconst* and *De Beghinselen des Waterwichts*, and the mathematical works. Page after page of the *Wisconstige Gedachtenissen* illustrates his ability in this respect, as we have already seen when discussing *Vande Deursichtighe* in Chapter Nine. Even in *De Thiende*, which could have been a work without illustrations, the calculations and proofs are supported by graphic language. Charles van den Heuvel (2005) also examines Stevin's logical order and pedagogy when he analyses *De Sterctenbouwing* and *Huysbou*.

Any discussion of Stevin's originality as a didactician has to mention his innovative use of thought experiments, such as those discussed in Chapter Six. A unique aspect here is that he succeeded in explaining unmanifest physical laws to the reader without prior knowledge – as with his *clootcrans*, his trademark 'wreath of spheres', for instance. Several of those who study the role and significance of thought experiments refer to Stevin as an innovator.

Chapter Eleven

An unfinished composition

De Spiegheling der Singconst – On the Theory of the Art of Singing
In the Wisconstige Gedachtenissen, *Stevin devoted a short passage to his thoughts on musical theory. In 1624 Isaac Beeckman was able to inspect Stevin's unpublished manuscripts,* De Spiegheling der Singconst (On the Theory of the Art of Singing). *The title is perhaps a little misleading, as 'singing' refers not to the human voice but to music in general. It is interesting to follow the further adventures of these manuscripts and their reception, and the exact significance of Stevin's ideas on music. His was a* mathematical *theory of the tuning of musical instruments, something that had already been discussed by the ancient Greeks. Only after Simon Stevin, and starting with Isaac Beeckman, would the* physics *of music be studied.*

1. Introduction

Stevin never published *De Spiegheling der Singconst*. Did he simply not get around to taking it to the printer? Did he consider the work to be incomplete?

Stevin never had any difficulty finding a printer. Even so, there were other works that remained unpublished at the time of his death in 1620, such as *Vande Molens* (*On Mills*), *Van den Handel der Cammen en Staven* (*On Cogs and Staves*) and *Burgherlicke Stoffen* (*Civic Matters*). These were published by his son Hendrick Stevin and/or copied down by Isaac Beeckman into his famous *Journal*. We may assume that only practical preoccupations had held up their publication until it was too late, or that perhaps Stevin felt they still required some finishing touches.

In the case of *De Spiegheling der Singconst*, however, it could be that other scruples prevented publication. Stevin had sent his manuscript to a number of musicians, including Abraham Verheyen, an organist at Nijmegen. Though courteous in his reply, Verheyen was critical of some of Stevin's ideas on music. Perhaps Stevin hesitated to publish before he had given sufficient consideration to Verheyen's remarks. He had originally intended to include *De Spiegheling der Singconst* in the *Wisconstige Gedachtenissen*, but he had obviously had second thoughts about it.

Through Beeckman the unpublished manuscript of the *Singconst* ended up in the collection of Constantijn Huygens, and is today preserved as part of his correspondence in the National Library of the Netherlands in The Hague (ms. KA XLVII). In fact, the *Singconst* comprises two manuscripts, a draft version in Stevin's own hand and an essentially different, more definitive-looking version in a fair copy by his son Hendrick. The draft is evidently no more than a first sketch. In some places entire paragraphs have been crossed out and have not always been replaced by new text. In Hendrick's fair copy small changes have been made, but no substantial alterations. However this version does have certain gaps, which were presumably to be filled in later with additional text. (In Bierens de Haan these are indicated by '... [sic] ...'. Incidentally, his transcription is not without errors. In fact it is surprising that to date, as far as we know, not a single page of Stevin's autograph text has appeared in facsimile.) It is clear that this text was not meant for the printer either. Perhaps the intention was to send it to some critical reader such as Verheyen.

1.1. In historical terms, Stevin's rational mathematical theory represents a transitional stage

In essence, Stevin's contribution to musical theory in *De Spiegheling der Singconst* consists of a mathematically accurate description of the system of equal temperament, in which all the successive tones in an octave relate to one another in the same way. Already in antiquity mathematical studies had been made of tone systems, most famously by Pythagoras.

Stevin's unpublished thoughts on music have been judged in various ways. Beeckman initially endorsed Stevin's mathematical insights into musical theory as he knew them from the short passage in the *Wisconstige Gedachtenissen*. But later he himself formulated an innovative theory of coincidence, with a physical concept of consonance that diverged from Stevin's ideas regarding the division of the octave.

Consonant tones are two separate sounds that, unlike dissonant tones, sound pleasing to the ear. This is not a very precise definition but it will do for our purposes here. Besides, what had previously been regarded as dissonance is now often perceived as consonance, so that these concepts are in any case rather vague.

Until the rediscovery and publication of *De Spiegheling der Singconst* by Bierens de Haan in 1884 only Descartes, Mersenne and Huygens (father and son) seem to have joined Verheyen and Beeckman in being aware of its existence. Descartes undoubtedly knew of it thanks to his friend Beeckman, who also wrote about it to Mersenne. Apart from Beeckman, until 1884 Christiaan Huygens was the only other scholar to have actually seen Stevin's original manuscripts. Beeckman and Huygens principally objected to Stevin's statement that the ratio of the fifth is given by 1.4983..., which is the twelfth root of 128, or can be mathematically expressed as $(128)^{1/12}$.

Just to be clear, it should be noted here that the interval C–G is an example of a fifth, while the ratio of the interval is the number that indicates by how much we must multiply the frequency of a certain tone to move from the lowest to the highest tone of the interval.

In 1635 Descartes referred to Stevin's 'error', though he offered no support for this statement. In the foreword to his *Harmonie universelle* (1636–37), Mersenne wrote that he thought Stevin's theory of music strange, but his direct knowledge of Stevin's ideas was limited to the aforementioned brief passage in the *Wisconstige Gedachtenissen* (which had been translated into French).

The basic biographies of Simon Stevin do not probe too deeply into the status of his musical theory. Given the opinions of Verheyen, Beeckman, Descartes, Mersenne and Huygens, it is important to look at exactly what Stevin said, both in the short passage in the *Wisconstige Gedachtenissen* and in his draft version of *De Spiegheling der Singconst*.

Recently there has been fresh interest in Stevin's treatise on music (Cohen, 1987; Rasch, 1998).

2. Stevin divides the octave into 12 equal intervals

De Spiegheling der Singconst is a mathematical approach to music. It belongs to a centuries-old tradition that stretches back to antiquity, when music was one of the liberal arts, part of the *quadrivium*, which also included arithmetic, geometry and astronomy. Of course, music is infinitely more than a mathematical discipline, just as man is more than a skeleton. Nonetheless, anatomy is of great importance and it is certainly important to learn the mathematical order that is concealed in musical sounds.

No doubt Stevin embarked on the *Singconst* because of the lessons he devised for Prince Maurice. But as we have come to expect by now, in this area too he was not content simply to keep to the pedagogic presentation of already existing knowledge. He contributed to the discussion of a problem that lay at the heart of musical performance, which indeed had been debated since Pythagoras – the tuning of notes in relation to each other. The basic question is this: how – that is to say, according to which system – do we tune a musical instrument? How do the pitches of succeeding tones relate to one another? Anyone familiar with musical theory knows the problem. With only a limited number of keys in an octave, it is basically impossible to get every possible consonance in its pure form (that is, free of beats). Therefore the method of tuning always depends on one's priorities in this respect. In the course of history, this has led to many different tone systems (or systems of tones). In Stevin's day several methods were already in use, with which the names of Pythagoras, Zarlino, Fogliano and Grammateus (among others) are connected (fig. 1).

When Isaac Beeckman visited Stevin's widow in 1624, he came across fragments of a manuscript of *De Spiegheling der Singconst*. He found two drafts of chapters and two appendices. One draft was in Stevin's own hand. The section of the manuscript that looked as if it had been copied out in preparation for the printer contained mostly basic theory and definitions. Thus neither of these drafts was complete. Apart from the pedagogic aspects of the *Singconst*, Stevin had

Figure 1: Caravaggio's *Lute Player*. The virtually insoluble problems in placing the frets on the neck of the lute that accompany a perfect tuning of intervals provided the initial reason for a practical compromise, which we nowadays call equal temperament. The first propositions in this direction were purely pragmatic. Stevin, on the other hand, developed a mathematical underpinning for this tone system.
The State Hermitage Museum, St Petersburg. © 1990. Photo SCALA, Florence.

provided a personal contribution in which he had put forward his opinion of how to place the frets – the bars or ridges – on the fingerboard of a stringed instrument, such as a viola da gamba or lute, in order to achieve the correct intervals. It is also striking that, like the Greeks, Stevin went to work in terms of mathematics rather than physics. He makes no mention of beat, a sort of tremolo or rapid fluctuation of intensity, created when two more or less equally high notes are sounded together. Yet beat is easy to observe.

As regards its structure, Stevin's draft of *De Spiegheling der Singconst* begins, like *De Beghinselen der Weeghconst*, with definitions, explanations and postulates. But gradually the division becomes freer, and larger sections are worked out with titles such as 'Comparison of Geometrical Ratio with Musical Ratio' and 'On the True Ratios of Natural Tones'. These longer chapters are not subsectioned, which also indicates that the *Singconst* was unfinished.

Stevin's premise is 'all whole-tones to be equal and likewise all semitones to be equal'. This is in contradiction to the tone systems most commonly used in his day (meantone temperament, for instance). Ever intent on the use of the Dutch language, in his definitions Stevin calls the intervals *eerste* (first or prime), *tweede* (second), *derde* (third), and so on. Major and minor intervals are defined in the usual way.

At this point Stevin interpolates a discussion 'On Ratio in General', in which once again he gives us his view on the semantic power of Dutch to make a word, such as *Evenredenheyt* – 'equirationality', express its meaning properly. He believed the Greeks had failed to fathom these concepts, opining that 'no very thorough understanding of "equirationality" was found among the Greeks and their successors.' Of course these reflections do not detract from Stevin's technical and mathematical discourse, any more than the animistic ideas of Kepler – the first to gain a mathematical understanding of the orbits of the planets – or the many notes on mysticism made by Newton detract from their discoveries in physics. It astonishes us today that these pioneers of modern science were so clearly influenced by ideas on metaphysics, mysticism and alchemy.

In composing the *Singconst* Stevin had the only correct division of the scale in view. In the process, he reopened a discussion that had been regarded as closed since the Middle Ages, since which time the reproduction of 'true' or 'natural' interval ratios as perceived from antiquity onwards had generally sufficed. Contributions by later theoreticians had been limited almost exclusively to the development of 'pragmatic compromises', which made it possible, for instance, to perfectly tune the most commonly used minor or major thirds with as little beat as possible in the fifths. Stevin, on the other hand, reopened the main debate concerning the essence of our tone system, and he did so from a completely new angle. He stated that there was a single 'just' division of the octave, to obtain which it was necessary to abandon simple fractions with integers (such as 3/2) as a ratio of the pitch of the highest and lowest notes in an interval, such as G and C in a fifth. Though in practical terms Stevin's division of the octave corresponded to equal temperament – the tone system most commonly used in western music today – he arrived at it via the formulation of a problem quite separate from historical musical evolution (fig. 2).

The most significant factor that drove him to search for his system seems to have been his inability to accept that ascending through a stack of twelve fifths is to reach almost, but not exactly, the same note as is reached when ascending through a stack of seven octaves. That the difference, known as a 'Pythagorean comma', was so small led him to conclude that failure to arrive at the same note was exclusively due to the impossibility of establishing the fifth accurately enough by ear alone. He expressed this as follows:

> *It is therefore obvious that no human hearing, however keen it may be, is able to fit two tones quite surely in their perfection. From this*

it follows that many such mistakes, each of which in itself is inappreciable, yet in combination produce an appreciable error...

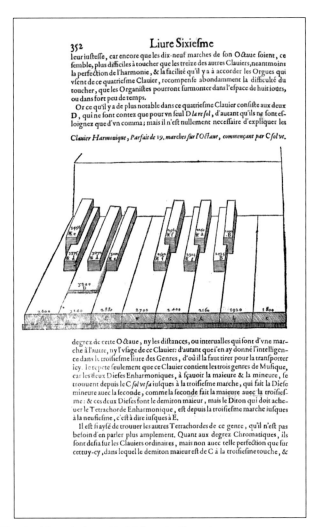

Figure 2: The Archicembalo, with 18 keys to the octave.
  'or there would have to be as many frets as there are ratios of pitch'. It was not only with fretted instruments that attempts to perfectly tune as many intervals as possible led to complications. It was the case with keyboard instruments as well. To Stevin this proved that the traditional theories of interval ratios were incorrect. From: F. Marin Mersenne, *Harmonie universelle*. Paris: Sebastien Cramoisy, 1636. 'des Instruments' (p. 352).
  © *Royal Library of Belgium (Brussels). Rare Books Department, Fétis 5347 C (ektachrome: JTD).*

2.1. Experiments had a different status in Stevin's day

It may strike us as remarkable that Stevin passed over the fact that the difference is invariable. From the distinction we have learned to make between systematic and random differences, it seems evident to us that such a difference cannot be caused by the normal tolerances in the tuning of perfect fifths. In that case the stack of twelve fifths would be smaller just as often as it was larger than the stack of seven octaves. In fact, it is better here (and also more in line with practice) to speak of the stacking of five ascending fifths and seven descending fourths. The result is the same, but the practical accuracy is greater. We should also remember that statistics and error estimation did not exist in Stevin's day. Nor was it usual practice to repeat an experiment that had already been carried out in order to verify the results. Significantly, Stevin refers to experiments which, when actually carried out, do not give the results he states. He also appears to have underestimated the accuracy with which intervals can be perfectly tuned, another indication of his lack of practical musical experience. This was also remarked on by Verheyen, who correctly wrote of the major third '...which can be established as perfectly by natural hearing as the octave'. In reality, it is possible to tune a major third by ear even more accurately than an octave, which in practice is one of the least critical intervals for small imperfections! Figure 7 shows an example of Verheyen's music.

A second factor that seems to have spurred Stevin on was evidently the practical problem of the placing of the frets on lutes and viola da gambas. In fact, this can only be solved by using equal temperament. Although in principle this does give imperfect intervals, the player, by making certain small corrections, can reduce their imperfection to a point where it is barely noticeable. This may have given Stevin the impression that equal temperament produces perfect intervals. Had he heard a harpsichord or an organ tuned in this way, however, he might have thought differently.

Moreover, throughout history, simplicity and elegance have been important criteria for theories and systems (provided, of course, that they explained the phenomena well). Mathematically seen, this simplicity is naturally a strong point of a system with 12 equal semitones in an octave. There is only one sort of semitone; all intervals can be related to one another in a simple way; an octave is exactly three major thirds; two major thirds are exactly a minor sixth, and so on. There are no more discrepancies between combinations of intervals that seemingly ought to add up to the same interval; four fifths give exactly the same tone as two octaves and a major third, and so forth. Finally, only one empirically determined interval ratio has to be used, the 2/1 of the octave.

Cohen has also pointed out, and probably rightly, that with his division of the octave, Stevin also tried to give a new explanation of consonance (in the sense of the perfection of intervals). It is striking that Stevin, contrary to what we have come to expect from him and, it is true, in a draft text which he never published, made rather haughty statements about the division of the scale. He called his division the only true one, branded other systems as aberrations, and declared:

> *Now someone might wonder, according to the ancient view, how the sweet sound of the fifth could consist of so unutterable, irrational and inconvenient a number. To this we might give a detailed answer, but as it is not our intention here to teach to the unutterable irrationality and inconvenience of such a misunderstanding the utterability, rationality, convenience and natural wonderful perfection of these numbers...*

The fierceness of these utterances seems hardly explicable as a simple rejection of the traditional explanation. It seems rather to be a repudiation of the philosophy underlying it, which held that particular properties must be attributed to certain numbers. Stevin, with his modern mathematical ideas, considered this way of thinking to be completely out of date. He saw no reason to ascribe a special meaning to a simple fraction. Science was not yet sufficiently far advanced for a physical explanation. Therefore Stevin sought his explanation in a mathematical system, which he thought would provide the true ratio between intervals.

Already in determining the whole-tone, ditone and so on, we see that Stevin used only ratios that are simple whole powers of the twelfth root of 2, and that he assigned no special place to the 'simple' ratios 3:2, 4:3 and so on. The essence of Stevin's contribution is that he divided an octave into 12 equal steps. Because the human ear seems to work logarithmically, to determine the next higher note the frequency must always be multiplied by the same factor. Now Stevin was a trendsetter in the matter of logarithms (see Chapter Three, box 1), and he was highly competent in working with roots. Given that he retained the octave and sought to divide it into 12 steps, the jump between two successive tones always corresponded to a factor of the twelfth root of 2. Before Stevin, simple rational numbers, ratios that led to the fifth, major third, and so on, were always chosen as a starting point. Stevin attached no importance at all to the rational character of ratios, however. His rejection was undoubtedly connected to the fact that the physics underlying beat-free intervals was still unknown in his day. Without this knowledge, there was no reason at all, from a theoretical point of view, to ascribe a special status to a simple rational ratio.

Prior to Stevin, Vincenzo Galilei (*c.* 1525–1591), father of Galileo Galilei, was – as far as is known – the only one to have worked out a system with equal semitones in any detail, namely in his *Dialogo della musica antica e della moderna* (1581). However, this system had originated as a practical compromise for the placing of the frets on the lute and had nothing to do with a coherent tone system. It had no mathematical model as its basis, only a semitone interval that had proved empirically to be a suitable standard. Stevin's semitone interval, on the other hand, was the result of a mathematically based system that had been comprehensively thought out. To Galilei, the equality of semitones was exclusively a technical necessity for lutes and viols; to Stevin it was a fundamental and universally valid property of our tone system (fig. 3).

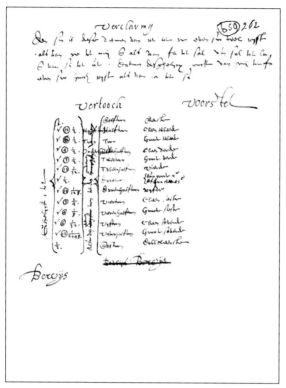

Figure 3: Stevin was the first to make a mathematically correct calculation of equal temperament. Compared with earlier pragmatic approaches, which accidentally gave more or less the same result, this was pioneering work. Perhaps the most extraordinary thing, however, is Stevin's contention that these were also the 'true' interval ratios. In his tone system Stevin uses only the twelfth root of 2.
The Hague, National Library of the Netherlands, KA XLVII, fol. 262 (ektachrome: JTD).

Box 1: Explanation of the most important ideas in the *Singconst*

Interval

In music this is the 'distance' between two notes. The name given to an interval is derived from the number of notes in a scale, the first note itself included. Thus an interval of five notes such as C–G (C=1, D=2, E=3, F=4,

G=5) or D–A (D=1, E=2, F=3, G=4, A=5) is called a fifth; an interval of three notes, such as C–E (C=1, D=2, E=3) or D–F(D=1, E=2, F=3) is called a third.

Semitone: the interval between an E and an F or a B and a C
Whole-tone: an interval such as C–D, or D–E, about the double of E–F
Major third: a third consisting of two whole-tones, for example C–E
Minor third: a third consisting of a whole-tone and a semitone, for example D–F and E–G.

Coincidence theory

A theory developed by Isaac Beeckman that endeavours to explain consonance on a physical basis. Simplified, the gist of it is that a tone is regarded as a regular series of pulses of air whose frequency determines the pitch. When the pulses of two tones coincide more often, the interval they form is more consonant (the bars represent the pulses):

We see that with the fifth C–G the pulses already coincide again after 2 and 3 pulses, respectively. With the major third C–E these numbers are 4 and 5, respectively and with the whole-tone C–D 8 and 9, respectively. When the coincidence only takes place at such great distances as in this last case, the consonance was perceived in Beeckman's time as dissonance.

It is clear that according to this theory, Stevin's interval ratios that could not be reduced to fractions would all be regarded as dissonance, including those of the traditional consonances!

Beat

The waves of tones whose frequencies do not entirely coincide (see coincidence theory) move periodically in the same and in the opposite directions. The periodic increase and decrease in volume this causes is called beat (the 'wah-wah-wah' effect). Consonances that have no beat are perceived as pure.

Temperament

Temperament is the 'tempering' of musical intervals (increasing or decreasing) away from the natural scale (that is deducible by physical laws) to fit them for practicable performance. If an instrument with a limited number of notes to the octave (such as a lute, viola da gamba or keyboard instrument) is to be playable, it cannot have too many notes per octave. Usually this number is limited to 12, as in the case of the piano and the guitar. But with this limited supply of notes the perfect tuning of all the consonances is impossible. Hence, from the sixteenth century onwards, various compromises were developed, whereby to have a larger number of usable consonances certain acceptable imperfections were deliberately introduced. These compromises were later called 'temperaments'. Which consonances these compromises would serve was determined by the requirements of the music and the technical particularities of the instrument in question.

Meantone temperament

In the sixteenth century, the third became more and more important in music, so a temperament was developed in which the eight frequently used thirds were made perfect at the expense of the four rarely used ones. Eleven of the twelve fifths are a little too small, but usable; the twelfth is far too large and unbearably out of tune (the 'wolf fifth').

Equal temperament

In the second half of the eighteenth century, the use of meantone temperament gave way to the system of tuning that is still in use today, known as 'equal temperament', in which the octave is divided into 12 equal steps, as Stevin specified. The consequence is that all the steps in a particular key have the same imperfection. Though this results in a poverty of colour and a reduced beauty of the intervals as such, it does allow the system to be used without restrictions.

3. 'Now in order to divide the monochord geometrically...' and the 'Arithmetical Division of the Monochord'

After the definition and discussion of ratio, in general, and with reference to *Singconst*, or as he himself writes, 'Comparison of Geometrical Ratio with Musical Ratio', Stevin makes a fairly comprehensive survey of what the Greeks had understood in this respect. He believed that the Greeks, who started from certain simple rational ratios for the intervals of the fifth, fourth and major third, used a wrongly divided scale.

By and large, Stevin's criticisms of the traditional systems are only valid if one accepts his premises as a norm. In particular, Stevin seemed to start from the tacit assumption that a closed system with a limited number of tones in the octave is in any case possible. Consequently, every discrepancy is ascribed to the system in question rather than to nature – in this respect Stevin was at his clearest in his criticism of the Ptolemaic system. For the same reason, he also baulked at interpreting as such certain tone systems that were designed as a compromise. He criticised Gioseffo Zarlino's tone systems for their arbitrariness, which would only have been justified if these systems had made any pretence of producing 'true' intervals (figs. 4.1 and 4.2).

Stevin dilates on the fact that the Greeks – according to him extremely intelligent as a rule – had failed to correctly grasp the idea of 'ratio'. He associated this shortcoming with their language, which lacked the semantic potential of Dutch. It is evident that the word *evenredenheyt* – 'equirationality' – concerns equal ratios; in Stevin's view this already makes it clear that the intervals in the octave must be obtained by its division into equal ratios.

Figure 4.1: Certain parts of the draft of *De Spiegheling der Singconst* survive in two autograph versions. We have fragments of what was apparently a first draft and a sort of a provisional fair copy to which minor corrections had been made. It can be seen from this page that much care had been taken with the fair copy.

The Hague, National Library of the Netherlands, KA XLVII, fol. 235^r (ektachrome: JTD).

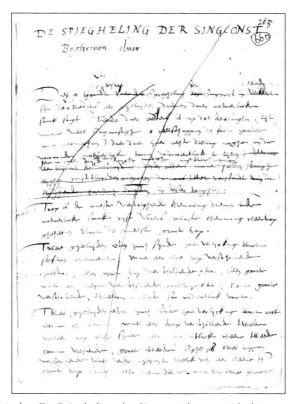

Figure 4.2: Maybe *De Spiegheling der Singconst* began with this page. It is the first folio of the surviving draft copy, with the title at the head of the page. The Hague, National Library of the Netherlands, KA XLVII, fol. 263ʳ (ektachrome: JTD).

In the section entitled 'Geometrical Division of the Monochord', geometry is used to determine precisely where the frets on a lute must be placed in order to achieve the abovementioned division of the scale (fig. 5):

> *Now in order to divide the monochord geometrically in such a way that we have the true perfect sounds of natural singing...*

Then follows what Stevin calls the 'Arithmetical Division of the Monochord', in which he calculates the (relative) positions of the frets numerically. In the accompanying table (see fig. 6), we see the results of his calculations (10,000 = the full length of the string). In this day and age, with a pocket calculator, it takes no great skill and no time at all to obtain the numbers Stevin worked out by hand. In his day, however, it was not only a very laborious process to figure out

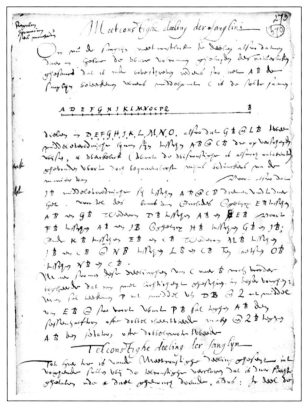

Figure 5: *'Meetconstighe deeling...'* or the 'Geometrical Division of the Monochord', a study of the placing of the frets on the neck of a lute. In some places it is clear to see that the text was not yet finished.
The Hague, National Library of the Netherlands, KA XLVII, fol. 273ʳ (ektachrome: JTD).

all these numbers, it was also necessary to develop certain special methods of computing. Stevin obtained his results here via a combination of square and cubic root extraction. He had previously published the method for doing this in *De Thiende* (see also Chapter Three, box 3) and in *Arithmetic*, parts of the *Wisconstige Gedachtenissen*.

Stevin's figures differ slightly here and there from the correct values, yet as a rule these are only errors produced by rounding off. We should not lose sight of the fact that the modern convention in which the number of decimals should give an exact picture of the minimum accuracy with which we have calculated the number in question, did not exist in Stevin's day. Given a sufficiently relative accuracy of the number concerned, an error of one or two units in the last decimal was evidently still acceptable. Even an accumulation of such errors was apparently not too disturbing when this condition was fulfilled. In practice, the

Figure 6: Stevin's calculations of pitches. Stevin states not only how interval ratios should look in theory, he also computes their values with more than sufficient accuracy for practical purposes. To do this he used methods of root extraction he had developed himself.
The Hague, National Library of the Netherlands, KA XLVII, fol. 274ʳ (ektachrome: JTD).

small differences in Stevin's tables are irrelevant. It would never be possible to place the frets with this kind of accuracy. In one case, however (the number 8,404, which should be 8,409), an error in copying is obvious: a carelessly written 9 can be taken for a 4. Because this number is easy to find if one divides 10,000 by $\sqrt{(\sqrt{2})}$, rounding-off errors of this magnitude can be excluded, and an error in calculation seems less likely. If one calculates the differences in 100ths of a semitone, or 'cents', the numbers Stevin computed are more than accurate for practical purposes; the largest difference amounts to 0.6 cents.

It was Stevin's habit to make a clear distinction between what he regarded as knowledge already acquired and subjects on which opinions still differed. Therefore, *De Spiegheling der Singconst* included an appendix that covered controversial issues. In 'Chapter on the Fourth' Stevin entered into the debate about whether or not the fourth was consonant. Also in the appendix is a chapter, which according to the table of contents bore the title *Bemollaris cantus is onnut*

Figure 7: If one tries to tune all the consonances in this example of music by Verheyen so that they are free of beat, small semitones are created that amount to less than 2/3 of the interval of the large semitone. This marked contrast was essential for the chromatic music of the time, which lost all its colour in an equal temperament. This example of music is not entirely correctly reproduced by Bierens de Haan.

The Hague, National Library of the Netherlands, KA XLVII, fol. 276^r (ektachrome: JTD).

onderscheijt... In it Stevin argues that one can avoid having to add a flat at the clef by transposing the music in question downwards by a fourth. This is only true, however, on condition that the pitch in the notation is regarded as purely relative. But that was no longer the case in Stevin's day. Certain 'standard pitches' did exist (Fokker's assertion in *The Principal Works of Simon Stevin*, (vol. V) that absolute pitch had no importance in Stevin's day is not correct in a general sense).

Finally, among the fragments of the *Singconst* that were left was yet another chapter 'in which is explained the Cause of the Imperfection that arises when Organs and Harpsichords are tuned'. Here, Stevin argues that in the customary tuning of keyboard instruments the imperfection of the fifth at G-sharp is false only because of inaccuracy in the tuning of the other fifths. However, this explanation rests on the erroneous assumption that the fifth at G-sharp '...at one time turns out slightly too large, at another time slightly too small, sometimes also good...' and on an underestimation of the accuracy with which fifths and fourths can be perfectly tuned. Given the practice of his day, Stevin's reasoning contains another weak point. He takes it for granted that keyboard instruments would be tuned on the basis of perfect fifths and fourths, but this was not the case at that time, as, for that matter, can also be inferred from Verheyen's reaction. It can be concluded from Stevin's argument that he was unaware, or only vaguely aware, of the practical side of tuning keyboard instruments.

4. Temperament in Stevin's time and the placing of the frets on the neck of a lute

In Stevin's day a clear distinction was made between the tuning of keyboard instruments (such as the organ, harpsichord and clavichord) and fretted string instruments (such as lutes and viola da gambas). In the case of the first group of instruments, their technical construction imposed no restriction at all on the initial choice of tuning; this could therefore be determined on aesthetic grounds, entirely to suit the music. In practice, given the importance of the third in contemporary music, the tuning generally chosen was meantone temperament, which had as its basis the greatest possible perfection of the most frequently played thirds. Limitations are inherent in the technical construction of fretted instruments, however. The intervals chosen for one string inevitably became the intervals for the other strings. This could lead to annoying discrepancies, which could only be resolved by making all semitone intervals the same length – which would automatically have led to an equal temperament. Clearly this had nothing to do with appreciation; there was simply no other choice.

4.1. Stevin's practical knowledge was probably confined to the lute

Because, broadly speaking, Stevin propagated a scientific system that, when applied in practice, would lead to this equal temperament, some historians believe that he was unfamiliar with the musical practice of his time, which such a tuning would not have served.

I think this conclusion is a little too radical. Stevin had read at least some part of the authoritative works on musical theory that were available then. But there were obviously certain gaps in his knowledge of musical praxis, including surely any practical knowledge of the tuning of keyboard instruments. It seems likely that he did not have a particularly good ear himself, otherwise he would never have underestimated its accuracy in tuning. Such gaps are only normal in someone who had not made such a complex matter as music his special study, and they are also not to be wondered at in one who was active and outstanding in so many other fields. It is also clear that Stevin himself had never carried out the experiments he mentions, or he would assuredly have established their inaccuracy.

One practical musical problem with which Stevin was evidently familiar was the placing of the frets on the fingerboard of a lute. Here his calculations could have had practical importance, for they produced significantly more accurate positions than did those of Vincenzo Galilei, for instance. But as *De Spiegheling der Singconst* was never published they remained unknown to the musical world. It was Mersenne who first published accurate calculations of equal temperament in his *Harmonie universelle* (1636–37), though his method was much less innovative than Stevin's.

5. What Verheyen had to say…

The Nijmegen-born organist Abraham Verheyen responded to a draft of the manuscript Stevin had sent him. One of the heads of his criticism was that, although the mathematical correctness of Stevin's calculations was beyond question, they did not produce the desired perfect intervals. Unlike Stevin, Verheyen knew from practical experience of tuning that the traditional interval ratios were the 'just' ones, even though he obviously lacked the physics that would give this correctness a sound theoretical basis. He well knew the accuracy with which the various intervals could be tuned by ear and knew, therefore, that the slight differences, which Stevin ascribed to margins of error, were in fact real.

Verheyen showed why a lute could not be tuned otherwise than in equal temperament, 'or there would have to be as many frets as there are ratios of pitch'. He gave four possible ways of tuning keyboard instruments, including the meantone temperament already mentioned. He also suggested a calculation of the intervals for this tuning. The figures are exact to the last decimal. Did he intend to show Stevin the scientist that when it came to calculation he could hold his own?

Verheyen drew Stevin's attention to the empirically incorrect basis of his system and then showed him how instruments were tuned in practice. It can easily be imagined that an intelligent and critical man like Stevin could not simply disregard these comments and observations, and that he was at least prepared to consider the possibility that in this field his lack of practical experience had led him astray. The postponing of publication of the *Singconst* was almost certainly partly the result of Verheyen's criticism of it.

6. The historical place of Stevin's tone system

Discussions on the *Singconst* by Dijksterhuis and Fokker have often been found uncritical. It has been said of both that they favour the mathematics and the physics, and that perhaps they are also a little too keen to add a musical feather to Stevin's already well-plumed cap. In any case, we should take a look at whether Stevin's intentions as regards the *Singconst* have been well understood in the past. H. Floris Cohen has gone into the matter (Cohen 1987).

Stevin was the first to calculate intervals that correspond to what we call equal temperament and he did so with considerable accuracy. That merit is due to him and is also ascribed to him by history. What strikes us here is that Stevin believed that this division of the octave was the 'true' tuning. It has more than once been said that Stevin was the first to propagate the general application of equal temperament and, in doing so, was ahead of his time. However, Cohen suggests that Stevin had something other than equal temperament in view. A temperament is a compromise that makes practice possible, whereas Stevin's system was certainly not produced to fulfil the need for a practical compromise, but rather to arrive at an understanding of the actual connection between tones. According to Cohen, Stevin hoped that by doing so he would be able to give a new explanation for the perfection of consonant intervals.

It cannot really be said that Stevin was much ahead of his time in propagating equal temperament. He did not realize that it was a compromise and thus he did not propagate deliberate musical choices. We even get the impression that he was unfamiliar with the principle of temperament and that, as regards keyboard instruments, he had no knowledge of the solutions employed in practice. In this regard, it is significant that he criticized Zarlino's suggestions for temperaments as if they were attempts at calculating true interval ratios (from which point of view Stevin's criticisms would have been justified). Evidently he regarded what he saw as normal tuning practice (the simple perfect tuning by ear of 11 of the 12 fifths) only as an imperfect form of perfect tuning, which owing to the inaccuracy of the human ear did not entirely succeed. He took the view that a mathematical definition of the correct notes must lead to an improvement of the desired result. (Mathematical ratios have been transposed into sound since antiquity with the aid of a monochord. This is a sound box with a single string and a moveable bridge whose position can be determined by mathematical calculations, as a result of which every interval ratio can be accurately transposed into sound (see fig. 8). The notes thus produced can then be copied on the instrument to be tuned.) That the mathematical determination of the correct tones would also produce an essentially different tuning seems to have escaped him however. It was Verheyen who drew this to his attention.

So, can it perhaps be said that Stevin was a Schoenberg *avant la lettre*? Schoenberg also abandoned the idea of compromise in favour of the division of the octave into 12 equal semitones (his so-called 12-note technique). The comparison does not hold water. The essential difference between Schoenberg and Stevin is that the former did not devise his system in order to find the true ratio of perfect intervals. He abandoned the traditional ideas of perfection and relationships between notes for a new kind of music that deliberately assumes the absolute equality of all 12 notes within the octave as a basis. Stevin, on the other hand, was seeking the exact opposite: he believed that he had found the true interval ratios, which we experience as pure. It was precisely this purity that was to him the crucial point, only he argued that this was achieved *de facto* when two notes correspond with an interval ratio that is to be traced back to the twelfth root of two. Whereas Schoenberg made a radical choice and deliberate intervention in the tone system, Stevin meant only (as the first to do so) to provide the correct mathematical description of the 'existing' tone system.

Finally there remains the question of whether, in our own day, Stevin has not been proved right insofar that we have become conditioned to regarding the interval ratios he calculated as the 'true' ones. We have grown up with equal temperament in the tuning of guitars, pianos, organs (more than 95 per cent of

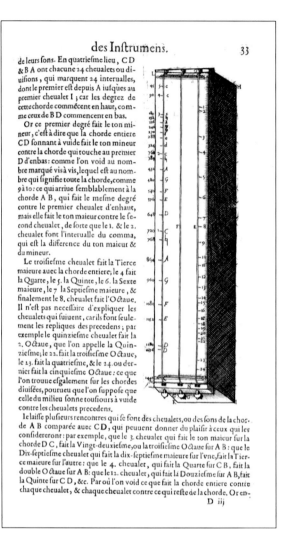

Figure 8: Until the arrival of electronics, the monochord, which was already known in antiquity, had been the link between mathematics and the consonances produced by calculated interval ratios. By moving the bridges, every ratio of length of both strings could be established. From: F. Marin Mersenne, *Harmonie universelle*. Paris: Sebastien Cramoisy, 1636. 'des Instruments' (p. 33).
© *Royal Library of Belgium (Brussels). Rare Books Department, Fétis 5347 C (ektachrome: JTD).*

organs are tuned in equal temperament despite the increasing tendency to tune historic organs and newly built 'historic' organs in unequal temperaments once more), harps and not forgetting the electronic instruments whose sounds assail us every day (often whether we want to hear them or not). Whether, as a result, we have come to perceive this as more correct than the truly perfect intervals is another question, however. Research has shown that musicians' tuning is a highly complex affair and that standard equal temperament tuning is certainly not the rule. We therefore cannot state that, in general, equal temperament tuning is the norm in modern musical praxis.

In his draft of the *Singconst*, we can see how Stevin worked out and defended the 'equal' division of the octave on a mathematical, and not a physical, basis. The actual sound that was the result of his calculations (never contemplated by him), equal temperament, had no place in the logical line of the musical evolution of his day. The greatest significance of Stevin's calculations seems to lie precisely in his innovative way of calculating, in his new view of numbers and in the rejection of the special meaning that some before him had ascribed to certain numbers. Our familiarity with decimals after the point, fractional powers, logarithms and so forth often makes us forget that, in part, we also have Stevin to thank for the development of these ideas.

Chapter Twelve

The resonance of Simon Stevin and his work

In this chapter, we shall take a look at the influence of Stevin's work on some of the eminent scholars of the sixteenth and seventeenth centuries, and glance at some later reactions up to the present day (see e.g. fig. 1).

1. Introduction

Simon Stevin occupies a unique place in the history of the Low Countries. Any number of squares and streets are named after him. In 1846, in his native town of Bruges, a bronze statue was unveiled amid great celebration (box 1). There are numerous Simon Stevin societies of the most diverse kind. In Belgium and the Netherlands there are Simon Stevin sailing clubs; there is a Simon Stevin study association for students in the department of mechanical engineering at the Technische Universiteit Eindhoven; several observatories bear his name; and a technical institute for architecture, health sciences and management in Bruges is named after him, as is the Simon Stevin Flemish Fortification Centre, which promotes the study, conservation and restoration of fortifications no longer in military use.

> ### Box 1: The statue of Simon Stevin in Bruges
>
> In Bruges, the town where he was born, Simon Stevin's statue dominates the Simon Stevinplein. The monument is 3m high, 1m wide and 2m deep. It was cast in bronze by the Belgian sculptor Louis Eugène Simonis and stands on a plinth of Belgian bluestone.
>
> In 1839 the provincial government of West Flanders and the Bruges corporation decided to put up a statue of the Bruges-born scholar. But the plan was brought to a halt by violent protests in the national parliament. When the house sat on 20 February 1845, the Walloon Catholic member Bartholémy Dumortier, a famed botanist, pointed out that it was not appropriate to erect a statue of a person who had not only fled his country to side with the enemy but had even taken service in that enemy's army. The press entered the fray. On 5 March 1845, an article by Dumortier appeared in *L'Observateur*. It ended with the following: 'Sir, your readers are undoubtedly educated people; well! I wager

not one in a thousand had heard of Simon Stevin before the question of erecting a statue to him arose, and I confess to my shame that I was one of them.'

Dumortier's outburst caused a tremendous to-do. A defender of Stevin was found in the liberal politician Sylvain Van de Weyer. Dumortier and Van de Weyer were political opponents at a moment when the split between Catholics and Liberals had brought about a severe political crisis. Their personal beliefs were diametrically at odds in every respect. Dumortier was an ultramontane Catholic; Van de Weyer a liberal to the core. In 1830 Dumortier had fought on the barricades against the Dutch troops; Van de Weyer had fled to safety in France. Dumortier was fervently anti-Dutch; Van de Weyer, as Belgian ambassador in London, had negotiated the treaty with the United Provinces, to Dumortier's great dissatisfaction. Public interest in the dispute was undoubtedly increased when in 1846 Van de Weyer was appointed head of a Liberal-Catholic coalition government and had to take Catholic resistance into account in his policy. Van de Weyer made a savage attack on Dumortier in a pamphlet under the pseudonym of J. Du Fan. The issue now was no longer Stevin, the historical figure, but rather the political significance of modern science and the liberal state. Van de Weyer maintained that Stevin had been driven from his homeland by Catholic repression, the same repression that now sought to ban the raising of the statue, and he warned of a new wave of Catholic repression, stating scathingly that 'they practice on the dead, before getting a grip on the living'.

Figure 1: Stevin's sailing chariot also inspired nineteenth-century artists. This lithograph by J.G. Canneel dates from the mid-nineteenth century. *Ghent, University Library (photo: University of Ghent).*

> The monument got the go-ahead. Its inauguration was planned for 26 July 1846. There was one small problem – the statue was not ready. The official ceremony still took place, though it was a plaster stand-in that was unveiled. A year later, the real statue was set on its plinth. To mark the event, the city organized great celebrations. A whole week of festivities was planned, including a competition for an ode to Stevin and a piece of prose. Ironically enough, only texts in French were accepted. Everyone in Bruges was involved in the celebrations. During the week of the festival there was free entry to a number of exhibitions; the city architect turned the market square into a historical showcase, there were dances, and a great procession was organized by and for the Bruges societies.

Stevin's influence on a whole variety of scientific organizations and bodies is still being felt today. There are Simon Stevin exchange programmes between eight departments of the history of science in the Netherlands, England and North America. Every year the Netherlands Technology Foundation STW (Stichting voor de Technische Wetenschappen) offers the prestigious title of 'Simon Stevin Master' to a leading researcher in the technological field (500,000 euros go with the title). The history section of the BBC's education webpage includes Simon Stevin in its list of 'Historic Figures'. One of the moon's craters has been named 'Stevinus'. Also worth mentioning is the Belgian mathematics and physics journal *Simon Stevin* (nowadays known as the *Tijdschrift van de Belgische Mathematische Vereniging*). Stevin's name regularly figures in the leading scientific journals *Nature* and *Scientific American*. On a list compiled by eminent authors, Simon Stevin is one of 10 historical figures whose work, vision and technological know-how paved the way for today's computers; his name is mentioned in the same breath as Pascal, Babbage, Hollerith, von Neumann, Zuse and so on. Stevin's *De Thiende* is on the famous list 'Printing and the Mind of Man', which highlights books that have had a significant impact on the evolution of Western civilization (box 2). It is one of the 65 books listed in the section devoted to the sixteenth century.

Box 2: Stevin's books in the antiquarian book world

The value of early editions of historical printed works is determined by their age, the repute of their author and printer, their illustrations, the number of surviving copies, the situation in which they are found, their provenance and so on.

The tenth-century Archimedes palimpsest, which was mentioned in Chapter One, box 1, is the oldest Archimedes manuscript known to survive.

It is our only source for a work, the *Method of Mechanical Theorems*, that had long been thought lost and it brings us near to one of the greatest scholars who ever lived. Though the manuscript is not in the best material condition, it is virtually beyond price; witness the amount that was paid for it when it was auctioned by Christie's in 1998. The designs and drawings of Leonardo da Vinci are also among the most highly valued manuscripts: in addition to their unique character and historical significance, they also possess an aesthetic component. Other historical scientific manuscripts and printed books that appeal to the imagination are the works of Copernicus, Vesalius and Mercator, for instance. The market value of scientific manuscripts and printed books is possibly assessed a little more objectively than that of a work of art, where taste plays a greater part. The factors that determine the exceptional sums paid for Van Gogh's sunflowers seem less quantifiable than those that determine the price of a printed copy of Huygens's *Horlogium...* or Mercator's *Great Atlas* of 1585. But in both the art and the antiquarian book market it ultimately comes down to supply and demand.

Below are a few of the prices that were paid for Stevin first editions, posthumous publications and translations. In parentheses is the year of auction when the book was sold.

- *De Beghinselen der Weeghconst. De Weeghdaet. De Beghinselen des Waterwichts*, Leiden, 1586: 14,000 guilders (1982)
- *Wisconstige Gedachtenissen*, Leiden, 1605-1608: 2,800 guilders (1971) // 5,400 DM (1981)
- *La Castrametation*, Rotterdam, 1618: 3,250 guilders (1981)
- *L' Arithmetique de Simon Stevin de Bruges*, Leiden, Plantin, 1585, bound with *La Pratique d'Arithmetique*, ibid., 1585: 4,200 DM (1984) // 13,000 DM (1995)
- *Les Œuvres Mathematiques de Simon Stevin de Bruges*, A. Girard, Leiden, 1634: 350 DM (1961) // 750 DM (1979) // 900 guilders (1984) // 1,600 DM (1985) // 2,100 DM, (1998)
- *Nouvelle Manière de Fortification par Escluses*, Rotterdam, 1618: 900 DM (1985)
- *La Castrametation*, Leiden, 1618, bound with *Nouvelle Manière de Fortification par Escluses*, ibid., 1618: 2,200 DM (1990) // 5,000 DM (1995)
- *Problematum Geometricorum*, Antwerp, 1583: 5,600 DM (1996)
- *Nieuwe Maniere van Sterctebou, door Spilsluysen*, Rotterdam, 1617: 55 guilders (1956) // 240 guilders (1967) // 1,400 guilders (1995)
- *De Sterctenbouwing*, Leiden, 1594: 55 guilders (1956) // 130 guilders (1960)
- *Castrametatio, Dat is Legermeting…*Rotterdam, 1617: 840 DM (1974) // 1,500 guilders (1975) // 2,900 guilders (1977)

> *De Beghinselen der Weeghconst, De Weeghdaet, De Beghinselen des Waterwichts* very rarely comes up for sale. According to data from Asher Rare Books (Berlin), a copy was sold at Christie's on 15 June 1998 for 46,000 US dollars. In early 2002 its price had risen by a considerable amount. In sales catalogues such as those published by Christie's *De Beghinselen der Weeghconst...* is described as 'The Single Most Important Work on Mechanics in the 16th Century'.
>
> These data come mainly from the *Jahrbuch der Auktionspreise für Bücher, Handschriften und Autographen* (Dr Ernst Hauswedell & Co – Stuttgart). It seems that especially the first editions of Stevin's works are rarely offered for sale.
>
> By way of comparison, *Traité de la lumière* (1690), a groundbreaking work by Christiaan Huygens in which he postulates the wave character of light, reached 57,500 US dollars when auctioned by Christie's on 15 June 1998. On 11 January 2001 the same work was sold by Sotheby's in New York for 115,750 US dollars.
>
> It can be said that in the world of rare and antiquarian books with its incunabula and historical atlases, the works of Simon Stevin are valued very highly among the printed books of the sixteenth and early seventeenth century. Stevin's inclusion on the famous list 'Printing and the Mind of Man' and the prices paid for his works state this plainly.

As part of the commemoration of the 400th anniversary of Simon Stevin's birth in 1948, the Academie voor Wetenschappen, Letteren en Schone Kunsten van België devoted all 112 pages of one of its bulletins to the Bruges scholar. In 1995 a major exhibition focusing on his work was held in the Bruges Municipal Library, which houses many unique historical manuscripts. In 1998, Ghent University's Central Library mounted an exhibition in which Stevin's works from the library's own collection were confronted with works by his contemporaries. The catalogues of these two exhibitions provide a good overview of what is currently known about the great scientist and inventor. On 11 and 12 December 1998, the authors of the present volume and Dr Charles van den Heuvel organized a symposium entitled *Simon Stevin Bruggelinck* at which specialists from the Netherlands, Denmark and Belgium discussed Stevin's importance to science and technology.

The last 50 years have seen the regular appearance of publications on Stevin. All cannot be listed here, but some of the most important have been mentioned. The work of Eduard Jan Dijksterhuis, one of the best-known Dutch historians of science, has already been mentioned. He and a number of his colleagues published an extremely extensive study of Stevin's collected works, *The Principal Works of Simon Stevin*, in which a considerable part of the Bruges mathematician's oeuvre is reproduced, with commentary and a modern English translation. In this work, and also in *Het Land van Stevin en Huygens* (1966),

Dirk Struik, a Dutch professor of mathematics at MIT (United States), gives his personal view of science in the Golden Age, drawing much attention to Stevin in the process. There are also extensive references to Stevin's works in *Geschiedenis van de wetenschappen in België van de Oudheid tot 1815*, particularly in the chapters by Paul Bockstaele and Patricia Radelet-De Grave (see also Geert Van Paemel).

The exhibition *Maurits, prins van Oranje*, held in the Rijksmuseum in Amsterdam from late 2000 to early 2001, shed much light on the relationship between Stevin and Prince Maurice. The catalogue that accompanied the exhibition examined the contacts between the two men in considerable detail. After 1590 most of Stevin's own publications dealt with topics that were important to Maurice or to the emerging Republic.

In October–November 2004, the Royal Library of Belgium held an exhibition entitled *Simon Stevin (1548–1620) – De Geboorte van een nieuwe wetenschap* ('Simon Stevin (1548–1620) – The Birth of a New Science'). It was accompanied by Dutch and French catalogues with the same title (2004 Brepols Publishers Turnhout), containing contributions by various authors.

Since the publication of the Dutch edition of the present volume, three new books on the work of Stevin have appeared: *Simon Stevin, De la vie civile 1590* (edited by Catherine Secretan, which describes and translates Stevin's *Vita politica*); *De Huysbou* by Charles Van den Heuvel (which presents a reconstruction of an unfinished treatise by Stevin on architecture, town planning and civil engineering); and *Simon Stevin, 'Het Burgherlick Leven'*, by Pim den Boer.

That Stevin was highly regarded by the renowned scientific figures of his own day is clear from the writings of Willebrord Snel, William Gilbert, Isaac Beeckman, Constantijn and Christiaan Huygens, and Adriaan van Roomen. The latter praised Stevin for his statics and his sailing chariot. Beeckman transcribed into his *Journal* a number of Stevin's manuscripts, which he had borrowed from the latter's widow. Hugo de Groot also showed great esteem for his friend Stevin's work. Later, admiring references to Stevin's achievements were made by no lesser scientific luminaries than Joseph-Louis Lagrange in his *Traité de Mécanique Analytique*, Ernst Mach in *Die Mechanik in ihrer Entwickelung: historisch-kritisch Dargestellt*, and Richard P. Feynman in his famous *Feynman Lectures on Physics*.

The young Newton was influenced by John Wilkins's *Mathematical Magick*, a work intended for the general reader that was much indebted to both Stevin and Galileo. It was not just Stevin's works on physics that were influential, however, as is evident from the references to his work on bookkeeping made by bookkeeper and teacher Richard Dafforne in *The Merchants Mirrour* (1635) (fig. 2). Bernard Forest de Bélidor, author of the standard work on civil engineering, *Architecture hydraulique...* [1737–1753], stated that modern hydraulics originated in the Netherlands and that Simon Stevin had been the first to write on innovations in sluice design.

Figure 2: Stevin's work on bookkeeping was very influential, as can be seen from this example of seventeenth-century English specialist literature. From Richard Dafforne, *The Merchants Mirrour...* 2nd edition, London, 1651, printed by J. L. for Nicolas Bourn.
Amsterdam, University Library, UBM: OF 80-140 (ektachrome: JTD).

2. Stevin's influence on some of his contemporaries

2.1. Adriaan van Roomen. 'Ad Staticem Stevinii quod attinet, non facile in latinam linguam vertetur...'

One of Stevin's most outstanding contemporaries was Adriaan van Roomen, who in the traditional manner of Renaissance scholars Latinized his name to Adrianus Romanus. He was born in Antwerp and began his studies at the Jesuit college in Cologne. Later he studied medicine at Leuven. In 1585, on one of his several trips to Rome, he came into contact with Christopher Clavius, one of the translators (1574) of Euclid's *Elements*. From 1586 to 1592 he was professor of mathematics and medicine – a combination that was still possible at that time –

at Leuven University. In 1593 he was appointed as the first professor of medicine at Wurzburg, and between 1603 and 1610 lived alternately in Leuven and the Bavarian town. In 1604 he was ordained as a priest. After 1610 he taught mathematics in Poland. Whatever else, Van Roomen was a much-travelled man. In addition to journeys to and stays in Rome, Heidelberg, Nuremberg (near Wiesbaden), Cracow, Frankfurt am Main, Bologna, Geneva and Prague, he also travelled throughout France.

Van Roomen's letters reveal just how closely he followed the scientific activities of his day. He corresponded with the cream of Renaissance scholars – Ludolph van Ceulen (see Chapter Two, box 4), the abovementioned mathematician and astronomer Christopher Clavius, Johannes Kepler, Justus Lipsius, Jan Moretus of the Officina Plantiniana and Frans van Ravelingen – both sons-in-law of Christopher Plantin – Francois Viète, the eminent French mathematician... the impressive list goes on. He met Kepler, Tycho Brahe and, of course, Justus Lipsius. Unfortunately, much of Van Roomen's correspondence has been lost. In his letters he writes of the square of the circle, the making of goniometric tables, the solution of algebraic equations and the Gregorian reformation of the calendar, with which Clavius was also closely involved. Writing to Clavius, Van Roomen mentions Stevin several times, calling Stevin and Van Ceulen the only mathematicians of the Low Countries. He makes more than one reference to the fact that Stevin published his works in *Nederduyts*. He cites the *Wegkunst* (*sic. De Beghinselen der Weeghconst*) and explains that it is difficult to translate into Latin some of the concepts Stevin introduced in Dutch:

> *Ad Staticem Stevinii quod attinet, non facile in latinam linguam vertetur propter terminos quibus utitur, qui cum vario modo sint compositi, non facile Latine reddentur v.g. Rechthefwicht, Scheef hefwicht etc. ubi quod syllabae tot dictiones, priorem ad verbum ita redderem recte-elevandum-pondus, alterum oblique- (sive ad obliquos angulos) elevandum-pondus...*
>
> ['The terminology and, particularly, the many compound words that Stevin uses in his statics make it difficult to translate the work into Latin. For example, consider *Rechthefwicht* [an upward force] or *Scheefhefwicht* [an oblique force], in which each syllable is a word: rendered literally the first would be *recte-elevandum-pondus* and the second *oblique-(sive ad obliquos angulos) elevandum-pondus*.']

This passage from a letter written to Clavius in 1597 (cited by Bockstaele in his study of Van Roomen) suggests that terms used in *De Beghinselen der Weeghconst*, such as *Rechthefwicht*, were indeed coined by Stevin.

Van Roomen was a gifted mathematician. He excelled in the precision of his argumentation, generally applying existing methods. He calculated π to 16 decimal places. He also wrote a commentary on the algebra of Al-Khwarizmi. Tragically, the only two known copies were lost in 1914 and 1944, when the University Library in Leuven was destroyed by fire.

2.2. William Gilbert and Stevin's *De Havenvinding*

Dijksterhuis regards five years as *anni mirabiles* for the new physics that was emerging:

1543: Copernicus's *De Revolutionibus*
1586: Stevin's *De Beghinselen der Weeghconst* and *De Beghinselen des Waterwichts*
1600: Gilbert's *De Magnete...*
1609: Kepler's *Astronomia Nova*
1638: Galileo's *Discorsi*

William Gilbert's *De Magnete, magneticisque corporibus, et de magno magnete tellure; Physiologia nova* was published in London in 1600. Gilbert, doctor and physicist, had won an excellent reputation as a practitioner of medicine; he was elected president of the Royal College of Physicians and was court physician to Elizabeth I.

The importance of his work in physics is primarily methodological. What he wrote about the nature of magnetism was already known to a large extent. He was, however, the first to make a clear distinction between electrical and magnetic effect. He was also the first, as far as we know, to refer to the earth as a giant lodestone, or magnet. Magnets were especially important to navigation, the magnetic compass having for centuries been an essential instrument for determining position at sea.

Gilbert introduced a robust 'empirical' component in his work. Like Stevin, he was one of those who have created a bridge between theory and praxis. Gilbert believed findings based on experiment to be the only basis for physics. Experiment meant something rather different to Gilbert than it does to us, however. Nowadays it goes without saying that experimental conclusions go hand in hand with the most precise possible mathematical registration of what is observed. It was not so in Gilbert's day. His approach was entirely qualitative. Thus he carried out no measurements but rather recorded what he saw when observing physical phenomena, acknowledging only the observation of nature as a valid method for physical research. His motto, *Nullius in verba* ('On the words of no one'), signifying his commitment to establishing the truth of scientific matters through experiment rather than through citation of authority, would later be adopted by the Royal Society of London. 'Authority' was no longer regarded as an argument in physics. Gilbert's belief in observation as the only arbitrator in science concurred with that of Vesalius, who is generally acknowledged as a pioneer of the empirical approach.

Galileo tells us that he owed his copy of *De Magnete* with its new ideas to 'a famed peripatetic philosopher who presented it to me, I think in order to purge his own library of the contagion thereof'. *De Magnete* claimed Galileo's attention, and Kepler's too, because the ideas expressed in it were closely in line with their search for new principles to support the heliocentric model of the universe.

Gilbert's attitude to the ancient Greeks differed somewhat from those of Stevin or Leonardo da Vinci, for example. Stevin's *Wijsentijt* betrays a certain nostalgia for the time of the great Greek thinkers. Gilbert, the empiricist, distanced himself from them. He said that the ancient Greek had doubtless acquired a very great deal of knowledge but that they would certainly happily accept the more recent findings.

De Magnete contains an introduction by Edward Wright, a fellow of Caius College, Cambridge, and an outstanding navigator, mathematician and cartographer. He published a description of the use of logarithmic tables and translated Napier's works from Latin. In his introduction to Gilbert's book he wrote:

> *Nor is there any doubt that those most learned men, Petrus Plantius ... and Simon Stevinius, a most eminent mathematician, will be not little rejoiced when first they set eyes on these your books and therein see their own λιμνευρετικήν... or method of finding ports so greatly and unexpectedly enlarged and developed ...*

Wright had helped Gilbert in the realization of *De Magnete*. The previous year, 1599, he had translated Stevin's *De Havenvinding* into English as *The Haven-finding Art*. Gilbert mentions this in a footnote:

> *Simon Stevinus, Portuum Investigandorum Ratio, Leyden 1599.* This was printed in English, the same year, by the celebrated mathematician, Edward Wright; it was afterwards attached to the third edition of his *Certaine errors in navigation, detected and corrected.*

Thus Gilbert makes a specific reference to Stevin's *De Havenvinding* barely a year after the appearance of Wright's English translation. Given his good contacts with Wright this is not surprising. Gilbert refuted certain observational details in Stevin's *De Havenvinding*, however he approved of Stevin's method of the inclination. Gilbert states that:

> *Stevinus (quoted by Hugo Grotius), in his Portuum Inveniendarum Ratione* [i.e. De Havenvinding, translated into Latin in 1599 by Hugo de Groot], *distinguishes variation according to meridians.*

Gilbert criticized some of the data on compass variations that Stevin had adopted from Portuguese mariners and others. He mentions Stevin's method thus:

> *But the method of finding the port on long voyages to distant parts by means of accurate knowledge of the variation is of great importance, if only fit instruments be at hand wherewith the deviation may positively be ascertained at sea.*

An interesting contradiction is that Gilbert, who so clearly regarded observation as the only source of knowledge, suddenly went off at an animistic tangent in his endeavours to arrive at a system for his empirical data. This illustrates how difficult it was for the new science to free itself of medieval traditions and convictions. Gilbert was not the only one in this. Kepler explored similar paths and Newton may well have spent more time on mysticism than on the *Principia* and his other work on physics. Stevin and Galileo were much more down-to-earth in this respect. Stevin's motto, *Wonder en is gheen wonder*, clearly indicates the pragmatic character of his research. Galileo was also thoroughly convinced of a new spirit when he abandoned 'what' and made 'how' the central question.

2.3. Willebrord Snel van Royen (Snellius), who translated Stevin's *Wisconstige Gedachtenissen* into Latin

Snellius believed that it was important to make Stevin's work available to the international community and therefore translated the *Wisconstige Gedachtenissen* into Latin: *Hypomnemata mathematica* came out in 1608. The translation was no easy task. In the first place, it was a voluminous work; moreover, Stevin had introduced many new words in statics and hydrostatics, which had no Latin equivalent. He had, it is true, given the Latin words for his Dutch terms here and there in the margin, but very sparingly.

Snellius also found it appropriate to indicate that it was his own idea to translate Stevin. As a rule the prefaces so full of Maurice's praises could well create the impression that it was the prince of Orange himself who had contributed so much to science. Hugo de Groot in particular excelled in composing glowing panegyrics to Maurice: in the case of the translation of *De Havenvinding*, for instance, he seems to imply that the translator was the prince, who 'let himself speak as though through an interpreter'. Snellius was more restrained in this respect.

2.3.1. Snellius's name is forever associated with the law of light refraction
Beeckman's *Journal* tells us how, in his lessons at Leiden, Snellius explained the heliocentric system with the aid of didactic material he had made with his own hands:

> *Willebrordus Snellius cùm explicaret motum trepidationis in Terrâ, sinistrâ tenuit globum ligneum loco Solis quiescentis, dextrâ autem alterum globum, oblongo ligno priori annexum, loco Terrae mobilis. Volvebat circa Solem ita ut poli Terræeandem plagem semper respicerent ad similitudinem Terrae Verae...*

> ['To explain the precessional motion of the earth, Willebrord Snel held in his left hand a wooden ball to represent the stationary sun and in his right a ball that was attached to the first by a long stick and which represented the moving earth. He made this turn around the

sun so that the poles of the earth were always orientated in the same direction, like the real earth...']

Snellius's lessons were undoubtedly inspired by his reading of Stevin's *Vanden Hemelloop*. Willebrord Snel (his father Rudolph was a mathematician and a man of letters) ranked among the great scholars, and his contributions to science are of historical importance. His name is given to the law of light refraction. The passage of light through lenses was of crucial importance for both the microscope and the telescope. Galileo and Christian Huygens both discovered new moons of planets through the newly invented telescope. In 1683 Antoni van Leeuwenhoek discovered microbes by means of a microscope. In the grinding of lenses – Huygens hand-ground his own lenses, and very finely too – Snellius's law was also vital. Kepler had given an impetus to the law of light refraction. It is possible that Descartes discovered it independently of Snellius, who formulated this law at the end of his life – in 1626 at the latest. Huygens saw the unpublished manuscript shortly after his death. In 1626 Descartes told Isaac Beeckman that he had derived the law of refraction in 1626. But Snellius empirically obtained an analysis of the law that bears his name; Descartes did not.

In Stevin's work we come across two references to light refraction. In the Summary of *Vande Deursichtighe*, which forms part three of the *Wisconstige Gedachtenissen*, he announces that his treatise is to consist of three books (see also Chapter Nine): *Vande Verschaeuwing* (perspective, also called 'scenography'), *Vande Beginselen der Spieghelschaeuwen* ('reflections in mirrors' or catoptrics), and *Vande Wanschaeuvving* (refraction). The book on refraction has not reached us. We know, however, that Stevin used the word *wanschaeuwing* for refraction: in the margin of the Summary he gives the Latin term, *refractione*. He speaks of *wanschaeuwing* a second time in his explanation of the Nova Zembla phenomenon.

Snellius, who was also highly adept at the technique of calculation, proposed the method of triangulation for determining the length of the meridian. The triangulation method itself was developed by the mathematician and mapmaker Gemma Frisius. The Flemish cartographers made frequent use of it in determining long distances.

Triangulation is the tracing and measurement of a series or network of triangles to determine the distances and relative positions of points spread over an area by measuring the length of one side of each triangle and deducing its angles and the length of the other two sides by observation from this baseline. Cartographers used church towers as observation points and vertices for their triangles. Snellius sought to determine the meridian – the circumference of the earth. The result of his first measurement, expressed in modern terms, was 38,500 kilometres.

Stevin also dealt with the problem of determining position and distance, chiefly in *Vande Zeylstreken* and *De Havenvinding*, though he was mainly concerned with the finding of longitude at sea.

Willebrord Snel made several journeys abroad, including trips to Prague and Wurzburg. Tycho Brahe, Johannes Kepler and Adriaan van Roomen were among the scientists he met on his travels. Kepler called him *geometrarum nostri seculi decus* – 'adornment to the geometry of our century'. We can assume that he and Stevin knew each other.

Willebrord's father, Rudolph Snel, was a professor at Leiden when Stevin was a student there. They were both members of a commission and Stevin had named him a guardian of his children although, for reasons unknown, Willebrord declined to act as such.

2.4. Isaac Beeckman and his Journal. References to and transcriptions of Stevin's work

Isaac Beeckman's name may be fairly unfamiliar to the general reader, yet in some of his ideas on physics he was ahead of his time. Beeckman was born in Middelburg (in the province of Zeeland). In 1607 he began his studies at Leiden, where Rudolph Snel was very likely one of his tutors, and Willebrord Snel may well have been a fellow student. In 1618 he graduated in medicine from Caen in France and, already in that same year, he defended the proposition that the rise of water in a pump is attributable not to the *horror vacui* but to air pressure. At the same time he postulated a version of the principle of inertia, though in this he had been preceded by Ibn al-Haytham, an Arab scholar born in 965. In 1618 in Breda, Beeckman came to know René Descartes. Subsequently, the two men worked together and derived a formulation of the law of fall.

The *locus classicus* for the proof of the law of free falling bodies is to be found in the Third Day of Galileo's *Discorsi* of 1638. There had, however, been several previous attempts to understand it. First, there was De Soto's work in Spain around 1540. Then there were the earlier attempts by Galileo himself, which certainly dated back to 1604. It is generally assumed that in 1618 Beeckman and Descartes carried out their work on the law of fall independently of Galileo. Nevertheless, there are remarkable similarities between Galileo's work and that of the two scholars in Breda. The history of the law's origin and related concepts is highly complex and entire volumes have been dedicated to it. It was only when Newton gave his masterly synthesis in mechanics that the law of fall was transparently interpreted. Before that the conceptual situation was distinctly muddy.

In his exploratory researches Galileo had to start with medieval traditional concepts which, with ups and downs, were transformed into what is now referred to as 'preclassical mechanics'. As regards the law of fall, De Soto and Descartes and Beeckman were in the same position.

Although Beeckman's scientific work is to be found only in his *Journal*, which was discovered in 1905 and published between 1939 and 1953, he influenced scientific development through his collaboration, first with Descartes and subsequently, with Mersenne and Pierre Gassendi. In his role as rector of the Latin school at Dordrecht, he was also influential. From his voluminous notebook-cum-diary it appears that Beeckman had constructed a very coherent mechanical worldview, though the fact that he neither published his insights nor produced completely worked out propositions meant that his influence did not spread as far as it might have done. His mastery of mathematics is not to be compared with, for instance, Stevin's. Beeckman's talent lay primarily in his intuitive insight. He saw the importance of coupling mathematical reasoning with experimental testing for the advancement of what was then referred to as 'natural philosophy'.

The *Journal* is a remarkable, disorganized document. Beeckman made notes on anything and everything. Here is half a page in Dutch, there a passage in Latin, interspersed with quotations in Greek.

Our interest here lies first and foremost in points of contact between Beeckman and Stevin. Beeckman was 32 when Stevin died. As far as we know, they were not personally acquainted, though Beeckman was certainly very familiar with Stevin's works. As a pupil of Rudolph Snel and a fellow-student of Willebrord Snel, he would undoubtedly have come into contact with them. On 15 June 1624, Beeckman visited Stevin's widow at Hazerswoude. He recorded in his *Journal* a list of Stevin's works that he had been allowed to inspect or borrow. Thanks to Beeckman we have extracts from the following unpublished Stevin manuscripts: *Huysbou, Spiegheling der Singconst, Cammen ende Staven, Watermolens ende Cleytrecking, Waterschueringh* and *Van de Crijchconst*. Beeckman's extensive transcribing has greatly contributed to the recording of Stevin's unpublished works. Later, Stevin's son Hendrick would publish the surviving manuscripts. He complained that his mother had dealt rather thoughtlessly with his father's manuscripts.

In his *Journal*, Beeckman makes many comments on the propositions and statements from Stevin's work. As a source he had the *Wisconstige Gedachtenissen* and his transcriptions of the unpublished manuscripts. In several places, he appeals to the *Beghinselen der Weeghconst* and *De Beghinselen des Waterwichts* to prove his own insights, and some of the diagrams he uses are virtual copies of those in Stevin's works – where he discusses the motion of bodies on an inclined plane or the pressure on the bottom of a water-filled vessel for instance, or where he too proposes a kind of 'Almighty', albeit with pulleys instead of the cogwheels Stevin had used (fig. 3).

Stevin's influence is also manifest in Beeckman's statement that 'to explain' is to reduce something wondrous to something that can be understood, which amounts to the same thing as Stevin's *Wonder en is gheen wonder*. Like Stevin, Beeckman rejected the idea of perpetual motion and he employed terms such as *staltwichtig* (in an application with pulleys) and *scheefwicht*, words which Stevin had been the first to use. Beeckman also copied into his *Journal* notes written by Descartes under the title *Parnassus*, from which it appears that Descartes knew Stevin's work on the force a liquid exerts on the wall of a vessel. Only occasionally does Beeckman differ with Stevin, for instance about elements of the *Singconst*; otherwise he cites Stevin as an obvious authority, certainly in the matter of statics, hydrostatics and bookkeeping. In his treatment of how pulleys work, he writes *Stevyn Tartagliae praefertur*... – 'Stevin is to be preferred to Tartaglia...' (*Journal*, 15–22 May 1633).

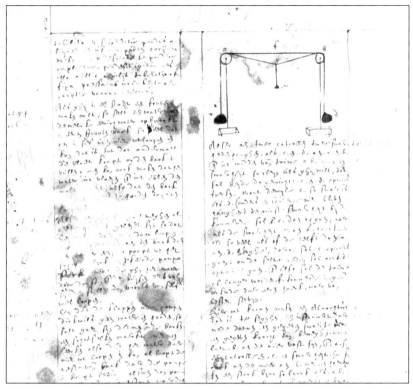

Figure 3: Inspired by Stevin's 'Almighty', Isaac Beeckman designed his own version: an *Almachtich van touwen*. From the *Journal tenu par Isaac Beeckman* (April 1614 – January 1615).
Middelburg, Zeeuwse Bibliotheek (ektachrome: JTD).

2.5. Christaan Huygens

It seems highly unlikely that a young Dutch mathematician and physicist would become the principal member of the group of distinguished scientists who made up the newly founded Académie Royale des Sciences in Paris, yet this is what Christiaan Huygens did. In the Netherlands, Christiaan's father, the eminent man of letters and diplomat Constantijn Huygens, is perhaps better known, but in global terms, Christiaan's fame is greater. Indeed, he is generally acknowledged as one of the greatest scholars in history.

2.5.1. A child prodigy studies De Beghinselen der Weeghconst

In the list of the year's 'textbooks' that were set for the fifteen-year-old Christiaan Huygens in 1644 by Jan Stampioen, the private tutor hired by his father, we find among others the following: Tycho Brahe, *Astronomiae instauratae progymnamata*; René Descartes, *Discours de la méthode*; Johannes Kepler, *Mathematici dioptrice*; Nicolaus Copernicus, *De revolutionibus orbium coelestium*; Apollonius Pergae, *De elementa conica*; Ptolemy, *Caelestium motuum*; Simon Stevin, *De Beghinselen der Weeghconst* and *Wisconstige Gedachtenissen*... Clearly, Master Stampioen meant to bring his exceptionally talented young pupil into speedy contact with the prevailing scientific thought. The list of well-chosen works also shows that Stevin was already regarded as one of the authoritative authors.

In his own work, Christiaan Huygens made crucial use of Stevin's composition law of forces. To calculate the shape of a rope or chain suspended by its ends (the 'chain–curve'), he studied a model that consisted of a set of identical spheres connected to each other by cords of equal length. The model is clearly reminiscent of Stevin's *clootcrans* or 'wreath of spheres'. In combining the weights of the spheres and the forces that operate between them he called on Stevin's composition parallelogram. As a result of this study, carried out when he was only seventeen, Christiaan Huygens realized that the contention by Galileo and others that the curve of a chain hanging under gravity would be a parabola did not correspond to reality. Even though the mathematical operation that Huygens set down in 1646 is not without flaw – it was only much later, in 1691, that he would arrive at a conclusive geometric argument – his early insight was nonetheless correct and interesting. In 1691 he was also the first to use the term *catena* for the curve – now known as the catenary – being sought (fig. 4).

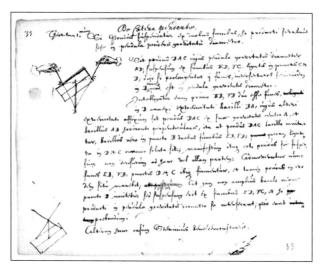

Figure 4: At an early age, Christiaan Huygens was already very familiar with Stevin's work, especially *De Beghinselen der Weeghconst* and *De Beghinselen des Waterwichts*. He refers to Stevin in this 1646 study of the chain–curve. 'De Catena pendente. Theorema 1.^{mum} Si, pondus suspendatur ex duobus funibus ...' ('On the chain–curve. Theorem 1. If a weight be suspended by two cords...') The last line reads: *'Alterum hunc casum Stevinius bene demonstravit'*. ('Stevin has demonstrated this second case very well.')
Leiden, University Library, HUG 17, fol. 23^v (ektachrome: JTD).

In 1657 Christiaan Huygens, in partnership with his clockmaker, patented a pendulum clock. It was based in part on his grasp of the properties of another famous curve, the cycloid. This is the curve traced by a point on a circle being rolled along a straight line. If a cycloidal curve is positioned like an inverted arch, and a ball bearing, say, is released from any point on it (except for the bottom), it will always take exactly the same time to reach the bottom, irrespective of where it starts from on the curve. This enabled Huygens to construct a pendulum whose oscillation took the same length of time no matter how large or small its swing was.

In working out the mathematical underpinning of the pendulum clock, Huygens called, at least implicitly, on Stevin's composition of forces. This is not surprising, since around the middle of the seventeenth century *De Beghinselen der Weeghconst* represented the pinnacle of knowledge in statics.

Huygens, who played the harpsichord, also came into contact with Stevin when he turned his mind to music. Like Stevin and other mathematicians he made studies of music that were inspired by and based on mathematics. Around 1662 Huygens was making notes – which have not survived in their entirety – on the natural dominant tone and how it could be framed in a mathematical order. He knew Stevin's *De Spiegheling der Singconst*. Indeed, the very survival of this unpublished work is partly due to the copy in the Huygens family library. Christiaan was also fascinated by Mersenne's *Harmonie universelle* (1636–37). He endeavoured, like Stevin before him, to work out and put into practice an optimal system of equal temperament (see Chapter Eleven). Huygens differed with Stevin, who stated that his mathematical values for the tempered musical intervals were the 'true' ones. Huygens was sensitive to the fact that in meantone tuning a cadence in 'E' is somewhat 'more plaintive and soft' than the cadences in 'D'. In the mathematical equal temperament such subtle nuances of tonal diversity are lost.

There is another point of contact between Huygens's and Stevin's work in hydrostatics. Stevin discovered the hydrostatic paradox, a contribution of the first magnitude (see Chapter Six). He was also the first to calculate the force exerted by a liquid on the wall of a vessel. For his part, Huygens – 66 years after Stevin's work – supplemented the classical studies of floating bodies made by Archimedes with *De iis quae liquido supernatant*. This work, which also incorporated the ideas of Torricelli, was published posthumously, more than a 100 years after Stevin's *De Beghinselen des Waterwichts*.

Huygens also worked on theories of collision and centrifugal force and on optics, but he is most firmly associated with the fundamental principle, named after him, that governs the propagation of light. He was outstanding for his skill in grinding lenses and making telescopes. In 1655 he discovered Titan, a satellite of Saturn. His creative work was on a par with that of Newton, Kepler, Leibniz and the Bernoullis, with whom he kept in touch. As regards mechanics and hydrostatics, one might refer to a line of thought running from Archimedes to Stevin to Huygens.

2.6. Nicolaes Witsen. 'I am more convinced by the discourse of the renowned Mr S. Stevin...'

Stevin was not only one of the most creative scientists of his age, but his patents and contributions to navigation, warfare, drainage mills and sluices, hydraulic engineering, town planning and architecture also show him to have been an extraordinarily inventive engineer and a pioneer in the undergirding of technological applications with mathematical and physical analyses and calculations (see box 3).

Box 3: Simon Stevin: an innovator and polymath

(The text in this box was translated by Dr Francis Wray, from: J.T. Devreese, in *"Simon Stevin (1548–1620) – De geboorte van de nieuwe wetenschap"*, Epiloog., page 149. Catalogue, 2004 Brepols Publishers, Turnhout, ISBN 90-5622-054-3 [in French: ISBN 2-503-51704-8].)

Although we have a reasonably good understanding of the methodology, depth and perspective of Stevin in his work on statics and hydrostatics, there remain several open questions. For example, was Stevin familiar with the early work on mechanics from the schools of Jordanus and of Paris? Why did he not pursue the study of dynamics after his pioneering experiments on falling bodies? Such questions address the larger and more important issue of the positioning of Stevin and his work.

For his time, Stevin's profile was that of an innovative man of science. However, he was not immediately accepted by traditional scholars. During the early years after his arrival in Leiden in 1581, Stevin dedicated himself to mathematics and mechanics, until in 1585 he became, *inter alia*, a hydraulic engineer. How did Stevin come to immerse himself so deeply in mechanics?

Around 1584, Justus Lipsius, a scholar of the then humanist orthodoxy, distanced himself from Stevin and described him as *'a mathematician only who knows no other sciences, hardly even the language* [Latin – Authors]*, and who is a technician rather than a theoretician'*. Was Stevin excluded from the world of traditional scholars? Did he try to transcend that world? Was the initially negative attitude of Lipsius towards Stevin nothing more than the antithesis between classical humanities and applied engineering? Did Stevin's illegitimacy, like that of Da Vinci, play a role in his career? Or did Stevin lack academic qualifications?

The historical significance of Stevin is intimately linked to the new philosophy that he promoted and applied in his science and technology – practice. According to Stevin, theory must go hand in hand with practice and experiment. We see how Stevin, rather than limiting himself to a deeper study of dynamics, progressively involved in a broader spectrum of disciplines: logic, the organization of the State, language, geography, cosmology, etc. Was this Stevin's way of distinguishing himself from traditional scholars?

Whereas in his early works Stevin had the objective of acquiring new knowledge, always combining *Spiegeling en Daet*, he later stressed the importance of reacquiring knowledge that had been known long ago, in the so-called *Wysentyt*, but which was lost in the mists of time. Was this partial reorientation of his standpoint towards that of the traditional scholars a way of his acquiring their status? Or was it a political move to make his own philosophy more acceptable to the establishment of classical scholars, including Hugo de Groot, who was himself greatly influenced by Stevin?

Through the use of references, Stevin constantly demonstrated his excellent knowledge of the writers of classical antiquity. Was this an attempt to impress classical scholars? This did not, however, prevent him from confidently demonstrating that ancient works have weak points and that he could go further and devise new solutions.

Whatever Stevin's initial handicaps were, his career comprised a series of successes. Even before 1590, he became the tutor and, for the rest of his life, confidant of Prince Maurice. He often had his tent next to the Prince's in army camps. The Prince showed off the sailing chariot invented by Stevin. He then entrusted him with the important task of creating a school of engineering at Leiden.

But how could Stevin, who was not of noble birth, even though supported by an influential and rich family, become so influential at the court of Maurice (Nassau)? Perhaps the well-documented incident with Emmius hints at Stevin's diplomatic skills. There have been many references to the attack on Stevin's heliocentric viewpoint in 1608 by Ubbo Emmius, who later became the first rector of the University of Groningen. This attack could potentially displease the Prince. Was this not the frustration and jealousy of a classical scholar towards the tutor and influential adviser at the court, who, nevertheless, was not a humanist scholar? It is striking that Emmius expressed his criticisms, which, after all, could indirectly damage the Prince, only by correspondence so as to avoid divulging them publicly. Does not Stevin display a superior sense of diplomacy when, in the *Wisconstige Gedachtenissen*, he comments exhaustively on his dialogues with Maurice, except in the chapter where he defends his heliocentric viewpoint?

It is not known exactly why Stevin did not become a university professor like Rudolf and Willebrord Snellius. Was this related to his studies? In the register of Leiden University one reads *'studuit artes apud Stochium'*. As the 'artes' are the preparatory studies, this suggests that Stevin had not studied at University before 1581. Undoubtedly, he was to a large extent self-taught. Neither is there any proof that he obtained any academic diploma at the end of his studies at Leiden University. At that time, was he already too preoccupied with his own research which would result in works of historical significance? Or did Stevin, because of his novel ideas, displease classical scholars? Or was his prominent position at the court of Maurice, with a much higher salary than that of a professor, with greater influence and at the centre of power in the Republic, just more important for him?

> In any case it is important to realise that the study of mechanics was the main activity of Stevin only for a limited period from 1581 to 1586.
> Literature: R. Vermij (2002); C. van den Heuvel (2000); G. Van Paemel (1995).

So how was Stevin's work received by technologists? We find interesting evidence on this point in the book *Aeloude en hedendaegsche scheepsbouw en bestier*, (*Ancient and Modern Shipbuilding and Sailing*), written by Nicolaes Witsen at the age of 20, though it appeared only 10 years later, in 1671.

Witsen was mayor of Amsterdam, a governor of the VOC (the Dutch East India Company) and a special envoy to London. He was, moreover, an exceptionally cultivated and well-travelled man. He was a Fellow of the Royal Society and had a predilection for cartography. It was he who arranged an incognito apprenticeship for Peter the Great at the Amsterdam wharves of the VOC. He was befriended by the tsar, whom he introduced to William III, prince of Orange. He even spent a year or so in Moscow, and wrote the first book to make Russia known in the Netherlands.

Simon Stevin is often mentioned in Witsen's splendidly illustrated book on shipbuilding. In the seventeenth chapter, Witsen deals with the mathematics of how much water rests against the side of a ship and the measurements of ships. In a style highly reminiscent of Stevin himself, with propositions, examples and notes, he determines the weight that operates on the bottom or side of a ship. He also examines the centre of gravity – 'how to know the weight of a ship, how far a ship will sink in the water' – and related practical matters. He describes tests with immersed vessels partially filled with lead balls. Several of his diagrams (such as those illustrated in fig. 5) seem to have walked out of Stevin's works, to which he refers thus: 'The renowned Mr S. Stevin brought me to this in his *Waterwichts*, which can never be too highly praised.' Indeed, Witsen's first proposition is an application of Stevin's hydrostatic paradox.

It is striking that Witsen adopted what later came to be called 'Pascal's law' from Stevin's *De Beghinselen des Waterwichts*, 'because a point in the water (…) is pressed equally from all sides; on this, see Stevin's *Waterwichts*…' In the 'explanation of the shape of many parts of ships', in which he also dealt with the force exerted by the sails on the mast, Witsen gave pride of place to Stevin's insights. Having discussed the theory of the lever as propounded by Aristotle, Archimedes and Guidobaldo del Monte he writes, 'But all these proofs seem to be somewhat contradictory, which is why I leave them behind. I am more convinced by the discourse of the renowned Mr S. Stevin, which follows.'

Witsen's book shows that Stevin's work had an impact in the seventeenth century and was highly esteemed, not only by scholars such as Snellius, Beeckman and Christiaan Huygens but equally by those who put into practice the technology Stevin described, of which shipbuilding was a high point.

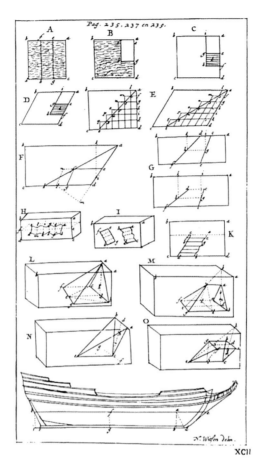

Figure 5: Nicolaes Witsen, an exceptionally cultivated and well-travelled man, Fellow of the Royal Society, member of the Dutch East India Company's governing body, mayor of Amsterdam and friend of Peter the Great, makes many references to Stevin's works in his splendid book on shipbuilding. From Nicolaes Witsen; *'Aeloude en hedendaegsche scheepsbouw en bestier,: ...'*. Printed at Amsterdam, by Casparus Commelijn, Broer and Jan Appelaer, 1671.
Archive JTD.

3. Later reactions of eminent scholars

From the eighteenth century up to the present day, many eminent scholars have commented on Stevin's work. We shall take a brief look at what has been said by three of them: Joseph Louis Lagrange, Ernst Mach and Richard Feynman.

3.1. Joseph Louis Lagrange. 'Cette démonstration de Stevin... trés ingénieuse...'

Joseph Louis Lagrange was one of the most brilliant and important mathematicians of the eighteenth century. He also made fundamental contributions to mechanics. His insights were so far-reaching that his name lives on today in core concepts of modern physics, such as the 'Lagrangian', and in fundamental theorems in higher mathematics.

The powerful mathematical structure that Lagrange devised for mechanics has even received a proper name, *Mécanique Analytique*, the title of his great treatise on the subject, published in 1788. In this masterly and enduring work, which William Rowan Hamilton described as a 'scientific poem' for the elegance of the analysis, Lagrange refers to Stevin's work. Where he studies the composition of forces he refers to Stevin's *clootcrans* and writes:

> *J'ai rapporté cette démonstration de Stevin, parce qu'elle est très ingénieuse et qu'elle est d'ailleurs peu connue. Au reste, Stevin déduit de cette théorie celle de l'équilibre entre trois puissances qui agissent sur un même point, et il trouve que cet équilibre a lieu lorsque les puissances sont parallèles et proportionnelles aux trois côtés d'un triangle rectiligne quelconque. (Voir les Éléments de Statique et les Additions à la Statique de cet auteur, dans les Hypomnemata mathematica, imprimés à Leide en 1605, et dans les Œuvres de Stevin, traduites en français, et imprimées en 1634 par les Elzevirs.)*

Thus Stevin's *clootcrans* proof made an impression on the man who already in 1766 had been referred to by Frederick the Great as 'the greatest mathematician in Europe'.

3.2. Ernst Mach. 'Die Stevin'sche Ableitung ist eine der werthvollsten Leitmuscheln in der Urgeschichte der Mechanik...'

Ernst Mach was a philosopher, positivist and physicist. In his famous and influential work *Die Mechanik in ihrer Entwickelung* (1883), he criticized Newton's concept of absolute time and space. His most lucidly written book influenced Einstein's formulation of the special theory of relativity. One of the starting points for Einstein's general relativity theory was what he called 'Mach's Principle'. Supersonic speeds are now measured in 'Mach' numbers.

Mach begins *Die Mechanik in ihrer Entwickelung* by presenting a clear overview of the history of statics and hydrostatics, which includes a thorough review of Stevin's work. Of the inclined plane Mach writes:

> *Stevin untersuchte zuerst die mechanischen Eigenschaften der schiefen Ebene und zwar auf eine ganz originelle Weise.*

And further:

> *Die Stevin'sche Ableitung ist eine der werthvollsten Leitmuscheln in der Urgeschichte der Mechanik und wirft ein wunderbares Licht auf den Bildungsprocess der Wissenschaft, auf die Entstehung derselben aus instinctiven Erkenntnissen.*

Also:

> *Die Betrachtung von Stevin erscheint uns so geistreich, weil das Resultat, zu welchem er gelangt, mehr zu enthalten scheint, als die Voraussetzung, von welcher er ausgeht.*

In his analysis of Stevin's *clootcrans* proof Mach draws his own diagrams. He goes more deeply into the parallelogram of forces 'nach dem Stevin'schen Princip'. He also discusses Stevin's contribution to hydrostatics. Mach made a thorough study of Stevin's work and, like Lagrange, was eloquent in his praise of it:

> *Die Stevin'schen Fictionen, z.B. jene der geschlossenen Kette auf dem Prisma, sind ebenfalls Beispiele eines solchen weiten Blickes. Es ist die an vielen Erfahrungen geschulte Vorstellung ... Für Stevin's feines Forschergefühl besteht in der Fiction ein Widerspruch, der weniger tiefen Denkern entgehen kann.*

It is particularly significant that Mach, who seems to have invented the word *Gedankenexperiment* (thought experiment) and who endeavoured to fathom the significance of thought experiments, always produced Stevin's *clootcrans* as his favourite example.

3.3. Richard Feynman. '...because the chain does not go around...'

Richard P. Feynman (Nobel Prize in Physics, 1965) is a name familiar not only to physicists but also to a wider public, thanks to his membership of the panel that investigated the Space Shuttle Challenger disaster. Feynman summed up his conclusions anent the tragedy in the following insightful words: 'For a successful technology, reality must take precedence over public relations, for Nature cannot be fooled.'

The three volumes of the *Feynman Lectures on Physics* (the 'little red books' of physics) provide a dazzling introduction to modern physics. When he comes to the composition law of forces, having already discussed two fine, albeit conventional, derivations, he writes:

> *Cleverness, however, is relative. It can be deduced in a way which is even more brilliant, discovered by Stevinus and inscribed on his tombstone'* [though the latter statement is erroneous].

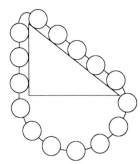

Figure 6: In *The Feynman Lectures on Physics* the inspirational physicist Richard Feynman speaks with great admiration of the inventiveness that underlay Stevin's *clootcrans* proof.

Then Feynman deals with the *clootcrans* (fig. 6):

> *explains that it has to be 3/5 of a pound* [obviously this problem involves specific numbers], *because the chain does not go around...*

That the *clootcrans* proof still appeals to the imagination is illustrated by the figure that was chosen to adorn the cover of the book of exercises that accompanied the *Feynman Lectures*.

Epilogue

Elizabeth Eisenstein writes in 'The Book of Nature Transformed' that the young Isaac Newton was stimulated by John Wilkins's *Mathematical Magick...* (1648), a popular scientific treatise that was in turn indebted to the works of Simon Stevin and the early tracts of Galileo.

On 5 February 1676 Newton wrote, 'If I have seen further it is by standing on the shoulders of giants'.

Simon Stevin was one of those giants.

Bibliography

An extremely detailed review of Simon Stevin's oeuvre can be found in the following works. The first is a commented study of Stevin's works. The following six reproduce his original works with a parallel English translation.

Dijksterhuis, E.J., *Simon Stevin*, Nijhoff: 's Gravenhage, 1943.

Dijksterhuis, E.J., (ed). *The Principal Works of Simon Stevin, vol. I, Mechanics*, N.V. Swets & Zeitlinger: Amsterdam, 1955.

Struik, D.J. (ed). *The Principal Works of Simon Stevin, vol. IIA, Mathematics*, N.V. Swets & Zeitlinger: Amsterdam, 1958.

Struik, D.J., (ed). *The Principal Works of Simon Stevin, vol. IIB, Mathematics*, N.V. Swets & Zeitlinger: Amsterdam, 1958.

Pannekoek, A & Crone, E. (eds). *The Principal Works of Simon Stevin, vol. III, Astronomy, Navigation*, N.V. Swets & Zeitlinger: Amsterdam, 1961.

Schukking, W.H. (ed). *The Principal Works of Simon Stevin, vol. IV, The Art of War*, N.V. Swets & Zeitlinger: Amsterdam, 1965.

Forbes, R.J., Fokker, A.D & Romein-Verschoor, A. (eds). *The Principal Works of Simon Stevin, vol. V, Engineering, Music, Civic Life*, N.V. Swets & Zeitlinger: Amsterdam, 1966.

These works naturally contain many bibliographical references. Below is a summary of works consulted in the preparation of the present volume, and of very recent literature, that do not appear in them.

Andersen, K., Stevin's theory of perspective: the origin of a Dutch academic approach to perspective. *Tractrix*, **2**, pp. 25–62, 1990.

Andriesse, C.D., *Titan kan niet slapen; Een biografie van Christiaan Huygens*, Contact: Amsterdam/Antwerp, 1994.

Arblaster, P., *Antwerp and the World. Richard Verstegan and the International Culture of Catholic Reformation*, Leuven University Press, 2004.

Arblaster, P., *A History of the Low Countries*, Palgrave Macmillan: Basingstoke, 2006.

Barbour, J.M., *Tuning and Temperament*, East Lansing: Michigan, 1951.

Beeckman, I., *Journal tenu par Isaac Beeckman de 1604 à 1634 publié avec une introduction et des notes par C. de Waard*, 1939–1953.

Beek, L., *Dutch Pioneers of Science*, Van Gorcum: Assen/Maastricht, 1985.

Beernaert, B & Schotte, B. (with contributions by D'hondt, J & Leenders, K.), *Over de Koe, het Hert, de Uil..., Zes huizen, Zes Verhalen*, Bruges Municipal Archive, Publicaties Levend Archief, 2000.

Bélidor, de, B.F., *Architecture hydraulique ou: L'art de conduire, d'elever et de menager les eaux pour les differens besoins de la vie*, Charles-Antoine Jombert: Paris, 1737–1753.

Bierens de Haan, D., *Bouwstoffen voor de geschiedenis der Wis-en natuurkundige Wetenschappen in de Nederlanden*. Reproduced from the *Verslagen en Mededelingen der Kon. Academie van Wetenschappen, Afd. Natuurkunde*, 2nd series, VIII, IX, X, XII, 1878.

Bierens de Haan, D., *Simon Stevin, 'Vande Spiegeling der Singkonst' et 'Vande Molens'*, Amsterdam, 1884.

Blockmans, W & Prevenier, W., *The Promised Lands. The Low Countries Under Burgundian Rule*, (trans. Elizabeth Fackelman), University of Pennsylvania Press: Philadelphia, PA, pp. 1369–1530, 1999.

Bockstaele, P.P., The correspondence of Adriaan van Roomen. *Lias III*, Holland University Press: Amsterdam, 1976.

Brandt, G., *Poezy* (3 vols.), Willem Barents: Amsterdam, 1725–1727.

Claes, F., *Simon Stevin als bron voor Kiliaan, Tijdschrift voor Nederlandse Taal- en Letterkunde*, 111, 55, 1995.

Cohen, H.F., Simon Stevin's Equal Division of the Octave. *Annals of Science*, **44**, pp. 471–488, 1987.

Coppens, T., *De vrouwen van Willem van Oranje*, fiction, De Fontein: Baarn, 1983.

Coppens, T., *Maurits, zoon van de Zwijger*, fiction, De Fontein: Baarn, 1984.

Dafforne, R., *The Merchants Mirrour*, R. Young for Nicolas Bourne: London, 1635.

Dalton, R., Through a looking glass. Did major artists use optical devices to plot points to help them paint? *Nature*, **412**, p. 860, 2001.

Damerow, P. et al., *Exploring the Limits of Preclassical Mechanics*, Springer-Verlag: Berlin, 1992.

Damisch, H., *L'origine de la perspective*, Flammarion: Paris, 1987.

Della Francesca, P., *De prospectiva pingendi* (ms Parmensis, Bibliotheek Palatine, Parma), 1576.

Den Boer, P & Fleurkens, C.G., *Simon Stevin. Het Burgherlick Leven*, Erven J. Bijleverd: Utrecht, 2001.

De Reu, M., Vanden Berghe, G & Van Hooydonk, G., *Simon Stevin*, exh. cat., Ghent, 1998.

De Vos, D., *Hans Memling; Het volledige œuvre*, Mercatorfonds Paribas: Antwerp, 1994.

Devreese, J.T., Simon Stevin Brugghelinck, "Spiegheling en Daet", *Spiegheling en Daet*, exh. cat., Bruges: De Biekorf, 9 December 1995 – 31 January 1996. *Ibid.*, exh. cat. accompanying the symposium and public lectures marking the 450th anniversary of Stevin's birth, 11–12 December 1998.

Devreese, J.T & Vanden Berghe G., Stevin, Flemish tutor to a Dutch prince. *The Low Countries,* **10**, 252–256, 2002.

de Vries, H. Vredeman, *Perspective...* Lugduni Batavorum, Henric. Hondius sculps. et excud. (Colophon: s'Graven-haghe, Beuckel Corneliszoon Nieulandt, for Hendrick Hondius (1604–1605)).

De Waal, P., *Van Paciolo tot Stevin, een bijdrage tot de leer van het boekhouden in de Nederlanden*, J. Romen en zonen: Roermond, 1927.

Dijksterhuis, E.J., Simon Stevin, *Simon Stevin,* **25**, pp. 1–21, 1947.

Dijksterhuis, E.J., *Simon Stevin: Science in the Netherlands around 1600*, Nijhoff: The Hague, 1970.

Dijsterhuis, E.J., *Clio's Stiefkind*, Bert Bakker: Amsterdam, 1990.

Donche, P., De familie Cant te Veurne in de 15de en 16de eeuw. *Vlaamse Stam,* **38,** 68–82, 2002.

Donche, P., Voorouders van de wiskundige Simon Stevin te Veurne. *Vlaamse Stam,* **38**, 178–199, 2002.

Feynman, R. P., Leighton, R.B & Sands M., *The Feynman Lectures on Physics*, Addison–Wesley: Palo-Alto, 1963.

Galilei, V., *Dialogo della musica antica e moderna*, Giorgio Marescotti: Florence, 1581.

Geschiedenis van de wetenschappen in België van de Oudheid tot 1815, Gemeentekrediet: Brussels, 1998.

Gilbert, W., *De Magnete*, Peter Short: London, 1600.

Haulotte, R & Stevelinck, E., *Luca Pacioli*, Instituut der Accountants: Brussels, 1994.

Hildebrandt, S. *et al.*, *Mathematics and Optimal Form*, Scientific American Books: New York, 1985.

Huygens, C., *Œuvres Complètes* (22 vols.), Martinus Nijhoff: The Hague, 1888–1950.

Israel, J.I., The Dutch Republic. Its Rise, Greatness, and Fall 1477–1806. *Oxford History of Early Modern Europe*, Clarendon Press: Oxford, 1998.

Kemp, M., Simon Stevin and Pieter Saenredam: A study of mathematics and vision in Dutch science and art'. *Art Bulletin*, **LXVIII (2)**, pp. 237–252, 1986.

Kemp, M., 'Saenredam's shapes'. *Nature*, **392**, p. 445, 1998.

Kemp, M., 'Journey into space'. *Nature*, **400**, p. 823, 1999.

Kool, M., De rekenkundige termen van Simon Stevin. *Scientiarum Historia,* **18**, pp. 91–107, 1992.

Kool, M., *Die conste vanden getale,* Verloren: Hilversum, 1999.

Kox, A.J & Chamalaun, M., *Van Stevin tot Lorentz*, Intermediair: Amsterdam, 1980.

Lagrange, J.L., *Mécanique Analytique*, Chez la Veuve Desaint: Paris, 1788.

Lenk, K & Kahn, P., To Show and Explain... *Visible Language* **26(3/4)**, pp. 272–281, 1991.

Logghe, V., Devreese, J. T., Meskens, A., van den Heuvel, C & Imhof, D., *Spiegheling en daet, Simon Stevin van Brugghe (1548–1620)*, exh. cat., Bruges: De Biekorf, 9 December 1995–31 January 1996.

Lombaerde, P., Het theoretische en praktische aandeel van Simon Stevin en Wenceslas Cobergher bij de heropbouw van Oostende na 1604. *Het Ingenieursblad*, **52(8)**, pp. 331–338, 1983.

Mach, E., *Die Mechanik in ihrer Entwickelung: historisch-kritisch Dargestellt*, F. A. Brockhaus: Leipzig, 1897.

Marijnissen, R.H., *Lof der mislukkeling* Pelckmans: Kapellen, 1997.

Mersenne, M., *Harmonie Universelle, contenant la theorie et la pratique de la musique*, Chez Sebastien Chamois: Paris, 1636–1637.

Mertens, R.A., Simon Stevin, *Nationaal Biografisch Woordenboek*, vol. 15, Koninklijke Academiën van België: Brussels, pp. 694–710, 1996.

Mertens, R.A., Simon Stevin: A European Scientist of the Renaissance, *Studia Europaea V*, pp. 91–122, 1998.

Morren, Th., Simon Stevin, *Eigen Haard*, pp. 52–56, 1899.

Norton, R., *Disme or The Art of Tenths*, S.S. for Hugh Astley: London, 1608.

Oomes, R.M.Th., Een onbekend werk van Simon Stevin, *Vereniging Jan Van Hout*, **7(1)**, 1996.

Pérez-Gómez, A & Pelletier, L., *Architectural Representation and the Perspective Hinge*, Cambridge: Massachusetts, 1999.

Printing and the Mind of Man, F.W. Bridges & Sons Ltd et al.: London, 1963.

Puype, J.P., *Simon Stevin en de Nederlandse militaire bevelstaal* (lecture given at the symposium *Simon Stevin, Brugghelinck*), Bruges, 11–12 December 1998.

Puype, J.P & Wiekart, A.A., *Van Maurits naar Munster*, Legermuseum: Delft, 1998.

Puype, J.P., Het Staatse leger en Prins Maurits, wegbereider van de moderne legers, *Armamentaria, Jaarboek Legermuseum*. Aflevering 35, Legermuseum: Delft, 2000.

Rasch, R., *Stevins Spiegheling der Singconst in 17de- en 20ste-eeuws perspectief* (lecture given at the symposium *Simon Stevin, Brugghelinck*), Bruges, 11–12 December 1998.

Rottier, H., *Rondreis door Middeleeuws Vlaanderen*, Davidsfonds: Leuven, 1996.

Sarton, G., Simon Stevin of Bruges (1548–1620). *Isis,* **21**, pp.241–303, 1934.

Sarton, G., The first explanation of decimal fractions and measures (1585). *Isis,* **23**, pp. 153–244, 1935.

Schlüter, L. and Vinken, P., *The Elsevier Non Solus imprint*, Elsevier Science: Amsterdam, 1997.

Schoutcct, A., De afkomst van Simon Stevin en diens werkkring in Vlaanderen, *Handelingen van het Genootschap 'Société d'Emulation' te Brugge,* **80**, pp. 137–146, 1937.

Simon Stevin (1548–1620). De geboorte van de nieuwe wetenschap, exh. cat., Brepols: Turnhout, 2004.

Simon Stevin (1548-1620). L'émergence de la nouvelle science, exh. cat., Brepols: Turnhout, 2004.
Simon Stevin De la vie civile 1590, (presentation and trans. Catherine Secretan), ENS Editions: Lyon, 2005.
Sinisgalli, R., *Per la storia della Prospettiva (1405–1605). Il contributo di Simon Stevin allo sviluppo scientifico della prospettiva artificiale ed i suoi precedenti storici* ('L'erma' di Bretschneider, Rome, 1978).
Smith, D.E., *History of Mathematics*, vol. II, Dover Publication Inc.: New York, 1958.
Struik, D.J., Simon Stevin and the Decimal Fractions. *The Mathematics Teacher,* **52**, pp. 474–478, 1959.
Struik, D.J., *Het Land van Stevin en Huygens*, Pegasus: Amsterdam, 1966.
Texts accompanying the exhibition on Leiden University in the Raadsgebouw on the Rapenburg, published as *Libelli introductorii ab Ricardo Vulpitio compositi* (anno MCMLXXV).
The New Encyclopaedia Britannica, William Benton *et al.*, 1977.
Receuil des pièces, affiches, journaux, etc., relatifs aux fêtes célébrées à Bruges à l'occasion de l'inauguration de la statue en bronze en 1846, Bruges: De Lay-De Muyttere, 1846.
Van Acker, J., De Veurnse voorouders en verwanten van Simon Stevin, *Liber Amicorum Roger Blondeau*, 1999.
van Berkel, K., *In het voetspoor van Stevin*, Boom Meppel: Amsterdam, 1985.
Vandamme, L., *De socio-professionele recrutering van de reformatie te Brugge, 1566–1567*, licence thesis History, K.U. Leuven, 1982.
van den Heuvel, C., *Papiere Bolwercken*, Dissertation, Groningen University (Drukkerij Vis Offset, Alphen aan de Rijn, 1991).
van den Heuvel, C., Stevins "Huysbou".... *Bulletin KNOB*, **93(1)**, p.1, 1993.
van den Heuvel, C., t Samenspreeckinghe betreffende de Architecture ende Schilderkunst, Schilders, Architecten en Wiskundigen over de uitbeelding van architectuur'. *Incontri, Rivista europea di studi italiani*, **1**, pp. 69–85, 1994a.
van den Heuvel, C. Architectuurmodellen en voorstellingen van een logische orde: Stevin en Ramus. *Feit & Fictie,* **2(1)**, pp. 95–111, autumn 1994b.
van den Heuvel, C., 'Van Waterschuyring tot Watermanagement. De betekenis van Stevins geschriften over de bouwkunst' (lecture given at the symposium *Simon Stevin, Brugghelinck*), Bruges, 11–12 December 1998.
van den Heuvel, C., Wisconstighe Ghedachtenissen. Maurits over de kunsten en wetenschappen in het werk van Stevin. *Maurits. Prins van Oranje*, ed. K. Zandvliet, Zwolle: Amsterdam, pp. 106–121, 2000.
van den Heuvel, C., Prive-lessen van Maurits (lecture given at the Rijksmuseum), Amsterdam, 20 February 2001.
van den Heuvel, C., *De Huysbou. History and Science and Scholarship in the Netherlands*, vol. 7, Royal Netherlands Academy of Arts and Sciences, 2005.
Van der Bauwhede, D & Goetinck, M., *Brugge in de Geuzentijd, Bijdragen tot de Geschiedenis van de Hervorming te Brugge en het Brugse Vrije tijdens de*

16de *eeuw* (Herdenking Oostvlaamse Synode (8 and 9 May 1582), Bruges, May 1982 (pub. Westvlaamse Gidsenkring v.z.w., 1982)).

van Duin, R.H.A & de Kaste, G., *Het Zuiderzeeprojekt in zakformaat,* 4th edn., 1995.

Van Hardeveld, I., *Lodewijk Mejer (1629–1681) als lexicograaf* (dissertation, Leiden University, 2000).

Van Paemel, G., *Een standbeeld voor Stevin: Wetenschap en cultuur in de Nederlanden.* Inaugural address, Katholieke Universiteit Nijmegen, 1995.

Van Rijckhuijsen, G.A., *Geslacht en Wapenboek van Gijsbert Ariensz. van Rijckhuysen, Bode met de Bussche te Leijden* (ms, Gemeentearchief Leiden).

Vermij, R., *The Calvinist Copernicans, The reception of the new astronomy in the Dutch Republic, 1575–1750,* Koninklijke Nederlandse Akademie van Wetenschappen: Amsterdam, 2002.

Woelderink, B., Het bezoek van Simon Stevin aan Dantzig in 1591. *Tijdschrift voor de geschiedenis der Geneeskunde, Natuurwetenschappen, Wiskunde en Techniek,* **3**, pp. 178–186, 1980.

Zandvliet, K., *Maurits, Prins van Oranje*, exh. cat., Rijksmuseum Amsterdam: Waanders uitgevers Zwolle, 2000.

Index of names

The name of Simon Stevin himself is not included in the index.

(1850) means 'was active in 1850'.

A

Abraham bar Hiyya Ha-Nasi (1070–1136) · 183
Acker, J. Van · see Van Acker, J.
Aken, Marthen Wentzel Van · see Van Aken, Marthen Wentzel
Albert of Austria (1559–1621) · xxxv, 29
Alberti, Leon Battista (1404–1472) · 7, 213, 214
Alexander of Parma · see Farnese, Alexander, duke of Parma
Al-Khwarizmi, Muhammad ibn-Musa (780–c. 850) · 60, 182, 186, 187, 195, 268
Alva, duke of (1507–1582) · xxxi, xxxii, xxxiii, 25, 29
Amstel van Mijnden, Johannes van · see van Amstel van Mijnden, Johannes
Anchemant, Heinric (1577) · 20
Andersen, K. (1990) · 219, 224, 225
Andriesse, C. D. (1993) · 173
Andriesz., Jan (1594–1633) · 43
Anne of Saxony (1544–1577) · 47
Anthonisz., Adriaan (c. 1543–1620) · 105
Apollonius of Perga (c. 262–c. 190 BCE) · 5, 6, 9
Archimedes (285–212 BCE) · 5, 6, 9, 10, 11, 12, 13, 14, 15, 131, 132, 133, 135, 139, 141, 150, 153, 155, 156, 162, 163, 165, 172, 173, 175, 176, 177, 263, 277, 280
Arethas, bishop of Caesarea (888) · 12
Aristotle (384–322 BCE) · 12, 139, 153, 154, 176, 280
Arminius, Jacobus (1560–1609) · 27
Astley, Hugh (1608) · 74

Aurelius, Cornelius (c. 1460–1531) · 27
Avicenna (c. 980–1037) · 203

B

Babbage, Charles (1791–1871) · 73, 263
Barbara of Nassau-Conroy (c. 1581) · 49
Bath, Adelar of (c. 1090–c. 1150) · 60
Beane, Clarcken van · see van Beane, Clarcken
Beane, Frans van · see van Beane, Frans
Becanus, Joannes Goropius (1518–1573) · 209
Beeckman, Isaac (1588–1637) · xxviii, 8, 34, 52, 84, 105, 106, 152, 153, 190, 224, 239, 240, 241, 248, 266, 271, 272, 273, 274, 275, 280
Beeckman, Jacob (1590–1629) · 52
Beernaert, B. (2000) · 19
Bélidor, Bernard Forest de (1693–1761) · 88, 266
Bellerus, Johannes (c. 1526–1595) · xxiii, 40
Benedetti, Giovanni Battista (1530–1590) · 7, 154, 176, 214, 231
Bernini, Gian Lorenzo (1598–1680) · 8
Bernoulli, Daniel (1700–1782) · 155, 174, 176, 178, 277
Bernoulli, Johannes (1667–1748) · 155, 178, 277
Beyer, Johann Hartmann (1563–1625) · 61
Beyts, Anthonis (1620) · 51
Bierens de Haan, David (1822–1895) · xxvii, 38, 84, 240, 254
Blaeu, Joannes (1596–1673) · 45
Bockstaele, P. (1976) · 212, 266, 268

Bombelli, Rafael (1526–1572) · 185, 186, 191, 195, 197, 198
Borelli, Giovanni Alfonso (1608–1679) · 14
Bouguer, Pierre (1698–1758) · 174
Bouwensz., Jan (1608) · xxv, 26
Boyle, Robert (1627–1691) · 155, 163
Brahe, Tycho (1546–1601) · 268, 272, 275
Brahmagupta (598–665) · 182
Brasser, Govert (1625) · 43
Bredero, Gerbrand Adriaenszoon (1585–1618) · 210
Briggs, Henry (1561–1630) · 62, 73
Bronchorst, Everhardus (1554–1627) · 37
Bruegel the Elder, Pieter (1525/30–1569) · 6, 8
Bruegel the Younger, Pieter (1564–1637/8) · 7
Brune, Jan de · see de Brune, Jan
Brunelleschi, Filippo (1377–1446) · 8, 213
Bürgi, Joost (1552–1632) · 62
Buridan, Jean (*c*. 1295–*c*. 1358) · 153
Byron, Ada, Lady Lovelace (1815–1852) · 73

C

Caesar, Julius (100–44 BCE) · 107, 129, 130
Calvin, John (1509–1564) · xxxi, 25, 133
Canneel, J. G. (1850) · 262
Cant, Adriaan, grandson of Lenaert de Cant · see de Cant, Adriaan
Cant, Lenaert · see de Cant, Lenaert
Cantimpré, Thomas of (1201–1272) · 203
Caravaggio (*c*. 1571–1610) · 242
Cardano, Girolamo (1501–1576) · 6, 7, 181, 184, 185, 195, 197, 198, 199
Caron, Emerentiana de · see Sayon-de Caron, Emerentiana
Caron, Hubrecht de · see de Caron, Hubrecht
Caron, Noël de · see de Caron, Noël
Casimir, Hendrik (1909–2000) · 152
Cauchy, Augustin Louis (1789–1857) · 135
Cervantes Saavedra, Miguel de (1547–1616) · 6
Ceulen, Ludolph van · see van Ceulen, Ludolph
Charlemagne (742–814) · 55
Charles V (1500–1558) · xxx, xxxi, 28, 70, 152
Christian IV of Denmark (1588–1648) · 46

Christoffels, Jan Ympijn (1543) · 123
Cicero, Marcus Tullius (106–43 BCE) · 5
Clavius, Christopher (1537/8–1612) · 11, 16, 61, 267, 268
Clusius, Carolus (1526–1609) · 37, 228
Cluyt, Dirck (1546–1598) · 37
Cobergher, Wenceslas (1557/61–1634) · 95
Cohen, H. F. (1987) · 241, 245, 256, 257
Comenius, Jan Amos (1592–1670) · 228
Commandino, Federigo (1509–1575) · 11, 13, 14, 132, 214
Copernicus, Nicolaus (1473–1543) · 4, 6, 7, 9, 28, 231, 232, 233, 234, 235, 236, 264, 269, 275
Courteaux, Willy (1971) · 74
Courtewille, Pieter de · see de Courtewille, Pieter
Craiy, Catharina (?–1673) · 24, 30, 49, 50, 52
Craiy, Jehan Carels (brother of Catharina) · 49
Craiy, Pieter (brother of Catharina) · 52
Croy, Philippe de · see de Croy, Philippe
Croy, Philippe III de, prince of Chimay (1526–1595) · 25, 26
Cusa, Nicholas of (1401–1464) · 8
Cusanus · see Cusa, Nicholas of

D

da Vinci, Leonardo (1452–1519) · 5, 7, 8, 9, 11, 132, 141, 175, 264, 270, 278
Dafforne, Richard (1635) · 266, 267
Dalton, R. (2001) · 214
Dante Alighieri (1265–1321) · 8
de Brune, Jan (1577) · 21, 23, 32
de Cant, Adriaan, grandson of Lenaert de Cant · 21, 23
de Cant, Lenaert · 21, 22, 23, 24
de Caron, Hubrecht · 19, 23
de Caron, Noël (?–1560) · 19, 23, 26, 28
de Courtewille, Pieter (1577) · 23
de Croy, Philippe (1668) · 91
De Decker, Ezechiel (1603/4–1646) · 62
de Fournier, Joachim (1562) · 19, 20, 32
De Gheyn, Jacob (1565–1629) · 46
de Graaf, Abraham (1676) · 224
de Groot, Hugo (1583–1645) · 16, 37, 97, 210, 212, 266, 270, 271, 279

Index of names   295

de Groot, Johan Cornets (1554–1640) · 29, 40, 41, 81, 86, 87, 114, 154
de Hubert, Andries (1620) · 51
de la Kethulle · see Ryhove, François de la Kethulle
de Montroial, Iehan · see Regiomontanus, Johannes
de Soto, Domingo (1495–1560) · 7, 152, 273
de Viry, François (1595) · 51
de Viry, Maurice (1595–1650) · 51, 52
de Vriese, Galiaen (1597) · 21
De Waal, P. (1927) · 124
de Waard, C. (1939) · xxviii, 30
Dechales, Claude F. M. (1621–1678) · 224, 227
Decker, Ezechiel De · see De Decker, Ezechiel
Descartes, René (1596–1650) · 8, 152, 153, 190, 192, 194, 197, 240, 241, 272, 273, 274, 275
Dijksterhuis, Eduard Jan (1892–1965) · xxiii, 17, 84, 86, 133, 151, 153, 170, 175, 176, 178, 195, 204, 205, 206, 209, 227, 256, 265, 269
Diophantos of Alexandria (around 250) · xxiv, 9, 195, 200, 227
Dodge, Mary Elizabeth Mapes (1831–1905) · 154
Dodoens, Rembert (1517–1585) · 2, 201, 228
Donche, P. (2002) · 22, 23
Dousa, Janus (1545–1604) · 28, 35, 36, 37
Drieux, Rémi (1519–1594) · 25
Dudley, Robert, earl of Leicester (1532/3–1588) · xxxv, 81
Dumortier, Bartholémy (1797–1878) · 261, 262
Dürer, Albrecht (1471–1528) · 8, 222, 223, 228
Dybvad, Christoffer (1572–1622) · 56

E

Egmond, Lamoraal, count of, (1522–1568) · xxxi, xxxii, 29
Einstein, Albert (1879–1955) · 177, 282
Eisenstein, E.L. (1979) · 285
El Greco, Dominikos Theotokopoulos (c. 1541–1613) · 5, 133

Elizabeth I of England (1533–1603) · xxxv, 269
Elsevier, Abraham (1592–1652) · xxvii, 40. 181,
Elsevier, Bonaventure (1583–1652) · xxvii, 40, 181
Elsevier, Isaac (1596–1651) · 39
Elsevier, Lodewijk (c. 1540–1617) · 35
Emmius, Ubbo (1547–1625) · 231, 279
Erasmus, Desiderius (1469–1536) · 4, 6, 27
Erathostenes (c. 276–c. 196) · 12
Euclid of Alexandria (c. 365–c. 300 BCE). xxiv, 6, 9, 12, 16, 182, 187, 196, 267
Euler, Leonhard (1707–1783) · 174, 176, 179, 199
Eutocius (480–540) · 10
Evens, Jacques (1597) · 21
Evertsz, Jan (1599) · 49

F

Falco, Charles M. (2001) · 214
Farnese, Alexander, duke of Parma (1545–1592) · xxxiv, 2, 26, 27, 29, 48
Ferrari, Lodovico (1522–1565) · 185, 195, 198
Ferreus, Scipio (1465–1526) 195
Feynman, Richard Phillips (1918–1988) · 266, 281, 283, 284
Fibonacci, Leonardo (1170–1250) · 60
Fior, Antonio (1526) · 183, 184, 195
Fogliano, Lodovico (?–c. 1538) · 241
Fokker, Adriaan Daniël (1887–1972) · 254, 256
Fonteyn, Thomas (1654) · 211
Fournier, Joachim de · see de Fournier, Joachim
Frederick the Great (1712–1786) · 282
Frisius, Gemma (1508–1555) · 231, 272
Fust, Johann (c. 1400–1466) · 2

G

Gailliaert, Johan (1554) · 25
Galilei, Galileo (1564–1642) · 8, 86, 131, 133, 135, 141, 151, 152, 153, 154, 155, 176, 179, 199, 231, 246, 266, 269, 271, 272, 273, 276, 285
Galilei, Vincenzo (c. 1525–1591) · 246, 255

Galle, Filips (1537–1612) · 184
Galle, Theodoor (1571–1633) · 3
Gassendi, Pierre (1592–1655) · 273
Gaurico, Luca (1476–1558) · 13
Gauss, Carl Friedrich (1777–1855) · 69, 135
Gechauff, Thomas (*c.* 1488–1551) · 13
Geeraerts, Jan (1577) · 19, 20, 21, 32
Gheyn, Jacob De · see De Gheyn, Jacob
Gilbert, William (1544–1603) · 4, 8, 231, 234, 266, 269, 270, 271
Giorgi, Giovanni (1871–1950) · 69
Giotto (Giotto di Bondone) (1266–1337) · 213
Girard, Albert (1595–1632) · xxii, xxvii, 97, 119, 181, 190, 194, 264
Gomarus, Franciscus (1563–1641) · 27
Graaf, Abraham de · see de Graaf, Abraham
Grammateus, Henricus (voor 1496–1525) · 241
Gravesande, Willem Jacob 's (1668–1742) · 224, 225
Gregory XIII, pope (1502–1585) · 7
Groenwegen, Juliaen van · see van Groenwegen, Juliaen
Groot, Hugo de · see de Groot, Hugo
Groot, Johan Cornets de · see de Groot, Johan Cornets
Gucht, Adriaen van der · see van der Gucht, Adriaen
Guicciardini, Francesco (1483–1540) · 6
Guidobaldo del Monte (1545–1607) · 213, 214, 224, 225, 280
Gustaaf Adolf (1594–1632) · 49
Gutenberg, Johannes (1400–1468) · 2
Gymnich, Jan (1541) · 201

H

Halifax, John (*c.* 1200–1256) · 60
Hamilton, Alexander (1757–1804) · 76
Haulotte, R. (1994) · 124
Heiberg, Johan Ludvig (1791–1860) · 13, 14
Helmholtz, Hermann Ludwig Ferdinand von · see von Helmholtz, Hermann Ludwig Ferdinand
Hendricz, Albert (1581) · xxiii, 40
Henry II of France (1519–1559) · 55, 119
Herigone, Pierre (1580–1643) · 224

Heuvel, C. Van den · see Van den Heuvel, C.
Hiero II of Syracuse (*c.* 306–215 BCE) · 150
Hockney, D. (2001) · 214
Hoecke, Gielis van den · see van den Hoecke, Gielis
Hogenberg, Frans (*c.* 1535–*c.* 1590) · 102
Hollerith, Herman (1860–1929) · 73, 263
Holstein, Ulric of (1602) · 46
Hooft, Pieter Cornelisz. (1581–1647) · 28, 210
Horne, count of (1518–1568) · xxxi, xxxii, 29
Hout, Jan van · see van Hout, Jan
Houten, Hendrik van · see van Houten, Hendrik
Houve, Jan van der · see van der Houve, Jan
Hubert, Andries de · see de Hubert, Andries
Hulthem, Karel van · see van Hulthem, Karel
Huygens, Christiaan (1629–1695) · 8, 131, 133, 153, 173, 177, 178, 212, 224, 240, 241, 264, 265, 266, 272, 275, 276, 277, 280
Huygens, Constantijn (1596–1687) · 106, 204, 239, 240, 266, 275

I

Ibn al-Haytham (Alhazen) (965–1039) · 273
Inghelbrecht, Pieter (1584) · 21, 23
Inghelbrechts, Jacques (1597) · 21
Isabella of Habsburg (1566–1633) · xxxv, 29
Isodorus of Miletus (6th century) · 12

J

Jacob of Cremona (1450) · 12
Jansz, Abraham (1609) · 50
Jefferson, Thomas (1743–1826) · 75, 76, 77
Johannes De Renterghem, Bartholomeus (1592) 124
John IV, duke of Brabant (1403–1427) · 4
Jordanus Nemorarius (1225–1260) · 132, 133, 141, 176, 278
Juliana of Stolberg (1506–1580) · 47

K

Kahn, P. (1991) · 228, 229
Kemp, M. (1986) · 225, 226
Kepler, Johannes (1571–1630) · 4, 7, 62, 153, 231, 243, 268, 269, 271, 272, 273, 275, 277
Kethulle, de la · see Ryhove, Frans Van den Kethulle, heer van-
Kiliaan, Cornelis (1530–1607) · 204, 205, 206, 207
Knuttel, W. P. C. (1889) · 113
Kool, M. (1992) · 58, 205, 206
Kues, Nikolaus von · see Cusa, Nicholas of

L

Lagrange, Joseph Louis (1736–1813) · 69, 143, 266, 281, 282, 283
Lambert, Johan Heinrich (1728–1777) · 225
Lambrechts, Joos (1562) · 201
Lammertijn, Passchier (c. 1562–1621) · 46
Langedijck, Gerritsz, Dirck (1620) · 51
Laplace, Pierre-Simon (1749–1827) · 69
Leeghwater, Jan Adriaenszoon (1575–1650) · 87
Leeuwen, Joanna van · see van Leeuwen, Joanna
Leeuwenhoek, Antoni van · see van Leeuwenhoek, Antoni
Leibniz, Gottfried Wilhelm Freiherr von (1646–1716) · 277
Leicester, earl of, · see Dudley, Robert, earl of Leicester
Lenk, K. (1991) · 228, 229
Leo of Thessaloniki (9th century) · 12
Lipsius, Justus (1547–1606) · 2, 28, 37, 100, 268, 278
Livius, Justus (1649) · 2, 28, 100, 268, 278
Lobelius, Matthias (1538–1616) · 228
Loe, Hendrik Van der · see Van der Loe, Hendrik
Loe, Jan Van der · see Van der Loe, Jan
Lombaerde, P. (1983) · 95
Louis of Nassau-Beverweerd (1604–1665) · 49
Lutte, Jan van der (also Lute, Jan vander) · see van der Lutte, Jan

M

Mach, Ernst (1838–1916) · 175, 176, 266, 281, 282, 283
Machiavelli, Niccolo (1469–1527) · 6,
Maerlant, Jacob van · see van Maerlant, Jacob
Magini, Giovanni Antonio (1555–1617) · 61
Margaret of Mechelen (c. 1581–1662) · 49, 52
Marnix van St. Aldegonde, Philip of (1540–1598) · 27
Marolois, Samuel (c. 1572–1627) · 224
Martin V, pope (Oddone Colonna) (1368-1431, elected in 1417) · 4
Martinus Smetius (1525–1578) · 28
Mary of Nassau (1593) · 86
Masterson, Thomas (1592) · 61
Maurice (1601-1617), bastard son of Maurice, prince of Orange · 49
Maurice, prince of Orange (1567-1625) · xxv, xxvii, xxxii, xxxv, xxxvi, 17, 29, 38, 42, 44, 46, 47, 48, 49, 51, 52, 57, 96, 98, 99, 100, 101, 102, 105, 111, 112, 128, 129, 152, 213, 215, 222, 223, 224, 227, 231, 233, 236, 241, 266, 271, 279
Maxwell, J. C. (1831–1879) · 69
Meetkerke, Adolf van · see van Meetkerke, Adolf
Meijde, Fop Pietersz. van der · see van der Meijde, Fop Pietersz.
Memling, Hans (c. 1435–1494) · 147, 148, 149
Menelaos (around 100 CE) · 6
Mennher, Valentin (1565) · 191
Mercator, Gerard (1512–1594) · 7, 97, 228, 231, 264
Mersenne, Marin (1588–1648) · 240, 241, 244, 255, 258, 273, 277
Mertens, R. (1998) · 212
Metsys, Quinten (1465/66–1530) · 124
Meurs, John of (Johannes de Muris) (c. 1290–c. 1351) · 60
Michelangelo Buonarotti (1475–1564) · 5, 8
Mierevelt, Michiel Janszen van · see van Mierevelt, Michiel Janszen
Montroial, Iehan de · see Regiomontanus, Johannes
More, Thomas (1478–1535) · 6

Moretus, Jan I (1543–1610) · 2, 268
Morren, Theo (1899) · 50
Morris, Robert (1733–1806) · 76
Mulerius, Nicolaus (1564–1630) · 27, 28, 132

N

Nansius, Franciscus (c. 1520–1595) · 28
Napier, John (1550–1617) · 62, 73, 270
Neumann, John von · see von Neumann, John
Newton, Isaac (1642–1727) · 8, 131, 133, 151, 152, 153, 224, 243, 266, 271, 273, 277, 282, 285
Norton, Robert (1608) · 56, 64, 74, 75, 76, 77

O

Ockham, William of (c. 1285–c. 1347) · 153
Oldenbarnevelt, Johan van · see van Oldenbarnevelt, Johan
Oomes, R. (1996) · 111, 113
Oresme, Nicole (1323–1382) · 153
Orliens, David van · see van Orliens, David
Overschie, Clement (1623) · 51

P

Pachtenbeke, Tanneken van · see van Pachtenbeke, Tanneken
Paciolo, Luca (1445–1517) · 122, 124, 183, 185, 195
Paemel, G. Van · see Van Paemel, G.
Palestrina, Giovanni Pierluigi da (1525/6–1594) · 133
Panhuysen, Bartholomeus (1620) · 51
Pannekoek, Anton (1873–1960) · 233
Pappus of Alexandria (around 300 CE) · 6, 12, 14, 133, 141
Pascal, Blaise (1623–1662) · 73, 131, 155, 157, 161, 176, 177, 263, 280
Pauw, Pieter (1564–1617) · 37
Pegolotti, Francesco Balducci (1315) · 120
Pelletier, L. (1999) · 225
Pérez-Gómez, A. (1999) · 225
Peter the Great (1672–1725) · 280, 281
Petrarch (Petrarca), Francesco (1304–1374) · 5, 6

Petri, Nicolaus · see Pietersz, Claes
Philip II of Spain (1527–1598) · xxx, xxxi, xxxii, xxxiii, xxxiv, xxxv, 2, 4, 26, 28, 36, 70, 71
Philip the Good (1396–1467) · 70
Philoponus, Joannes (6th century) · 154
Piero della Francesca (c. 1420–1492) · 7, 213, 214
Pietersz, Claes (1605) · 123
Plancius, Petrus (Plaetevoet, Petrus) (1552–1622) · 97, 98
Plantin, Christopher (c. 1520–1589) · xxiii, xxiv, xxv, 2, 30, 36, 39, 40, 43, 55, 56, 57, 97, 118, 204, 205, 207, 264, 268
Plato (c. 427–c.347 BCE) · 5
Plato of Tivoli (c. 1125) · 183
Polybius (c. 200–c. 118 BCE) · 100
Poort, Cathelyne (also Catelijne, Kathelyne) van der (also vander) · van der Poort, Cathelyne
Poort, Fransois van der · see van der Poort, Fransois
Poort, Hubrecht van der (also Huybrecht vander) · see van der Poort, Hubrecht
Poort, Margriete van der · see van der Poort, Margriete
Poort, Marie van der (also vander) · see van der Poort, Marie
Ptolemy I of Egypt (c. 366–283 BCE) · 150
Ptolemy, Claudius (c. 100–na 160) · 6, 9, 59, 231, 232, 233, 235, 275
Puype, J.P. (2000) · 102
Pythagoras of Samos (570–480 BCE) · 9, 240, 241

R

Rabelais, François (1494–1553) · 6
Raffaello Santi (1483–1520) · 53, 57, 311
Raphelengius, Franciscus · see van Ravelingen, Frans
Rasch, R. (1998) · 241
Ravelingen, Frans van · see van Ravelingen, Frans
Regiomontanus, Johannes (1436–1476) · 12, 59, 60, 61
Renterghem, Bartholomeus De · see Johannes De Renterghem, Bartholomeus

Reynier Pietersz. van Twisk (1598) · 99
Rijckhuijsen, Gijsbert Ariens. van · see van Rijckhuijsen, Gijsbert Ariens.
Roomen, Adriaan van · see van Roomen, Adriaan
Roosenboom, Alida · 53
Roosenboom, Catharina · 53
Roosenboom, Geertruyt · 53
Roosenboom, Johan · 53
Roosenboom, Karel · 53
Roosenboom, Maria · 53
Roosenboom, Simon · 53
Rubens, Jan (1530–1587) · 47
Rubens, Pieter Paul (1577–1640) · 5, 47
Ryhove, François de la Kethulle (1578) · 25

S

Saenredam, Pieter (1597–1665) · 220, 226,
Sarton, George (1884–1956) · 60
Sayon, Antoon · 26,
Sayon, Jacob · 25, 27, 52, 111,
Sayon, Joost (ook Joos) · 19, 20, 21, 25, 26, 32, 111
Sayon, Vincent · 25, 26, 27, 111
Sayon-de Caron, Emerentiana · 19, 23, 25, 26
Scaliger, Josephus Justus (1540–1609) · 37, 99
Schoenberg, Arnold (1874–1951) · 257
Schöffer, Peter (c. 1425–1503) · 2
Schooten, Frans Van, - the Elder · see Van Schooten, Frans, - the Elder
Schooten, Frans Van, the Younger · Van Schooten, Frans, the Younger
Schouteet, A. (1937) · 17, 19, 23
Shakespeare, William (1564–1616) · 6, 8, 73, 74, 75, 77
Simonis, Louis Eugène (1810–1882) · 261
Sinisgalli, Rocco (1978) · 220, 221
Snellius Rudolf (1546–1613) · 279
Snellius, Willebrord (1580–1626) · 51, 266, 271, 272, 273, 274, 279
Socrates (470–399 BCE) · 5
Soto, Domingo de · see de Soto, Domingo
Speckle, Daniel (1536–1589) · 102
Spinola, Ambrosio (1569–1630) · 48
Splinter, Jan Claesz. (c. 1610) · 50
Stadius, Joannes (1527–1579) · 233

Stampioen, Jan Jansz., the Elder (c. 1620) · 51
Stampioen, Jan Jansz., the Younger (1610–1690) · 275, 276
Stephanus (888) · 12
Stevelinck, E. (1994) · 124
Stevin, Adriaan (?–1534) · 21, 22, 24
Stevin, Anthuenis (also Antoon) · 19, 20
Stevin, Francisca · 23, 24
Stevin, Frederik (1612–?) · 30, 51, 52
Stevin, Hendrick (1613–1670) · xxi, xxvii, xxviii, 30, 43, 51, 52, 53, 79, 81, 84, 88, 91, 92, 93, 95, 100, 105, 107, 112, 127, 128, 203, 239, 240, 274
Stevin, Jan (1538) · 22, 24
Stevin, Levina · 30, 51, 53
Stevin, Magdalena · 21, 22, 23, 24, 28
Stevin, Pieter (?–1540) · 21, 23, 24
Stevin, Pieter (?–1556/7) · 24
Stevin, Simon Anthonis (cf. marriage announcement) · 17, 23
Stevin, Susanna · 30, 33, 51, 53
Stevin-Cant, Magdalena · 21, 22, 23, 24
Stevin-Courteville, Francisca · 23, 24, 122
Stevin-de Visch, Francine · 22, 24
Stochius, Nicholaas (1581) · 34
Stradanus, Johannes (1523–1605) · 3
Struik, Dirk (1894–2000) · 266

T

Tacitus, Cornelius (c. 55–c. 117) · 37
Tartaglia, Niccolo (1500–1557) · 6, 7, 13, 102, 181, 183, 184, 185, 195, 197, 274
Taylor, Brook (1685–1731) · 224, 225
Thabit ibn Qurra (around 836-901) · 14
Thales of Miletus (625–547 BCE) · 6, 9, 135
Theon of Alexandria (c. 335–c. 405) · 12
Thiessen van den Boom, Karl (1629) · 46
Thomson, Joseph John (1856–1940) · 69
Torricelli, Evangelista (1608–1647) · 133, 151, 155, 176, 178, 277
Trenchant, Jean (1558) · 119, 120
Trojanus, Curtius (1565) · 13

U

Uccello, Paolo (1397–1475) · 213

V

Valckesteyn, Magdalena van · see van Valckesteyn, Magdalena
Van Acker, J. (1999) · 22
Van Aken, Marthen Wentzel (1587) · 122
van Amstel van Mijnden, Johannes (1600) · 35
van Beane, Clarcken (1584) · 21
van Beane, Frans (1584) · 19, 21, 23
van Ceulen, Ludolph (1540–1610) · 38, 99, 122, 268
Van de Weyer, S. (1845) · 262
Van den Heuvel, C. (1991) · 105–106, 107, 225, 237, 265, 266, 280
van den Hoecke, Gielis (1537) · 206
van den Vondel, Joost (1587–1679) · 210
van der Gucht, Adriaen (1569) · 206
van der Houve, Jan (1577) · 20
Van der Loe, Hendrik (1552) · 2
Van der Loe, Jan (1552) · 2
van der Lutte, Jan (1584) (also van der Lute) · 19, 21, 23
van der Meijde, Fop Pietersz. (1591) 41, 42
van der Poort, Cathelyne · 17, 19, 20, 21, 25, 26
van der Poort, François · 21, 23
van der Poort, Hubrecht · 21, 23
van der Poort, Margriete · 21
van der Poort, Marie · 21, 23
van Eyck, Jan (1390–1441) · 8, 213, 214
Van Gogh, Vincent (1853–1890) · 264
van Groenwegen, Juliaen (1752) · 33, 53
van Hout, Jan (1542–1609) · 113
van Houten, Hendrik (1705) · 225
van Hulthem, Karel (1764–1832) · 203
van Leeuwen, Joanna (1642) · 52
van Leeuwenhoek, Antoni (1632–1723) · 272
van Maerlant, Jacob (c. 1235–c. 1300) · 203
van Meetkerke, Adolf (1611) · 28
van Mierevelt, Michiel Janszen (1567–1641) · 47
van Oldenbarnevelt, Johan (1547–1619) · 47
van Orliens, David (1600) · 105
van Pachtenbeke, Tanneken (1597) · 21
Van Paemel, G. (1995) · 266, 280
van Ravelingen, Christoffel (1599) · xxv, 97, 268
van Ravelingen, Frans (1539–1597) · xxiv, xxv, 2, 39, 40, 43, 44, 45, 131, 268
van Rijckhuijsen, Gijsbert Ariens. (1760) · 33, 49, 53
van Roomen, Adriaan (1561–1615) · 266, 267, 268, 272
Van Schooten, Frans, - the Elder (1581–1645) · 190
Van Schooten, Frans, the Younger (1615–1660) · 224
van Valckesteyn, Magdalena (1606) · 28
Van Varenbraken, Christianus (1532) · 206
van Waesberghe, Jan (1617) · xxvii, 46
van Warmenhuysen van Didolff Duyren, Johan Adriaensz. (1619) · 51
van Wender, Gerrit Beunen (1642) · 52
Vandamme, Ludo (1982) · 25, 26
vander Lute, Jan · see van der Lutte, Jan
vander Poort, Catelijne · see van der Poort, Cathelyne
vander Poort, Huybrecht · see van der Poort, Hubrecht
vander Poort, Kathelyne · see van der Poort, Cathelyne
vander Poort, Marie · see van der Poort, Marie
Varenbraken, Christianus Van · see Van Varenbraken, Christianus
Vaseur, Jacop (1551) · 19
Vauban, Sébastien Le Prestre de (1633–1707) · 105
Verheyen, Abraham (1600) · 239, 240, 241, 245, 254, 256, 257
Vermij, R. (2002) · 280
Verplancken, I. H. (1753) · 33
Vesalius, Andreas (1514–1564) · 228, 264, 269
Viète, François (1540–1603) · 185, 186, 192, 193, 194, 197, 268
Viruly, Paulus (1593) · 86
Viry, François de · see de Viry, François
Viry, Maurice de · see de Viry, Maurice
Vitruvius, Pollio Marcus (c. 90 BCE–c. 20 BCE) · 107
Vlietentoorn, Petrus (1635) · 53
Vlietentoorn, Simon, Cornelis and Catharina, children of Petrus

Vlietentoorn and Susanna Stevin (after 1635) · 53
Vollenhove, Joannes (1631–1708) · 53, 210
von Helmholtz, Hermann Ludwig Ferdinand (1821–1894) · 133
von Kues, Nikolaus · see Cusa, Nicholas of
von Neumann, John (1903–1957) · 73, 263
Vondel, Joost van den · see van den Vondel, Joost
Vredeman de Vries, Jan (1527–1604) · 348, 349, 350
Vriese, Galiaen de · see de Vriese, Galiaen
Vulcanius, Bonaventura (1537/8–1614) · 27

W

Waal, P. De · see De Waal, P.
Waard, C. de · see de Waard, C.
Waesberghe, Jan van · see van Waesberghe, Jan
Warmenhuysen van Didolff Duyren, Johan Adriaensz. van · see van Warmenhuysen van Didolff Duyren, Johan Adriaensz.
Washington, George (1732–1799) · 76
Weber, Wilhelm Eduard (1804–1891) · 69
Wender, Gerrit Beunen van · see van Wender, Gerrit Beunen
Weyer, S. Van de · see Van de Weyer, S.
Wijts, Jacques (c. 1580–1643) · 28, 49
Wijts, Jean (father of Jacques) · 28
Wilkins, John (1614–1672) · 266, 285
Willem Lodewijk of Nassau (1560–1620) · 48, 49

William III, prince of Orange (1650–1702) · 280
William of Moerbeke (c. 1215–c. 1286) · 10, 11, 12, 13
William of Nassau-de-Leck (1602–1627), bastard son of Maurice, prince of Orange · 49
William of Orange (1533–1584) · xxxi, xxxii, xxxiii, xxxiv, xxxv, 29, 36
William the Silent
see William of Orange · 29, 35, 42, 86
Winckelman, Bernaert (1567) 19
Witsen, Nicolaes (1641–1717) · 278, 280, 281
Woelderink, B. (1980) · 41
Woudanus, J. C. (1610) · 39
Wright, Edward (1558–1615) · 97, 270

X

Xenophon (c. 430–353 BCE) · 100
Xylander (Wilhelm Holzmann) (1532–1576) · 200

Y

Ympijn · see Christoffels, Jan Ympijn
Yperman, Jan (Jehan) (1310) · 322

Z

Zandvliet, K · 127
Zarlino, Gioseffe (1517–1590) · 241, 250, 257
Zuse, Konrad (1910–1995) · 263

Index of subjects

A

administration · xxvii, 17, 57, 87, 101, 102, 111, 125, 126, 127, 128, 129
afcomst (arithmetical calculation) · 206
agonic lines · 98
Album Amicorum · 35
Album Studiosorum · 37
algebra · xxiv, 40, 112, 131, 181, 182, 183, 185, 187, 189, 191, 196, 197, 199, 200, 224, 268
algebra, rule of · 199, 200
Almachtich ('Almighty') · 81, 150, 275
Almagest · 232
Almighty (*Almachtich*) · 81, 150, 275
Amsterdam · xxxv, xxxvi, 47, 68, 87, 95, 96, 97, 99, 113, 118, 123, 210, 266, 267, 280, 281
ancestry of Simon Stevin · 24
Ancient Greeks · 6, 9, 12, 14, 16, 59, 100, 108, 139, 141, 175, 179, 182, 198, 209, 239, 242, 243, 249, 250, 270
annual accounts (bookkeeping) · 126
Antiquity (Classical) · 1, 5, 6, 8, 9, 14, 59, 129, 131, 141, 172, 209, 240, 241, 243, 257, 258, 279
Antwerp · xxiii, xxx, xxxi, xxxiv, xxxvi, 2, 6, 29, 30, 33, 34, 39, 40, 56, 58, 68, 71, 96, 109, 111, 112, 114, 118, 120, 123, 124, 184, 201, 205, 209, 264, 267
apparent weight (*staltwicht*) · 138, 141, 142, 143, 209, 229
Appendice Algebraique · xxv, 29, 40, 198
Arabic · 5, 12, 14, 39, 60, 62, 63, 205
Archimedes palimpsest · 13, 14, 15, 263
Archimedes-Stevin tradition · 165, 166, 176
architecture · 1, 8, 14, 88, 105, 106, 107, 134, 177, 204, 225, 261, 266, 278

arithmetic · xxiv, 38, 62, 63, 65, 73, 74, 112, 120, 183, 199, 200, 203, 204, 205, 206, 207, 241, 249, 251, 252
Arithmetica (Joost Bürgi) · 62
Arithmetica Logarithmica (Henry Briggs) · 62
Arithmetica, ... edel conste (Gielis van den Hoecke) · 206
Arithmetica, Summa de (Luca Pacioli) · 122, 183
Arithmeticae, Practica (Girolamo Cardano) · 185
Arithmeticke, First booke of (Thomas Masterson) · 61
Arithmétique (Jean Trenchant) · 119, 120
Arithmetique, L' (Simon Stevin) · xxiv, 29, 40, 55, 57, 61, 118, 119, 181, 182, 186, 188, 189, 191, 193, 194, 195, 197, 199, 200, 210, 227, 264
armhuys (poorhouse) · 107
Arminians (Remonstrants) · 48
army · xxx, xxxi, xxxiii, xxxiv, xxxv, xxxvi, 28, 29, 43, 46, 47, 48, 49, 51, 79, 100, 101, 102, 105, 127, 261, 279
army administration · 101, 127
army camps, marking out of · xxvii, 46, 100, 101, 279
arts (faculty) · 4, 37
Astrolabium · 61
astrology · 203
astronomers · 16, 58, 60, 183, 232, 235, 268
astronomical calculations · 60, 67
astronomy · 8, 9, 28, 34, 203, 204, 207, 227, 230, 231, 241
axiom · 135, 139, 155, 162, 163, 166, 175, 215
azimuthal quadrant · 98, 99
Azores · 98

B

Babylonians · 9, 182
balance (weighing) · 134, 138, 142, 145, 159, 160, 161
balance sheet · 114, 115
baroulkos · 151
base number · 60, 187
bastions · 4, 103, 104, 107, 109, 237
Battle of Nieuwpoort · xxxv, 29, 48
beat · 241, 242, 243, 246, 248, 254
Beghinselen der Weeghconst, De · xxiv, xxv, 9, 14, 29, 40, 82, 91, 131, 132, 134, 135, 136, 137, 140, 141, 143, 144, 145, 147, 151, 152, 156, 162, 171, 177, 202, 204, 208, 209, 210, 227, 228, 229, 236, 237, 242, 264, 265, 268, 269, 274, 275, 276, 277
Beghinselen des Waterwichts, De · xxv, 9, 40, 91, 131, 132, 135, 151, 152, 154, 155, 156, 157, 158, 159, 160, 162, 163, 165, 168, 169, 170, 171, 173, 175, 176, 177, 179, 204, 227, 228, 229, 236, 237, 264, 265, 269, 274, 276, 277, 280
bit and bridle · 152
bookkeeping · 17, 40, 57, 111, 112, 113, 114, 115, 116, 117, 118, 122, 124, 125, 126, 127, 129, 130, 204, 207, 266, 267, 274
Bouckhouding, Van de Vorstelicke · xxvi, 17, 112, 115, 124
brake wheel (water mill) · 84
branches (company) · 118, 126
bridle · 49, 151, 152
Bruges · xxx, xxxi, xxxiv, xxxvi, 3, 17, 18, 19, 20, 21, 23, 24, 25, 26, 27, 28, 29, 32, 33, 34, 40, 41, 42, 43, 52, 56, 80, 91, 94, 106, 107, 112, 113, 118, 128, 131, 132, 134, 136, 137, 140, 141, 144, 146, 147, 148, 149, 156, 158, 160, 161, 165, 168, 171, 174, 181, 195, 202, 203, 206, 208, 216, 220, 225, 227, 230, 261, 263, 264, 265
Brugghelinck · xxiii, 17, 118
bulwarks · 94, 103
Burgherlick Leven, Het · xxv, 29, 43, 44, 266
Burgherlicke Stoffen · xxvii, 100, 102, 106, 107, 108, 239
Byzantium · 12

C

Calais · 91, 94
Calvinism · xxxi, xxxii, xxxiv, 2, 4, 25, 44, 122
canal lock · 89
cassebouck (bookkeeping) · 125
Castrametatio, Dat is Legermeting · xxvii, 30, 46, 79, 99, 100, 264
catenary · 276
Catrolwicht, vant · xxvi, 155, 152
chain-curve · 276
chariston · 150, 151
chief steward · 126, 127
China · 1
Classical Antiquity · 1, 5, 6, 8, 9, 14, 59, 129, 131, 141, 172, 209, 240, 241, 243, 257, 258, 279
classical scholars · 9, 133, 209, 231, 261, 264, 267, 268, 272, 273, 275, 278, 279, 280, 281
clootcrans · 52, 82, 133, 136, 137, 138, 140, 141, 142, 173, 175, 177, 178, 179, 237, 276, 282, 283, 284
codex A · 12, 13, 14
codex C · 12, 14
codex DD · 12, 14
codices · 9, 12, 14
coincidence theory · 248
comma · 61, 62
compagnie · xxiii, 28
compass · 96, 97, 98, 235, 269, 270
compass variation(s) · 98, 270
composition law of forces · 137, 144, 145, 146, 152, 178, 276, 286
composition parallelogram · 276
conics · 6
consonance · 240, 241, 245, 248, 249, 254, 258
Constantinople · 12, 13
conversations · 112
Cossic notation · 191, 192, 193
Council of State · xxxi, xxxiii, 42, 47, 51, 57
Counter-Remonstrants · 48
crane · 147, 148, 149
creditors · 116, 117, 118
credits · 59, 115, 116, 117, 125, 126, 127
crown wheel (water mill) · 81, 82, 84
cycloid · 277

D

daelwicht · 140
daet · 9, 131, 136, 159, 162, 179, 279
debits · 115, 116, 117, 124, 125, 127
debtors · 116, 117, 118
Decarithmia 56
decimal system · 55, 56, 59, 60, 61, 62, 63, 69, 73, 75, 199, 206, 252
defensive systems · 44
Delft · xxiii, xxxiv, 29, 37, 40, 41, 43, 81, 100, 113, 114, 153, 154
Deursichtighe, Vande · xxvi, 213, 215, 216, 218, 220, 221, 223, 225, 226, 237, 272
Deventer · 48, 91
Dialectike ofte Bewysconst · xxiv, 29, 30, 40, 210
didactics · 145, 147, 163, 213, 227, 230, 233, 235, 236
dispense · 127
dispensier · 127
Douai · 4
draining · 82, 85, 87
dredging · 81, 82, 83, 94
dredging machine · 79
dredging net · 83
dredging vessel · 81
Dutch language · xxiv, xxx, 27, 34, 58, 72, 201, 203, 204, 207, 209, 210, 211, 236, 243
dynamics · 133, 151, 152, 153, 176, 278

E

Ebbenvloet, Vande Spiegheling der · xxvi, 177
economist 111
Eertclootschrift, Vant · xxvi, 33, 96
Egypt · 150
Eighty Years' War · xxxii, 105
Elbing · 91
Elements (Euclid) · 12, 16, 187, 267
empirical ephemerides · 233
engineering school · 44, 105
engineers · xxxv, 8, 11, 49, 56, 81, 91, 95, 105, 209, 229, 278
equal temperament · 240, 242, 243, 245, 247, 249, 254, 255, 256, 257, 258, 259, 277

equation, algebraic · 40, 182, 183, 184, 185, 186, 187, 194, 195, 196, 197, 198, 199, 268
equation, cubic · 183, 184, 185, 194, 197, 198
equation, higher degree · 193, 195, 196
equation, quadratic · 182, 183, 187, 194, 195, 196, 197
equation, quartic · 182, 183, 185, 194, 195, 198
Europe · xxix, xxxvi, 1, 2, 3, 4, 6, 12, 14, 60, 71, 73, 76, 84, 111, 119, 183, 201, 213, 282
example(s) · 16, 43, 58, 59, 62, 64, 66, 68, 72, 82, 91, 114, 115, 116, 117, 118, 121, 122, 125, 137, 139, 142, 161, 162, 165, 169, 170, 171, 176, 182, 185, 186, 191, 193, 194, 195, 198, 199, 200, 206, 207, 208, 212, 213, 219, 220, 221, 222, 223, 227, 229, 230, 233, 235, 236, 237, 240, 245, 248, 254, 267, 268, 270, 278, 280, 283
explanation · xxiv, 64, 66, 83, 94, 113, 121, 122, 125, 157, 166, 209, 222, 224, 229, 230, 231, 242, 245, 246, 247, 254, 257, 272, 280
exponential notation · 191, 193
extraordinary finance 127
eye · 83, 127, 213, 215, 216, 217, 218, 219, 221, 222, 226

F

fairs · 38, 119
falling bodies · 29, 152, 153, 154, 273, 278
fifth(s) · 243, 245, 249, 254, 257
finance, extraordinary 112
fixed bodies · 143
fixed Earth · 232, 233
flameng · 56
Flanders, county of · xxx, xxxiv, xxxv, 6, 21, 25, 26, 32, 47, 73, 94, 124, 261
floats (water mill) · 84
fortification · 38, 43, 44, 45, 46, 49, 79, 95, 102, 105, 124, 127
fortification(s) · xxv, xxvii, xxxv, 8, 17, 38, 90, 95, 99, 105, 261
fortification-building · 38, 44, 49
fractions · xxiv, 40, 55, 56, 59, 60, 61, 62, 63, 64, 65, 66, 199, 243
freely falling bodies · 152

French Revolution · 48, 68, 69, 73
frets (string instruments) · 242, 244, 245, 246, 251, 252, 253, 255, 256
function (mathematical) · 196, 199

G

gaugers · 57, 58, 71
Gdańsk · 41, 42, 91, 92
geocentric · 227, 232, 233, 235, 236
geocentric system · 231, 233
geometric perspective · xxvi, 107, 163, 182, 183, 188, 189, 194, 196, 197, 198, 225, 228, 229, 249
geometrical numbers · 187, 188, 189
geometry · 40, 42, 135, 145, 178, 181, 183, 189, 190, 196, 198, 199, 214, 219, 225, 241
Ghemengde stoffen, Van de · xxvi, 17, 105
glass (theory of perspective) · 214, 214, 215, 216, 217, 218, 219, 222, 223
Golden Compass · xxiii, 99, 118
golden rule · 193
Golden Shield, The · 19, 25, 26
Gomarists (Counter-Remonstrants) · 48
grand parti, Le · 119, 120
great-circle sailing · 96
Greek · 2, 4, 5, 6, 9, 10, 11, 12, 13, 14, 16, 27, 36, 39, 59, 69, 74, 100, 107, 108, 131, 135, 139, 141, 155, 157, 169, 171, 175, 179, 182, 198, 200, 204, 207, 208, 209, 213, 231, 239, 242, 243, 249, 270, 274
Groningen · xxx, xxxv, 27, 28, 48, 231, 279
guardians · 19, 20, 26, 32, 51, 273

H

Haarlem · xxxii, 154, 155
harbour(s) · 41, 80, 82, 83, 88, 89, 92, 93, 94, 95, 97, 98, 99, 140, 151, 237
Havenvinding, De, Byvough van de · xxv, 29, 25, 96, 97, 98, 269, 270, 271, 272
heavenly motions (the movement of the heavenly bodies) · xxvi, 227, 231, 236
heliocentric · 228, 231, 232, 233, 235, 236, 269, 271, 279
heliocentric cosmology · 231
heliocentric system · 231, 233, 271
Hemelloop, Vanden · xxvi, 227, 230, 231, 232, 233, 235, 236, 272

house-building · xxviii, 107, 108
Huguenots · 119
humanism · 3, 4, 5, 36
humanities · 5, 6, 36, 37, 278
Huysbou · xxviii, 105, 106, 237, 274
hydraulic engineering · 88, 95, 278
hydraulic press · 159, 161, 169, 176
hydraulics · 88, 124, 266
hydrostatic paradox · 133, 156, 157, 158, 159, 160, 163, 165, 168, 169, 176, 177, 212, 229, 277, 280
hydrostatics (see also *Beghinselen des Waterwichts, De*) · xxv, 9, 40, 131, 132, 133, 135, 155, 156, 157, 162, 163, 166, 169, 172, 173, 175, 176, 177, 204, 207, 271, 274, 277, 282
Hypomnemata Mathematica · 271, 282

I

iconoclasm · xxxiv, xxxvi, 25, 26
impoldering · 87, 95
inclined plane · 133, 136, 138, 140, 141, 142, 144, 145, 154, 274, 282
information graphics · 228, 229
inquisition · 6, 231
instruments · 8, 44, 80, 84, 99, 132, 134, 136, 145, 147, 150, 152, 222, 223, 224, 241, 242, 244, 248, 249, 255, 256, 257, 269
integer · 56, 62, 63, 65, 66, 67, 111, 243
interest · 38, 44, 111, 112, 118, 119, 120, 121, 122, 123
interest tables · xxiii, xxiv, 118, 119, 120, 121, 122, 123
interest, calculation of · 120, 121,
interval (musical) · 241, 242, 244, 245, 246, 247, 248, 249, 250, 253, 256, 257, 258, 259, 277
interval ratio (musical) · 241, 243, 245, 246, 247, 248, 257, 258
inverse perspective · 214, 221, 222
Istanbul · 12, 14

J

journael (bookkeeping) · 125
journal (water mill) · 84, 87
Journal tenu par Isaac Beeckman · xxviii, 84, 239, 271, 273, 274, 275

K

keyboard instruments · 244, 249, 254, 255, 256, 257

L

land reclamation · 87
Latin · xxx, 2, 4, 6, 10, 12, 13, 14, 16, 25, 27, 28, 33, 34, 36, 37, 38, 39, 60, 69, 97, 98, 133, 155, 181, 183, 185, 192, 200, 201, 204, 207, 208, 209, 212, 220, 225, 268, 270, 271, 272, 273, 274
law of falling bodies · 152, 153, 154, 273
law of light refraction · 271, 272
Legermeting · xxvii, 46, 99
Leiden · xxiv, xxv, xxvii, xxviii, xxxii, 2, 4, 24, 27, 28, 32, 33, 34, 35, 36, 37, 38, 39, 40, 42, 43, 47, 49, 51, 52, 53, 55, 81, 91, 94, 97, 105, 112, 113, 118, 131, 195, 224, 271, 273, 278, 279
lenses · 214, 272, 277
Leuven · 4, 37, 40, 203, 231, 267, 268
lever · 134, 136, 139, 140, 145, 147, 148, 149, 150, 152, 280
lever, law of the · 132
lifting weight · 140, 143, 209
light refraction · 271, 272
liquids (hydrostatics) · 155, 157, 162, 163, 166, 169, 170, 175, 176
loans · 32, 119, 125
Lochtwicht, Vant · 151
lock · 83, 89, 90, 91, 237
lock gates · 89
longitude · 38, 97, 98, 272
Low Countries · xxix, xxx, xxxi, xxxii, xxxiii, xxxiv, xxxv, xxxvi, 2, 4, 25, 28, 29, 70, 84, 88, 89, 123, 155, 207, 211, 214, 228, 231, 268
lowering weight · 140
loxodrome · 96
loxodromic sailing · 96
lune(s) · 98
Lutheranism · 122

M

magnetic pole · 97,
majority, attainment of · 32
management of lands · 112, 113

manuscripts · xxvii, 9, 12, 13, 14, 15, 28, 39, 52, 105, 106, 107, 203, 239, 240, 264, 265, 266, 274
Materiae Politicae · xxvii, 43, 52, 79, 100, 102, 105, 107, 128
meantone temperament · 243, 249, 255, 256
measurers of tapestry · 58
measuring (standardization, etc.) · xxvi, 38, 55, 57, 67, 69, 98, 272
mechanics · 7, 9, 14, 40, 111, 124, 133, 139, 143, 153, 155, 175, 265, 273, 278, 280, 282
Meetdaet, Vande · xxvi, 161
memoriael (bookkeeping) 125
mercantile bookkeeping · xxvi, 113, 115, 122, 124, 126
mercantile practice · 111
merchants · 23, 42, 57, 67, 72, 96, 111, 119, 120, 122, 123, 193, 205, 266
meridian(s) · 69, 96, 98, 270, 272
metacentre · 174,
Middle Ages · 1, 5, 8, 12, 68, 130, 150, 213, 228, 243
Middle Dutch · 202, 203
military operations · 112, 113, 127
military science · 99, 100, 207
Modern Dutch · 201
molengang · 87, 88
Molens, Vande · xxvii, 84, 131, 239
monetary system(s) · 70, 71, 76
money-changers · 121
money-lenders · 121, 124
money-masters · 58, 67, 72, 193
monochord · 249, 251, 252, 257, 258
moving Earth · 232, 233, 234, 271
music (theory) 8, 207, 239, 240, 241, 243, 245, 247, 249, 254, 255, 256, 257, 259, 277

N

Nassau · xxv, xxvii, xxxi, xxxii, xxxv, xxxvi, 42, 47, 49, 57, 86, 100, 105, 127, 128, 129, 279
navigation · 8, 38, 45, 49, 69, 96, 97, 204, 207, 269, 278
Nederduyts · 38, 268
neologism(s) · 163, 203, 204, 206, 207

Netherlands (Northern) · xxxiv, xxxv, xxxvi, 4, 27, 40, 42, 43, 46, 95, 122
Nieuwe Inventie van Rekeninghe van Compaignie · xxiii, 29, 40, 111, 113, 114, 210
Nieuwe Maniere van Sterctebou, door Spilsluysen · xxvii, 30, 46, 79, 88, 90, 91, 99, 264
Nieuwpoort (Battle of) · 29, 48
numbers · xxiv, 38, 56, 58, 59, 60, 61, 62, 63, 64, 65, 66, 74, 89, 122, 176, 182, 185, 187, 188, 189, 191, 192, 193, 194, 195, 197, 203, 204, 206, 207, 222, 224, 225, 228, 233, 240, 241, 246, 247, 248, 249, 252, 253, 261, 263, 265, 266
numbers, decimal · 56, 58, 61, 62, 63, 64, 65, 66, 73
numbers, notion of · 182
numbers, whole · xxiv, 60, 61, 63, 65, 74, 122
numerals, Arabic · 60, 63, 205
numerals, Roman · 59, 60, 205
numerical notation · 59

O

oblique lifting weights · 143
octave · 240, 241, 243, 244, 245, 246, 249, 250, 256, 257, 259
Œuvres Mathematiques, Les · xxii, xxvii, 97, 119, 182, 264
Officina Plantiniana · 1, 2, 268
oncostbouck (bookkeeping) · 125
orphans' court, Bruges · 19

P

palimpsest · 13, 14, 15, 263
pamphlet 581 · 113, 114
papyrus · 9
paraboloid · 11, 172, 173
paradigm (for the Renaissance) · 5, 231
paradigm (heliocentricity) · 358
paradox (hydrostatic) · 133, 156, 157, 158, 159, 160, 162, 163, 165, 168, 169, 176, 177, 212, 229, 277, 280
parallelepiped · 166, 170
parallelogram of forces · 136, 140, 141, 142, 143, 144, 179, 283
paraphrased equations · 200

patent(s) · 29, 41, 46, 79, 80, 81, 82, 83, 84, 87, 132, 237, 277, 278
payments · 21, 86, 119, 121, 122, 127
pendulum clock · 277
perpetual motion · 133, 134, 137, 138, 139, 162, 175, 177, 178, 274
perspective · xxvi, 213, 214, 215, 216, 217, 218, 219, 220, 221, 222, 223, 224, 225, 226, 228, 272, 278
petition · 26, 51, 263
pintle (water mill) · 84
pit wheel (water mill) · 84
pivoted sluice-lock · xxvii, 46
polder · 82, 85, 86, 87, 88
polder water · 82, 85, 86, 87, 88
Polyglot · 36, 38
polynomial(s) · 182, 191, 192, 193, 199
poorhouse · 107
postulate · 125, 135, 139, 151, 154, 155, 162, 173, 175, 177, 178, 217, 231, 242, 265, 273
pragmatic approach (Stevin's) · 135, 247
Pratique d' Arithmetique, La · xxiv, 264
praxis (*daet*) · 8, 9, 157, 162, 255, 259, 269
precessional motions · 234, 271
pressure · xxvi, 84, 89, 133, 151, 152, 155, 157, 159, 161, 166, 169, 170, 171, 176, 177, 273, 274
princely bookkeeping · xxvi, 17, 112, 122, 127
principal (bookkeeping) · 112, 113, 115
Principal Works of Simon Stevin, The · 233, 254, 265
principle of inertia · 152, 273
printing · xxiii, xxiv, xxv, 1, 2, 3, 6, 8, 38, 39, 40, 55, 58, 131, 263, 265
privileges · 3, 4, 57, 195
Problematum Geometricorum · xxiii, 16, 29, 40, 181, 210, 264
Pythagorean comma · 243

R

Reformation · 4, 5, 6, 201, 268
regional awareness · 201
Rekenkamer of Holland (Audit Office) · 81
Religious Peace · 26
Remonstrants · 48
Renaissance · 1, 3, 5, 6, 7, 8, 9, 11, 107, 130, 169, 175, 176, 201, 209, 213, 214, 267, 268

retaining walls · 90
rhumb · 96
root extraction · 55, 56, 64, 252, 253
roots · 6, 64, 66, 67, 76, 182, 184, 185, 187, 189, 196, 197, 209, 246
rule of algebra · 199, 200

S

sailing chariot · 29, 45, 46, 79, 262, 266, 279
sailing tracks · 96
Sapientiae immarcescibilis · 4
scales (weighing) · 229
Schiedam · 91
schultbouck (bookkeeping) · 125
scientific revolution · 1, 7, 8
scoop wheel (water mill) · 84, 85, 87, 171
scoop wheel race (water mill) · 84
scoop wheel shaft (water mill) · 84
scouring · 88, 89, 90, 91, 93, 94, 95
semantic neologism(s) · 203, 204, 206, 207
sexagesimal system · 56
sheet piling · 91
shipping · xxxii, 92, 162, 172
Singconst, De Spiegheling der · xxvii, xxviii, 190, 239, 240, 241, 242, 243, 247, 249, 250, 251, 253, 254, 255, 256, 259, 274, 277
sluice(s) · 46, 79, 88, 89, 90, 91, 92, 93, 94, 95, 155, 169, 177, 237, 266, 278
snaphaen · 145, 147, 229
solid body · 139, 162, 163, 166, 167
spabijhou ('spade-axe-pick') · 102, 103
spade-axe-pick · 103
Spain · xxx, xxxi, xxxii, xxxiv, xxxv, xxxvi, 2, 5, 6, 48, 71, 73, 273
spheres (falling bodies) · 52, 53, 82, 133, 136, 137, 138, 139, 141, 142, 153, 154, 173, 177, 178, 237, 276
spherical trigonometry · 96
Spiegeling en Daet · 279
Spieghelschaeuwen, Vande Beginselen der · xxvi, 215, 272
staltwicht (apparent weight) · 138, 141, 142, 209, 274
States Army · 28, 29, 79, 100, 101, 105
States General · xxxi, xxxiii, xxxiv, 28, 42, 46, 51, 71, 79, 81, 96, 105, 128

States of Holland · xxxii, xxxiii, xxxv, 35, 36, 41, 47, 79, 83, 99
statics · 40, 131, 132, 133, 135, 139, 145, 152, 176, 204, 207, 266, 268, 271, 274, 277, 278, 282
Stercktenbouwing, De · xxv, 29, 44, 45, 99, 104, 237, 264
stereometers · 58
stewards · 51, 126, 127
Stofroersel des Eertcloots, Vant · xxvi, 33
string instruments · 255
studium generale · 3, 4
studium lovaniense · 4
surveyor(s) · 51, 58, 71
symmetry · 100, 106, 107, 108, 109, 178

T

tables (goniometric) · 59, 61, 268
tables (logarithm) · 62, 246, 259, 270
tables of credit and debit · 115, 117
tables of interest · xxiii, xxiv, 112, 118, 119, 120, 122
tables of variation · 98, 99
Tafelen van Interest · xxiii, 29, 40, 111, 118, 119, 122, 123, 181, 210
Tauwicht, Van het · xxvi, 136, 142, 143, 151
temperament · 240, 242, 243, 245, 247, 248, 249, 254, 255, 256, 257, 258, 259, 277
terrestrial magnetism · 97
theorem · 11, 13, 135, 162, 197, 200, 214, 215, 217, 218, 219, 264, 276, 282
Thiende, De · xxiv, 29, 40, 55, 56, 57, 58, 59, 61, 62, 64, 66, 68, 72, 73, 74, 75, 76, 77, 122, 181, 191, 199, 206, 210, 237, 252, 263
third(s) · 14, 28, 56, 61, 64, 65, 67, 71, 73, 80, 86, 89, 115, 117, 118, 122, 152, 157, 161, 172, 186, 188, 191, 192, 193, 197, 198, 204, 209, 215, 230, 231, 233, 234, 243, 245, 246, 248, 249, 255, 270, 273
thought experiment · 133, 138, 141, 156, 162, 168, 177, 178, 237, 283
thrust journal (water mill) · 84, 87
tide · 89, 91, 93, 94, 95, 154, 177
tides, theory of · 177
Toomprang, Vande · xxvi, 151, 152
Topswaerheyt,Vande vlietende · xxvi, 151, 152, 172, 174

town planning · 8, 33, 105, 106, 107, 266, 278
translation(s) of Stevin's works · xxii, xxiii, 55, 56, 62, 64, 74, 76, 77, 97, 98, 100, 118, 133, 155, 181, 200, 203, 212, 225, 228, 264, 265, 270, 271
tresorier · 126, 127
tuning systems · 239, 241, 242, 245, 249, 254, 255, 256, 257, 259, 277
tuning, (equal temperament) · 240, 242, 243, 245, 247, 249, 254, 255, 256, 257, 258, 259, 277
tuning, (meantone temperament) · 243, 249, 255, 256
Twelve Years' Truce · xxxvi, 30, 48

U

uniform standards of measures and money · 55
unit · 59, 61, 63, 64, 68, 69, 70, 71, 73, 76, 165, 171, 176, 177, 189
unit of commencement · 63, 71, 73, 192
unit of measurement · 71
United East India Company · xxxvi, 29, 96
United Provinces · xxvii, xxxv, 23, 28, 44, 51, 57, 88, 262
universities · 1, 3, 4, 27, 34, 36, 39, 183
University of Leiden · 28, 35, 36, 97

V

Verenigde Oost-Indische Compagnie (VOC) · xxxvi, 29, 96, 280
vernacular · xxx, 6, 201, 203, 210, 211
Verrechting van Domeine · xxvii, 128, 129
Verschaeuwing, Vande · xxvi, 213, 214, 215, 217, 221, 222, 224, 225, 225, 227, 228, 272
vertical sluice · 89
Vierschaar · 36
visible language · 227, 228, 237
Vita Politica · xxv, 29, 31, 43, 266
vlackvat ('surface vessel') · 163

W

Wadden Islands · 95
wallower (water mill) · 84
Wanschaeuvving, Vande · xxvi, 215, 272

warfare · xxvii, xxviii, 8, 48, 49, 79, 96, 99, 100, 107, 112, 278
water gate (water mill) · 84
water mills · xxviii, 81, 84, 152, 155, 169, 171, 177
waterhol · 173
Watertrecking, Vande · 151
Weeghconst (Anhang) · 151
Weeghconst, Byvough der · xxvi, 136, 142, 143, 151, 169, 174
Weeghconst, De Beghinselen der · xxiv, 9, 14, 29, 40, 82, 91, 131, 132, 134, 135, 136, 137, 140, 141, 143, 144, 145, 147, 151, 152, 156, 162, 171, 177, 202, 204, 208, 209, 210, 227, 228, 229, 236, 237, 242, 264, 265, 268, 269, 275, 276, 277
Weeghdaet, De · xxv, 9, 40, 81, 131, 134, 136, 145, 146, 147, 148, 149, 151, 153, 229, 230, 264, 265
Weereltschrift, Vant · xxvi, 33, 230, 232, 234, 235
weighing · xxiv, xxv, xxvi, 9, 131, 132, 133, 136, 142, 145, 151, 172, 229
weights and measures · 55, 57, 67, 68, 70
wheel pit (water mill) · 84
Wieringen · 96
windlass · 89, 134, 143, 145, 147, 148, 149, 151
windshaft (water mill) · 84
wine gaugers · 58, 71
Wisconstich Filosofisch Bedryf · xxviii, 52, 84, 88, 91, 93, 95
Wisconstige Gedachtenissen · xxv, 16, 17, 18, 30, 33, 46, 49, 51, 59, 96, 105, 112, 114, 115, 117, 118, 129, 136, 174, 181, 215, 216, 218, 220, 223, 225, 227, 230, 232, 234, 235, 237, 239, 240, 241, 252, 264, 271, 272, 274, 276, 279
WNT (Woordenboek der Nederlandse Taal) · 203, 204, 207, 235
wreath of spheres · 52, 53, 82, 133, 136, 137, 138, 139, 141, 177, 178, 237, 276
Wysentyt · 130, 279

Z

Zeylstreken, Vande · xxvi, 96, 97, 272
Zuiderzee, impoldering of · 95
Zutphen · xxv, xxx, xxxii, 48, 91

WITPRESS ...for scientists by scientists

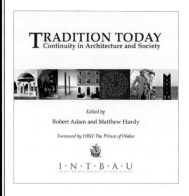

Tradition Today

Continuity in Architecture & Society

Edited by: **R. ADAM**, Robert Adam Architects, UK and **M. HARDY**, INTBAU, London, UK

In January 2002, after a two year gestation period, the International Network for Traditional Buildings, Architecture and Urbanism (INTBAU) was launched. To celebrate the launch, a conference was held to debate the place of tradition in modern society. While INTBAU was specifically concerned with building and urbanism, if tradition was indeed relevant then it must have a place throughout society. The conference forms the basis of this book.

It is an important feature of traditions that they adapt and change. So, while change accelerates so should the adaptation of traditions. If we rely on tradition for the transmission of culture, then the adaptation of traditions is a matter of importance to all of us. If change occurs without the transmission of culture, then culture itself dies; culture cannot be created anew every day. The evolutionary nature of tradition is something often ignored by supporters and opponents alike. It is important that history – that which measures our distance from the past – is not confused with tradition – the past living through us.

The papers presented in this book discuss these points and many others are a fascinating miscellany. With contributions ranging from the practical to the academic these papers can leave no doubt about the continued role and significance of tradition, the passion of those who understand its relevance and the dangers inherent in its denial.

ISBN: 978-1-84564-066-8   2008   160pp   £29.50/US$59.00/€44.50

Find us at
http://www.witpress.com

Save 10% when you order from our encrypted
ordering service on the web using your credit card.

WITPRESS ...for scientists by scientists

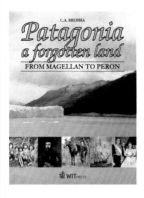

Patagonia, a Forgotten Land
From Magellan to Peron
C.A. BREBBIA, Wessex Institute of Technology, UK

This book describes the history of Patagonia from its discovery by Magellan to recent times. Since its early exploration Patagonia has been associated with conditions of extreme hardship and suffering. Men and ships were lost in the dangerous waters of the Straits of Tierra del Fuego, giving rise to tales of mysterious cities populated by the shipwrecked sailors of the many failed expeditions.

Early Spanish attempts to colonize Patagonia also ended in failure and the region remained largely uninhabited until the arrival of the Welsh in 1865. Their peaceful coexistence with the natives ended abruptly when the Argentine Army entered Patagonia and took over the Indian lands, which were promptly distributed to new settlers.

As a new and anarchic society, Patagonia could not fail to attract its share of desperados and adventurers, the most notorious of which are described in the book, including gold prospectors, hunters and bandits such as Butch Cassidy and the Sundance Kid.

The volume also relates the anarchist's struggles that took place in Patagonia at the beginning of the 1900s and the failed attempt by Peron's government to convert Argentina to a nuclear power.

The book conveys the image of Patagonia as still a largely unknown and forbidding place. Five hundred years of recorded history have not dispelled the image of Patagonia being a new frontier.

ISBN: 978-1-84564-061-3   2007   384pp   £33.00/US$59.00/€49.50

WIT*PRESS* ...for scientists by scientists

The New Forest
A Personal View by C.A. Brebbia

C. A. BREBBIA, Wessex Institute of Technology, UK

This is a new edition of a very successful book offering a personal view of the New Forest which stems from the author's many years of residence and research activities there. This has provided him with a deep appreciation of its unique rural charm and rich history. Its difference from many other National Parks is that it is home to many people and this has given the Forest a more dynamic environment.

    The narrative starts with a brief history of the area. There follows a description of the author's favourite places such as Lyndhurst, the so called capital of the Forest; the charm of Minstead and its unique church; Fritham, the wilderness of the northern Forest; Burley, the home of witches and smugglers; the extravagant Rhinefield House; old Brockenhurst and its churches; New Park, the site of many escapades of Charles II; the glory that was Christchurch Priory; the magnificence of Highcliffe Castle; Lymington, the ancient port of the Forest; Hurst Castle and its links to the unfortunate Charles I; the story of the ghost who designed Sway Tower; Boldre and its beautiful church; Beaulieu Abbey and how it was before its destruction; the ancient shipyard at Buckler's Hard and finally Ashurst Lodge from its early associations with saltpetre manufacturing to its current use as the home of the Wessex Institute of Technology.

    This new edition contains a substantial number of photographs taken by Pier Paolo Strona and Keith Godwin as well as other illustrations, including those of some famous characters associated with the New Forest.

ISBN: 978-1-84564-145-0   2008   128pp   £27.00/US$54.00/€40.50

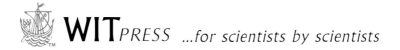 ...for scientists by scientists

Integral Equations and their Applications

M. RAHMAN, Dalhousie University, Canada

For many years, the subject of functional equations has held a prominent place in the attention of mathematicians. In more recent years this attention has been directed to a particular kind of functional equation, an integral equation, wherein the unknown function occurs under the integral sign. The study of this kind of equation is sometimes referred to as the inversion of a definite integral.

While scientists and engineers can already choose from a number of books on integral equations, this new book encompasses recent developments including some preliminary backgrounds of formulations of integral equations governing the physical situation of the problems. It also contains elegant analytical and numerical methods, and an important topic of the variational principles. Primarily intended for senior undergraduate students and first year postgraduate students of engineering and science courses, students of mathematical and physical sciences will also find many sections of direct relevance.

The book contains eight chapters, pedagogically organized. This book is specially designed for those who wish to understand integral equations without having extensive mathematical background. Some knowledge of integral calculus, ordinary differential equations, partial differential equations, Laplace transforms, Fourier transforms, Hilbert transforms, analytic functions of complex variables and contour integrations are expected on the part of the reader.

ISBN: 978-1-84564-101-6   2007   384pp   £126.00/US$252.00/€189.00

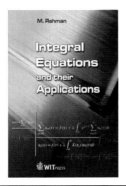

WITPress
Ashurst Lodge, Ashurst, Southampton,
SO40 7AA, UK.
Tel: 44 (0) 238 029 3223
Fax: 44 (0) 238 029 2853
E-Mail: witpress@witpress.com

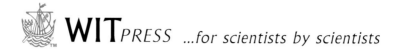

Advanced Vector Analysis for Scientists and Engineers

Edited by: **M. RAHMAN**, *Dalhousie University, Canada*

Vector analysis is one of the most useful branches of mathematics. It is a highly scientific field that is used in practical problems arising in engineering and applied sciences. Based on notes gathered throughout the many years of teaching vector calculus, the main purpose of the book is to illustrate the application of vector calculus to physical problems. The theory is explained elegantly and clearly and there is an abundance of solved problems to manifest the application of the theory. The beauty of this book is the richness of practical applications. There are nine chapters each of which contains ample exercises at the end. A bibliography list is also included for ready reference. The book concludes with two appendices. Appendix A contains answers to some selected exercises, and Appendix B contains some useful vector formulas at a glance. This book is suitable for a one-semester course for senior undergraduates and junior graduate students in science and engineering. It is also suitable for the scientists and engineers working on practical problems.

ISBN: 978-1-84564-093-4   2007   320pp   £95.00/US$165.00/€142.50

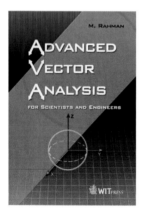

WIT eLibrary

Home of the Transactions of the Wessex Institute, the WIT electronic-library provides the international scientific community with immediate and permanent access to individual papers presented at WIT conferences. Visitors to the WIT eLibrary can freely browse and search abstracts of all papers in the collection before

Visit the WIT eLibrary at